Pollution in Tropical Aquatic Systems

Editors
Des W. Connell, Ph.D.
Division of Environmental Sciences
Griffith University
and
Government Chemical Laboratory
Brisbane, Australia
Queensland

Darryl W. Hawker, Ph.D.
Division of Environmental Sciences
Griffith University
Brisbane, Australia
Queensland

CRC Press, Inc.
Boca Raton Ann Arbor London

Library of Congress Cataloging-in-Publication Data

Pollution in tropical aquatic systems / editors, Des W. Connell,
Darryl W. Hawker.
p. cm.
Includes bibliographical references and index.
ISBN 0-8493-6581-3
1. Water—Pollution—Environmental aspects—Tropics. 2. Aquatic ecology—Tropics. 3. Water quality management—Tropics.
I. Connell, D. W. II. Hawker, Darryl W.
QH545.W3P65 1991
574.5′263′0913—dc20 91-29858
CIP

Direct all inquiries to CRC Press, Inc., 2000 Corporate Blvd., N.W., Boca Raton, Florida 33431.

International Standard Book Number 0-8493-6581-3

Library of Congress Card Number 91-29858
Printed in the United States of America 1 2 3 4 5 6 7 8 9 0

PREFACE

The tropical regions of the world are the location of a major proportion of the human population. The population of these regions is expanding at a rapid rate. Associated with this population expansion is the rapid growth of many diverse industries as well as tourism. These expanding activities produce large volumes of industrial waste water and sewage. Thus, tropical aquatic areas are being subjected to increasing stress from pollution.

Tropical aquatic areas contain significant resources which need to be protected so as to provide for the needs of the local population as well as nature conservation. It is well known that tropical systems are highly diverse and usually among the richest on the planet. These systems contain rich resources in terms of fisheries as well as providing a basis for a tourism industry.

Scientific knowledge is needed to provide a clear basis for pollution control in these areas. While temperate areas have a reasonable knowledge base, this information cannot be extrapolated to tropical areas with confidence.

We have identified a need to draw together knowledge of physicochemical and biological aspects of pollution in tropical aquatic systems. This book results from this and we hope will assist in providing management strategies to protect these systems from pollution effects.

In organising the book we have, as far as possible, attempted to cover the range of topics important in understanding pollution in tropical areas. Authors who are expert in their particular fields have been invited to contribute. We recognise that many topics remain uncovered but we hope this book will serve to assist in identifying these and stimulate interest in this area.

D. W. Connell and D. W. Hawker
Brisbane, Australia, 1991

THE EDITORS

Des W. Connell, Ph.D., is a Reader in environmental chemistry in the Division of Environmental Sciences at Griffith University and Director of the Government Chemical Laboratory in Brisbane, Australia. His broad interests are in water pollution, environmental chemistry, and ecotoxicology and he has published several books on these topics. His research interests are concerned with the behaviour of xenobiotic chemicals in the environment. He has been a member of the Great Barrier Reef Marine Park Authority Consultative Committee and the Water Quality Council of Queensland and is currently a member of the Water Quality Advisory Committee of the Great Barrier Reef Marine Park Authority. Before joining Griffith University, he was Marine Studies Coordinator of the Westernport Bay Environmental Study and Director of the Gippsland Lakes Environmental Study with the Victorian Ministry for Conservation. Dr. Connell received B.Sc., M.Sc., and Ph.D. degrees in Organic Chemistry from the University of Queensland.

Darryl W. Hawker, Ph.D., is currently Lecturer in Environmental Chemistry within the Division of Environmental Sciences, Griffith University. He received his Bachelor of Science Degree with First Class Honours in 1979 and Ph.D. in 1983 from the University of Queensland. Subsequently, he held postdoctoral research positions at Griffith University.

To date Dr. Hawker is the author or co-author of over 40 publications. He is an active member of the Environmental Chemistry Division of the Royal Australian Chemistry Institute. His current research interests include the environmental distribution and behviour of organic and organometallic compounds.

CONTRIBUTORS

Kevin G. Boto, Ph.D.
Australian Institute of Marine Science
Townsville, Queensland, Australia

Stephen L. Coles, Ph.D.
Research Scientist
Research Institute
King Fahd University of Petroleum and Minerals
Dharan, Saudi Arabia
Hawaii Institute of Marine Biology
Kaneohe, Hawaii

Des W. Connell, Ph.D.
Division of Environmental Sciences
Griffith University
Brisbane, Queensland, Australia

Iver W. Duedall, Ph.D.
Department of Oceanography, Ocean Engineering, and Environmental Science
Florida Institute of Technology
Melbourne, Florida

Miles J. Furnas, Ph.D.
Tropical Oceanography Program
Australian Institute of Marine Science
Townsville, Queensland, Australia

Darryl W. Hawker, Ph.D.
Division of Environmental Sciences
Griffith University
Brisbane, Queensland, Australia

Douglas A. Holdway, Ph.D.
Chief Environmental Toxicologist
Key Centre for Applied and Nutritional Toxicology
Royal Melbourne Institute of Technology
Melbourne, Victoria, Australia

Nicholas Holmes
Centre for Coastal Management
University of New England-Northern Rivers
Lismore, NSW, Australia

Paul L. Jokiel, Ph.D.
Hawaii Institute of Marine Biology
Kaneohe, Hawaii

Anne R. McNeill
State Water Laboratory
Rural Water Commission
Armadale, Victoria, Australia

P. Saenger
Centre for Coastal Management
Unviersity of New England-Northern Rivers
Lismore, NSW, Australia

Chih-Shin Shieh, Ph.D.
Department of Oceanography, Ocean Engineering, and Environmental Science
Florida Institute of Technology
Melbourne, Florida

Anitra Thorhaug, Ph.D.
Research Professor/Research Scientist
Sponsored Research Division
Florida International University
Miami, Florida

Eric Wolanski
Senior Principal Research Scientist
Australian Institute of Marine Science
Townsville, Queensland, Australia

TABLE OF CONTENTS

Section I: Physical, Chemical, and Hydrodynamic Features in Tropical Systems

Chapter 1

HYDRODYNAMICS OF TROPICAL COASTAL MARINE SYSTEMS

Eric Wolanski

TABLE OF CONTENTS

I. INTRODUCTION

This chapter addresses only coral reefs, mangrove swamps, and muddy coastal environments because these environments are common on a tropical coast.

Tropical coastal waters cover an area similar in size to temperate coastal waters, but until recently their hydrodynamics have been much less studied for various political, social, and logistical reasons. During the last few years, however, there have been a large number of scientific studies and associated publications on hydrodynamics of tropical coastal waters, principally concentrating on coral reefs and, to a lesser degree, on mangrove swamps.

Because the water circulation is determined by physical laws of motion and, hence, is theoretically predictable, the dynamic processes in tropical waters should apply also to temperate systems of similar topography and hydrodynamic forcing. For instance, tidal eddies in the lee of a coral reef in tropical waters are dynamically similar to tidal eddies in the lee of a headland in temperate waters. However, some processes, such as the existence of an evaporation-driven salinity maximum zone in mangrove-fringed tidal rivers in the dry season, are particular to tropical estuaries.

Similarly, because the topography and tidal forcings are similar, the water circulation in a tidal creek-mangrove swamp system in the tropics should be similar to that in a tidal creek-Spartina salt marsh system in temperate countries. However, until recently neither island wakes nor the hydrodynamics of Spartina salt marshes were the subject of much research in temperate countries. Fluid mud processes in turbid estuaries should be similar in both tropical and temperate estuaries, although not identical if only because of the influence of temperature on colloidal kinetics. Because mud dynamics have been extensively studied in temperate countries, only briefly some of their properties will be outlined here.

II. CORAL REEFS

As pointed out in the review of coral reef hydrodynamics by Hamner and Wolanski,[1] the physical and chemical characteristics of water masses in coral reefs reflect not only those of the surrounding offshore waters, but also local reef-driven influences. Coral reefs not only modify the water circulation, but also the chemicals and the living and nonliving organisms that are transported over and around the reef. Reefs modify the water circulation not only in their immediate surroundings, but also that in offshore waters many reef diameters away from the reef. In some cases, shown below, reefs can even modify the basin scale oceanic circulation.

A. LARGE-SCALE REEF-INDUCED DYNAMICS

When the reef comes close to the surface, only a minimal fraction of the water advected toward the reef by the prevailing current actually flows over the reef. In areas where reefs are densely scattered, the flow is restricted to narrow passages and the tidal currents can be exceptionally strong. Such is the case, for instance, in the Ribbon Reefs of the Great Barrier Reef and in Torres Strait. As a result of bottom friction acting on such strong tidal currents, little of the tidal wave energy can propagate through a wide and shallow reef barrier. Large-scale hydrodynamic consequences of this blocking effect can be dramatic. Indeed, only 30% of the semidiurnal wave energy is able to propagate through the Torres Strait. As a result of nonlinear coupling between tidal and low-frequency motions, long-term transport through the strait is negligible for oceanic budget computations.[2] Thus, as a result of the presence of coral reefs, two connecting oceanic basins, the Arafura Sea and the Coral Sea, are largely isolated from one another.

Another example of large-scale blocking by a reef barrier occurs in the southern region of the Great Barrier Reef near Broad Sound, where Hamon[3] reported a large amplification

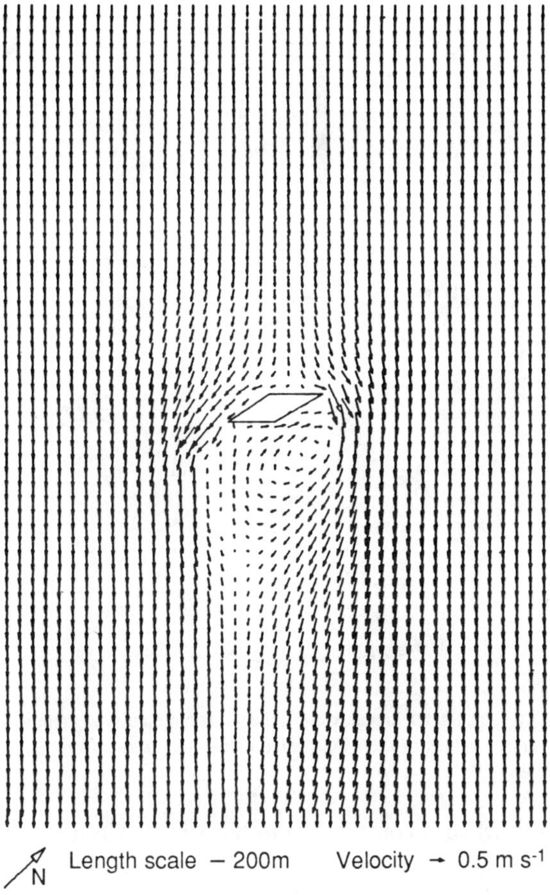

FIGURE 1. Numerically predicted depth-averaged circulation around Rattray Island after eddy spin-up. (From Wolanski, E., *J. Geophys. Res.*, 93, 1335, 1988. With permission.)

of the semidiurnal tide. Over 150 years ago, Flinders,[4] and more recently Middleton et al.,[5] explained this phenomenon as being due to the blocking of the oceanic tidal forcing by a dense matrix of reefs along the shelf break. The incoming oceanic tide can only enter the shelf through openings to the north and south. This creates two waves propagating from both north and south in the wave guide formed by the shore on one side and the reef on the other, and meeting in the middle near Broad Sound.

B. FLOWS AROUND REEFS

1. Barotropic Flows

Barotropic flows occur when the salinity and temperature gradients are too small to affect the flow field. There is evidence of blocking effects upstream of a coral reef in shallow water under steady or unsteady flows. Field studies and two-dimensional (depth averaged) models[6-14] suggest that the upstream flow is blocked only very close to the reef, a stagnation point being present. The width of the boundary layer and boundary layer dynamics are still unknown.[15,16] Figure 1 illustrates the numerically predicted depth-averaged circulation around Rattray Island, a 1.5-km-long island in shallow (mean depth = 20 m) coastal waters with strong (up to 0.7 m s^{-1}) semidiurnal tidal currents located in the central region of the Great Barrier Reef. It can be seen that, on the upstream side, the bulk of the water is deflected

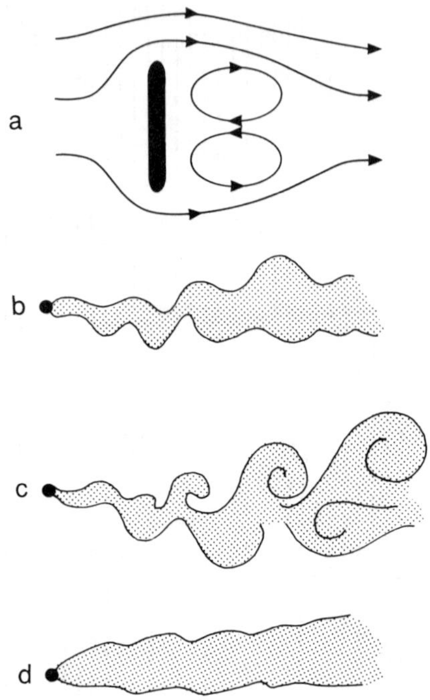

FIGURE 2. Sketch in plan view of the water circulation in shallow waters for various values of the island wake parameter, drawn from aerial photographs.[13,18,23]

around the island only when it comes close to the island. Upstream thus, the flow around islands or reefs resembles the flow around solid obstacles in laboratory experiments at very high values of the Reynolds number (Re >500). Drogues placed at the separation point have been observed to stall for hours, and zooplankton biomass in this area were found to be 20 times larger than further upstream.[17]

Downstream of reefs and islands in coastal waters, eddies of size comparable to the island diameter[13,18] have been observed (Figures 1 and 2a). Although visually the shape of these eddies resembles that of flow around solid obstacles in laboratory experiments at low values of the Reynolds number (Re = 10–50), there is no dynamic similarity between these two eddies because in the laboratory the velocities in the eddies are typically 1% those outside the eddy, while in coastal waters these velocities are of comparable magnitude. The most extensive studies of such eddies have been those of Wolanski et al.,[13] who deployed 26 current meters to map the eddy, and of Geyer and Signell,[9] who mapped the eddy using a Doppler profiling current meter mounted on a vessel criss-crossing the eddy. The first method offers truly synoptic views of the circulation sampled at 26 points, but does not resolve the vertical distribution, while the second technique filters out some temporal features, but solves the three-dimensional flow field. Figure 1 shows as an example the numerically predicted eddy at flood tide in the lee of Rattray Island. This numerical prediction was undertaken using a two-dimensional (depth-averaged) model extensively verified against the data from the 26 current meters moored around Rattray Island.[8] This model has since been further verified for tidal jets generated by tidal flows in reef passages[19] and for rotating tidal flows around coral reefs.[12,14]

In shallow waters when bottom friction effects are important, the eddy size can be predicted[13,20-22] using an island wake parameter, P, which is the product of Re and $(H/W)^2$, where H is the depth and W the width. In coastal waters, $(H/W) << 1$ but $Re >> 1$. For Rattray Island, as in Figures 1 and 2a, $P \sim 1$. For larger values of P, e.g., for stronger

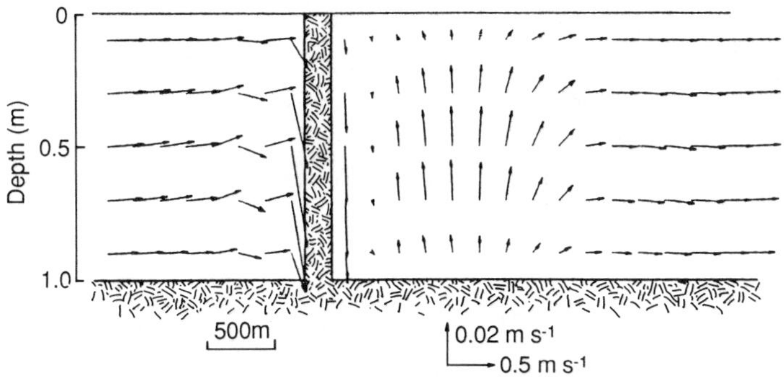

FIGURE 3. Numerically predicted[24] three-dimensional circulation around Rattray Island at flood tide after eddy spin-up in a vertical plane along mean stream axis. Note the zones of upwelling and downwelling. (Adapted from Deleersnijder et al.[24])

currents or smaller islands, evidence of flow instabilities in these eddies are apparent[23] with the formation of meanders very far downstream (Figure 2b). For even stronger currents or smaller islands, small vortices are formed when the meanders become dynamically unstable and overturn (Figure 2c). For very strong currents and small islands, long slab-line turbulent wakes with no apparent organized structure are observed (Figure 2d). This evidence suggests that vorticity plays a role in dynamics of flows around islands and reefs, vorticity being advected into the lee of the island at the separation point as a result of the velocity shear along solid boundaries, and this vorticity being dissipated further downstream by friction.[13,23] If the island, hence the eddy, is too small, bottom friction in the eddy is insufficient to dissipate this vorticity, and the remaining vorticity introduces flow instabilities further downstream.

Only recently has a numerical model been proposed for the three-dimensional barotropic circulation around coral reefs and islands.[24] Figure 3 shows the numerically predicted three-dimensional circulation at flood tide around Rattray island in a vertical plane parallel to the main stream and bisecting the island. Note in Figure 3 the strong downwelling on the upstream side of the island along the surface of the island or reef. Note also the strong upwelling in the center of the eddy in the lee of the island. In shallow water, the existence of a strong upwelling in the center of the eddy was observed by Wolanski et al.,[13] who argued that in shallow water, bottom friction plays a key role in the dynamics of the eddy, the eddy width being much larger than the depth. As a result, a three-dimensional circulation develops in an eddy in shallow coastal waters, similar to that in a tea cup (see the sketch in Figure 4), with convergence of bottom water at the eddy center in a benthic boundary layer, upwelling in the middle and downwelling at the edges. Such secondary circulation explains the trapping and aggregation of floating particles, such as coral eggs, along the eddy boundaries in shallow waters in the lee of coral reefs.[1,25]

For coral reefs in deeper waters (20 m $<$ depth) on the continental shelf, the secondary, bottom friction-driven, three-dimensional circulation may become dynamically less important for unsteady tidal flows. In this case, the effects of nonlinear inertia control the generation of such topographically controlled eddies, and the presence of an eddy can then be explained strictly from two-dimensional dynamics.[6-8,12,14] However, visual observations reveal, at least in calm weather conditions, that the three-dimensional circulation is still present in the eddy, though this secondary circulation may not have time to develop fully, since floating material usually ends up trapped along topographic fronts marking the eddy boundaries.[1,13,17,25-28]

Modeling of flows around reefs for predictive and management purposes is theoretically

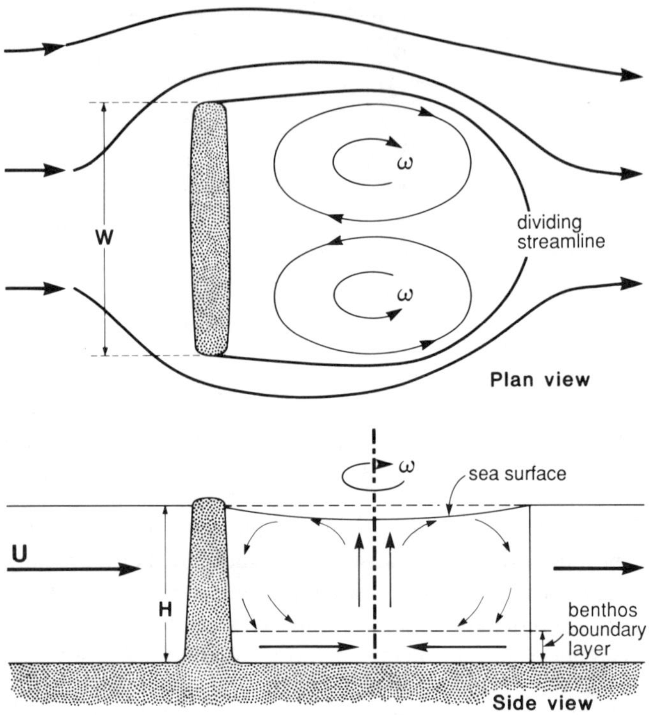

FIGURE 4. Sketch of the internal circulation in an island wake in shallow waters. The vertical scale is exaggerated, as typical depths about 20 m, whereas horizontal scales are of the order of 1 km. (From Wolanski, E. and Hamner, W. M., *Science,* 241, 177, 1988. With permission.)

feasible and economically practical using depth-averaged two-dimensional models.[7,8,12,14,19] Black[6] and Signell and Geyer[29] used such models to predict the fate of nonbuoyant larvae. Such modeling efforts should, however, be viewed with caution and may be unreliable for coral reefs in a reef matrix such as the Great Barrier Reef. This is because modelers have so far concentrated on simulating the water circulation around an individual reef, assuming the reef is hydrodynamically isolated from all surrounding reefs, i.e., assuming smooth uniform undisturbed upstream flows in the far-field of the reef. This assumption is quite unrealistic in a reef matrix system, even if the reefs are several reef diameters away one from the other, such as the Great Barrier Reef. Indeed, recent field studies[12,14] revealed a large horizontal shear of horizontal currents in a reef matrix system, even several reef diameters away from a reef. Numerical experiments reveal that the near-reef and lagoon currents are quite sensitive to this far-field heterogeneity.[12]

Dight et al.[30] attempted to simplify even more the numerical problems of dispersion in a reef matrix, by taking the mesh size of their numerical model to be of the order of a reef width, and by calculating separately and independently, and later adding linearly, the currents due to tides, wind, and oceanic forcing. This approach is unrealiable for at least three reasons. First, this approach neglects the important nonlinear interactions in shallow water between tidal and low-frequency currents.[31,32] Second, this approach neglects reef-scale (their sub-grid scale) recirculating flows around reefs, so that one must parameterise explicitly the poorly understood effects of reef-scale currents on mixing and trapping processes which are unnecessarily removed from the model. Third, the interreefal circulation is incorrectly modeled since this circulation is greatly influenced[12,14] by eddies and shear layers shed from reefs, and those are not included in the model.

FIGURE 5. Steady, barotropic, depth-averaged circulation over the central region of the Great Barrier Reef continental shelf under the influence of a 10-m s^{-1} southeasterly tradewind and of the southward East Australian Current generating an oceanic longshore pressure gradient. Note the southward current offshore and the northward current inshore. The zone of zero flow marks the offshore extent of the coastal boundary layer. This figure, the result of an ongoing study, was kindly provided by Mr. Brian King.

A more promising approach is that of King,[106] who models at small scales a wide area of the Great Barrier Reef continental shelf, from the shore to the shelf break. This approach enables an accurate representation through the open boundaries of the combined influences of tides, wind, and oceanic forcing on longshore and cross-shelf pressure gradients. This technique necessitates extensive computer power, but the results are very encouraging because the whole reef matrix is included. Figure 5 illustrates the results of such an application, illustrating the power of the model to calculate both the small-scale water circulation near a reef, the current variability in interreefal waters, and the blocking effect of the reef matrix on the prevailing oceanic-driven current. The model also reveals the existence of a coastal boundary layer that may trap land-derived contaminants in coastal waters (see later). Alternatively, one can use the results of this model, since it calculates the interreefal current variability, as open boundary forcing for a small-scale model of the circulation around an individual reef. Physical restrictions in computer power and sheer economics will determine how small the mesh size can ultimately be in such numerical simulations.

The constraints of economics and computer power will become even more limiting when three-dimensional hydrodynamic models are applied to a reef-studded continental shelf.

2. Baroclinic Flows

Baroclinic flows occur when the density stratification, controlled by the temperature and salinity, become sufficiently large to modify the water circulation. Flows around reefs in stratified water are exceedingly complex, just as they are also complex in laboratory experiments.[33,34] Eddies and meanders are shed and they are generally unsteady, if not unstable, although Johannes[35] reported oral evidence from indigenous fishermen of a stable eddy pair shed behind Tobi Island, a small island in deep water in the Carolina Islands group. Biologically important implications of flows around reefs in stratified waters is that they result in localized zones where upwelling is very strong.[36-38] This upwelling can be rapid and strong enough to stun fishes thermally.[39] No simple predictive models are yet available for such small-scale flows.

C. FLUSHING OF CORAL REEF LAGOONS

The flushing of a reef lagoon and the residence time of nutrients and pollutants within the lagoon have important ecological and engineering implications. Such studies were pioneered by von Arx[40] at Bikini for very practical applications, namely, any possible trapping of radioactive fallout. At Bikini, lagoon flushing was due primarily to the continuous inflow into the lagoon generated by breaking ocean waves on the reef crest. Atkinson et al.[41] undertook the most extensive study to date of lagoon flushing at Enewetak Atoll where they were able to parameterise the relative contribution of wave overtopping and of tidal and wind-induced currents. For such large and partially open lagoons residence time estimates range between 1 and 4 months. The residence time in closed atoll lagoons can be as high as 4 years.[42] Smaller shallow reef lagoons, a few kilometers in diameter and no more than 20 m deep, such as Bowden Reef in the Great Barrier reef, are completely flushed out in a day following strong winds.[1,43] Kinsey[44] pointed out that, for lagoons with entrances at the leeward side, the earlier concept of a reef lagoon flushed by flow over the reef crest is not applicable because the bulk of the water flushing the lagoon comes from the leeward side of the reef through passages into the lagoon.

One important feature of eddies generated offshore by tidal flows around coral reefs, is that these eddies occasionally intrude in coral reef lagoons.[12,45] Figure 6 shows numerical predictions of such an eddy generated northwest and offshore of Bowden Reef and, some time after generation, intruding in the lagoon. The intrusion of the eddy in the lagoon results in rapid flushing of the lagoon and in the formation of a barotropic friction-driven front where presettlement reef fish and zooplankton are most abundant.[45] Also, the eddy, by scooping up a large body of offshore water and bringing it into the lagoon, enabled the reef to become a larger target for presettlement fish than the reef size would suggest.

Nearshore barrier reefs and fringing reefs with or without a narrow lagoon between the reef and the shore are flushed very rapidly, typically in a day or so,[46,47] because of the strong longshore current generated in the narrow lagoons by the wave- and wind-induced inflowing current over the whole length of the reef crest. Two-dimensional depth-averaged models are very useful in estimating flushing rates in such simple systems.

Flows through the reef substrate can be important also in determining the flushing of lagoons and the currents over the reef flat if the substrate is particularly porous.[48,49] The coral substrate commonly comprised of connecting caves and the parameterisation of the flow through this substrate is difficult. This is the same unresolved problem hydrologists face in determining groundwater flow in limestone with numerous caves. Thus, it is not surprising that reported speeds of currents through a coral substrate range from microns per second to tens of centimeters per second.[48-53]

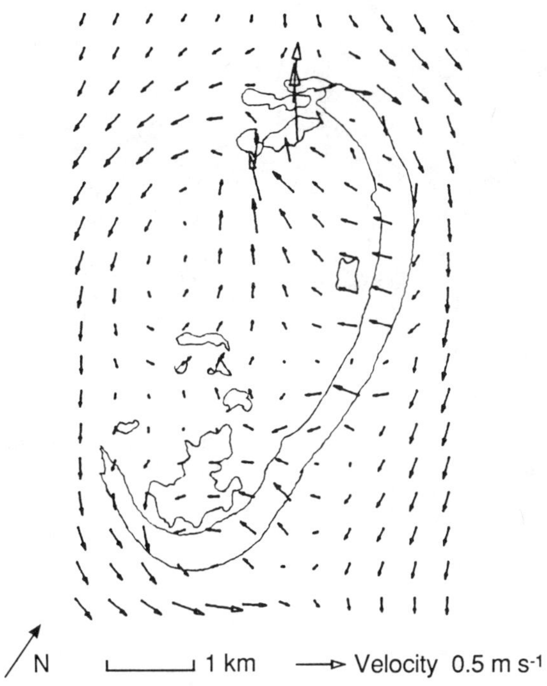

FIGURE 6. Synoptic distribution of the depth-averaged circulation around Bowden Reef, Great Barrier Reef, following spin-up of a tidal eddy offshore from Bowden Reef and intrusion of this eddy in the lagoon. (Adapted from Kingsford et al.[45])

D. TROPICAL CYCLONES

There exists a wide body of literature on storm surges generated by tropical cyclones (hurricanes) or other storms. Two-dimensional, barotropic, numerical models are now used routinely as an engineering predictive tool for such storm surges in the North Sea, the Gulf of Mexico, Bangladesh, and Australia, to name but a few. Of more interest here are the effects of waves in cyclones near coral reefs about which little is known, although there are numerous reports of cyclone-induced damage to living corals. Kjerfve et al.[54] have studied wave refraction patterns of waves offshore from a fringing reef in such storms. The damage these waves do to the coral reefs is related to the angle of attack of the waves.[55] The reef crest is an efficient breakwater,[56] but for lagoons with entrances at the leeward side, waves approaching from the back are able to propagate into the lagoon, causing much damage.[55]

E. UPWELLING

The 30-year-old view of coral reefs as ''closed'' ecosystems is not justified.[1] Certainly nutrients are tightly recycled within coral reefs, but there is also a nonnegligible inflow and an outflow of nutrients. Upwelling in many cases may be responsible for the nutrient inflow. Upwelling mechanisms on reefs are many and varied.

As sketched in Figure 7, the water that flows over the reef crest is drawn both from the surface layer, mostly due to wind forcing and wave breaking, and from deeper waters due to vertical entrainment into the grooves of a typical reef system.[57] Since the surface waters are generally oligotrophic, nutrient pumping along the boundary is biologically important. There are no models to predict the ratio of surface to deep water that crosses a reef crest. This problem is important as it determines the nutrient inflow on the reef crest.

Flows through the coral substrate are most important in deep oceanic atolls due to the endo-upwelling, sketched in Figure 8, whereby deep nutrient-rich waters are drawn to the

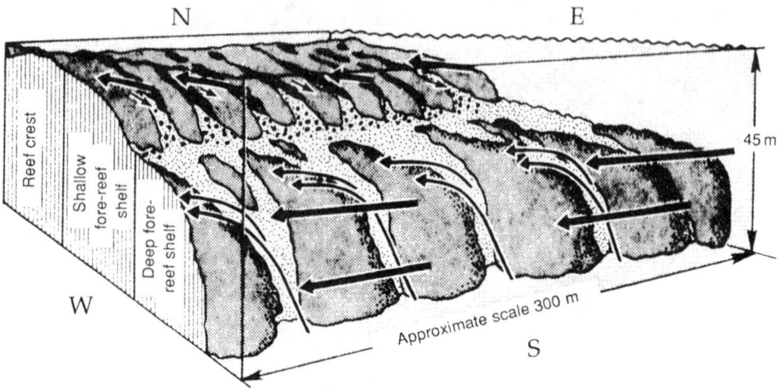

FIGURE 7. Upwelling in the spur-and-grooves system of a reef flat. (From Roberts, H. H., Murray, S. P., and Suhayda, J. N., *J. Mar. Res.*, 33, 233, 1975. With permission.)

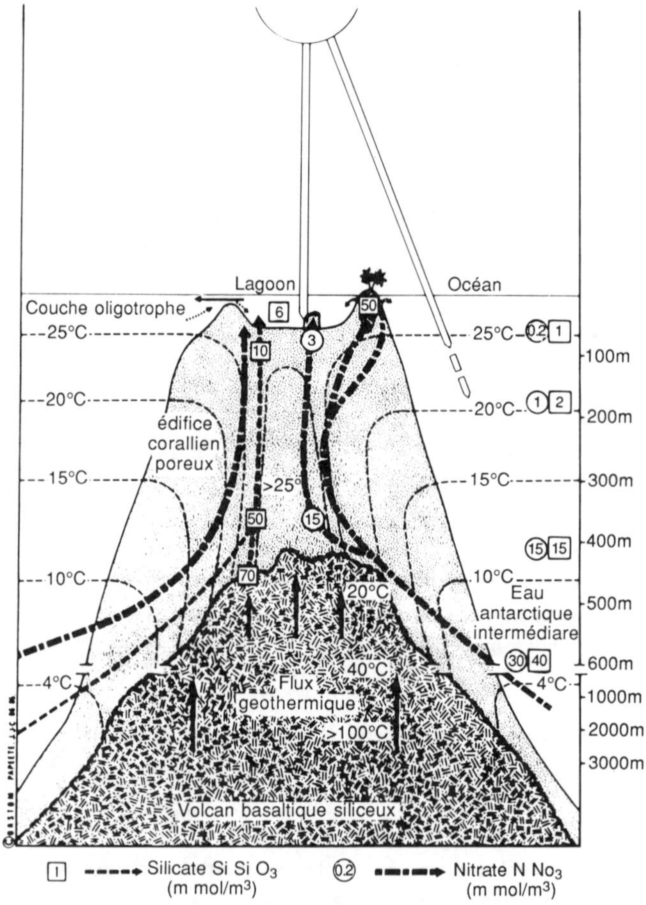

FIGURE 8. Endo-upwelling in a deep oceanic atoll. (From Rougerie, F. and Wauthy, B., *Oceanolog. Acta*, 9, 133, 1986. With permission.)

surface through the substrate, possibly enriching the reef and surface waters.[52,53] Waste discharged in the coral substrate in zones of endo-upwelling would ultimately find its way back to the surface.

In the central region of the Great Barrier Reef, shelf-scale upwelling occurs on the continental slope and is driven by wind fluctuations and events in the Coral Sea.[58,59] This upwelling controls phytoplankton dynamics over the shelf,[60] but it remains unclear how much it contributes to coral reef growth. There are also large-amplitude internal tides around reefs on the continental slope near a rugged topography due to reefs, and these may result locally in nutrient-rich water from below the thermocline, spilling onto the outer shelf.[38] Strong steady oceanic currents may also interact with a sloping reef to generate upwelling[36,37] and may explain the upwelling near the southern lagoon of New Caledonia.[61]

As sketched in Figure 9a, strong flood tidal currents through reef passages may also generate upwelling by a Bernouilli effect, also called tidal suction.[19,62,63] In the Ribbon Reefs area of the Great Barrier Reef, the upwelled nutrients are not readily available to the reef but instead are advected by a tidal jet effect (Figure 9b), toward *Halimeda* meadows several kilometers inshore.[19] This mechanism may be sufficient to sustain the nutrient needs of the *Halimeda* meadows. At ebb tides (Figure 9c) the tidal jet, injecting shelf water into the ocean through a reef passage, vertically entrains deep oceanic water. This entrainment enriches the surrounding outer coral reefs in nutrient and sustains pelagic fisheries concentrated exactly in front of such passages.[19]

If a barrier, occasionally interrupted by reef passages and separating a lagoon on one side from the ocean on the other side, inhibits sufficiently the exchange of water such that large ($\approx$0.1 m) sea level differences exist on either side of the reef, then geostrophic upwelling in response to changing sea levels may lift cold, oceanic, deep water in the passages, and from there, this water can be advected in the lagoon.[64]

Another upwelling mechanism is the enhanced boundary mixing as the currents are forced to flow around an island or a reef and reach maximum velocity near the two tips of the island. This mechanism, sketched in Figure 10, has been documented for continental shelf islands by Simpson et al.[65] and is particularly apparent if there is a strong stratification.

F. MODELING THE FATE OF CONTAMINANTS IN REEFAL WATERS

As discussed earlier, some two-dimensional (depth-averaged) models are able to reproduce several features of the depth-averaged water circulation near reefs. Nevertheless, these models are unsuitable and should not be used to predict the fate in reefal waters of many buoyant constituents of scientific and practical interest including coral eggs, zooplankton, sewage, and oil.

This is because, in calm weather at least, complex small-scale three-dimensional topographically driven circulation patterns not only inhibit diffusion of these constituents, but, instead, aggregate them into slicks.[1,25] Figure 11, for instance, sketches the internal circulation processes aggregating floating coral eggs near a reef crest, forming a coral slick, and preventing diffusion. Warmer lagoon water gently flows out of the lagoon over the reef crest and is deflected by the longshore currents along the reef in deeper water. The three-dimensional water circulation in the buoyant plume accumulates floating coral eggs along the outer edge of the plume, forming a coral egg slick.[25] An identical mechanism generates a foam line in coastal water at the outer edge of a river plume. Another important aggregating mechanism forming coral egg slicks near reefs is the process of boundary mixing[25] (Figure 12). Phillips et al.[66] argued that in the stratified waters of Chesapeake Bay, plankton trapped in the density-interface may be prevented by the secondary currents, generated by boundary mixing, from impinging on the sloping boundaries. A similar mechanism may exist near coral reefs.[16]

Another important effect controlling diffusion at small scales in reefal waters is the

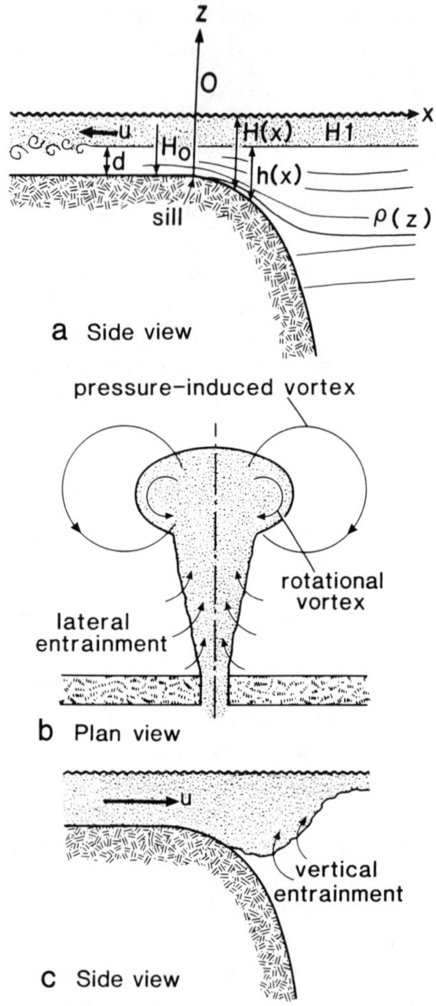

FIGURE 9. (a) Side view of upwelling by tidal suction at flood tide as a result of strong tidal currents through reef passages; (b) plan view of the tidal jet emanating from a reef passage; (c) side view of the tidal jet at ebb tide and the turbulent entrainment in the jet. (From Wolanski, E., Drew, E., Abel, K. M., and O'Brien, J., *Estuarine Coastal Shelf Sci.*, 26, 169, 1988. With permission.)

presence of topographically controlled free shear layers, occasionally less than 10 m wide.[8] These layers need explicit parameterization[11] in numerical models, as these models generally have a much larger mesh size of at least 100 m.

In windy conditions, field studies demonstrate that these secondary three-dimensional flows appear to be less important.[45] Nevertheless, buoyant constituents near coral reefs still end up aggregated, this time into Langmuir cells. These cells are more readily visible in reefal waters than in the open sea because of the shelter from waves offered by the reef.

The demonstration[25] that small scale three-dimensional circulation patterns are very efficient in aggregating coral eggs in calm weather throws open the question of how to model the fate of buoyant constituents near reefs, such as coral eggs, zooplankton, oil, and sewage. Certainly, it is inappropriate to use depth-averaged numerical models as was proposed by Sammarco and Andrews.[67] The answer may be in the use of small-scale three-dimensional

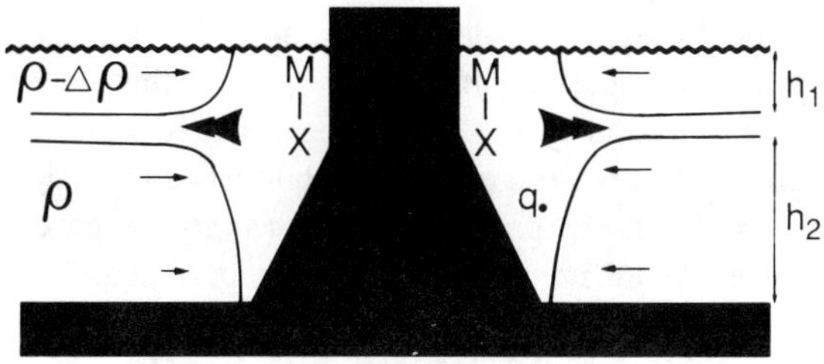

FIGURE 10. Side view of the internal circulation genereated by boundary mixing around an island. A two-layer stratified ocean is assumed, with a difference Δδ of density δ, the two layers being separated by a sharp interface, the pycnocline. Boundary mixing creates a mixed layer around the slopes of the island. Water in this mixed layer has intermediate density and by buoyancy spreads horizontally away from the island along the pycnoline. (From Simpson, J. H., Tett, P. B., Argote–Espinoza, M. L., Edwards, A., Jones, K. J., and Savidge, G., *Cont. Shelf Res.*, 1, 15, 1982. With permission.)

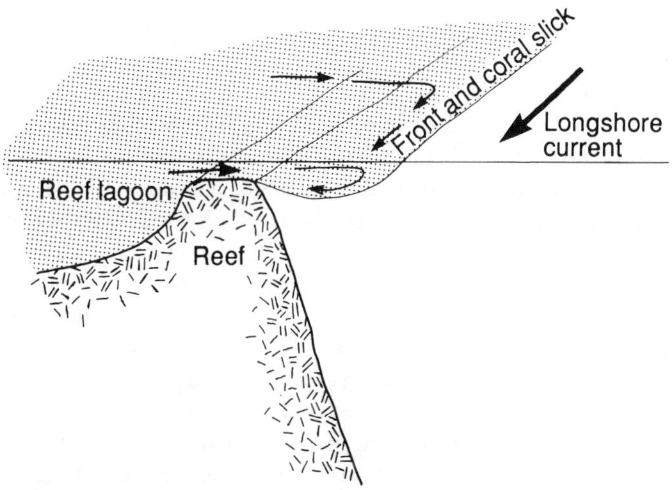

FIGURE 11. Sketch of the water circulation near a reef crest generating aggregation of buoyant particles such as coral eggs. (From Wolanski, E. and Hamner, W. M., *Science*, 241, 177, 1988. With permission.)

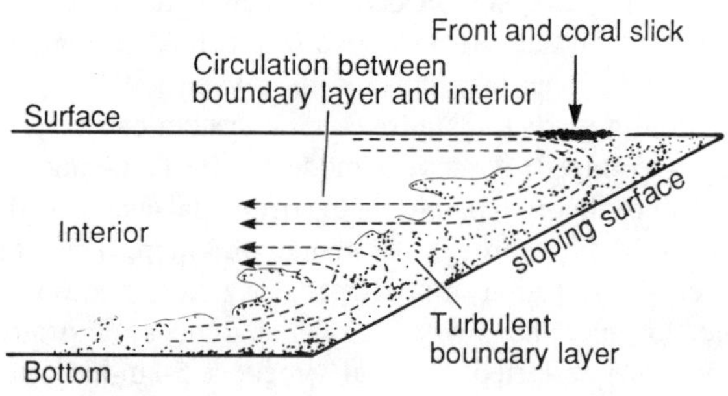

FIGURE 12. Side view of the internal circulation generated by boundary mixing along the sloping surface of a coral reef, aggregating coral eggs. (From Wolanski, E. and Hamner, W. M., *Science*, 241, 177, 1988. With permission.)

models, or even in the use of pseudo two-dimensional models with a negative eddy diffusion coefficient. In view of such complexity, it is not surprising that no reliable models of the fate of buoyant constituents around reefs have yet been put forward and verified. Complex small-scale three-dimensional currents aggregating buoyant material constitute an area of reef research that is presently largely ignored because it is so complex, yet that prominently stands above all the rest in scientific and engineering importance.

III. MANGROVE SWAMPS

Mangrove swamps are often seen as a haven for mosquitoes and sand flies, and therefore are seen by some as an ideal site for a number of human activities, such as waste dumping, land reclamation, aquaculture ponds, and dredging for navigation channels or marinas. Yet mangroves are a highly productive marine ecosystem and an important forestry resource; mangrove swamps serve also as a nursery ground for economically important marine species such as prawns (Penaeus spp.); mangrove swamps also export inorganic nutrients and mangrove litter to coastal waters where primary and secondary productivity is enhanced.[68-78] Mangrove swamps maintain deep water channels in tidal creeks draining the swamps,[79] and these channels are important for fisheries and navigation. The most vigorous mangrove stands are often associated with riverine conditions but only in areas where detritus accumulation is low,[80] and this enhanced growth may be controlled by groundwater.[81]

A. TIDAL DYNAMICS

Mangrove-fringed tidal rivers and creeks are estuaries, i.e., they permit saltwater intrusion and tidal propagation inland. Depending on physical constraints, such as the freshwater discharge at the tidal head of the estuaries, the length and depth of the estuary, the area covered by mangroves, and the tidal range, mangrove swamps may or may not play a significant role in the tidal dynamics and the mixing processes in the estuary. If the area of mangroves is small, the dynamic role of mangroves is negligible, and the estuary type can be any of the classic estuarine types described by Pritchard[82] including completely flushed estuaries where saltwater is found only at the mouth, the salt wedge estuary type, the partially mixed type, and the vertically well-mixed type. These classic types of estuaries are not discussed further, as the emphasis here is on the dynamic role of mangroves.

When mangroves cover a large area, they profoundly influence the tidal hydrodynamics. In the absence of a large freshwater runoff, strong tidal currents in mangrove creeks usually maintain vertical homogeneity or allow only a weak stratification in temperature and salinity. Under such conditions, buoyancy-induced baroclinic currents are weak, and flushing is primarily determined by tidal dynamics. The lateral velocity shear in the estuary, normally already high for wide and shallow channels with an uneven bathymetry,[83] becomes even higher as a result of the high vegetation density in mangrove swamps that prevents the occurence of strong currents in the swamps. As a result, tidal currents in the tidal channels can be typically ten times stronger than those in the swamps.[79]

The water circulation in a creek-mangrove creek system can be reliably modeled by a two-dimensional (depth-averaged) numerical mode.[79] The dominant dynamic effects turn out to be the highly nonlinear hydrodynamics in the deep tidal channels, the role of vegetation in enhancing friction for the flow through the mangrove swamp, and at rising tide, the loss of momentum by energy dissipation around mangrove trees for water, leaving the high velocity tidal channel to enter the low-velocity mangrove swamp area. The model was verified against obsevations collected in Coral Creek, a 5-km-long tidal creek-mangrove swamp system with zero freshwater inflow, located in northern Australia. The predicted synoptic water circulation in the mangrove swamp surrounding Coral Creek is shown at flood tide (Figure 13a) and ebb tide (Figure 13b), respectively. Figure 13 illustrates the tidal

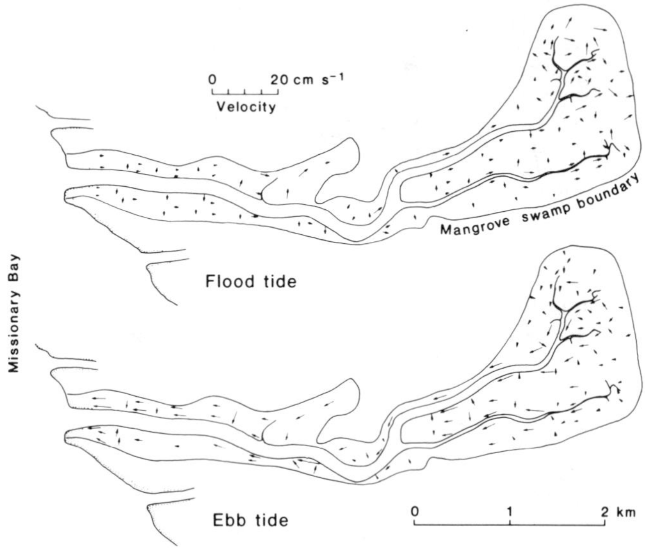

FIGURE 13. Peak tidal currents in the mangrove swamp of Coral Creek at flood tide and ebb tide. (Adapted from Wolanski et al.[79])

asymmetry of the currents in mangrove swamps, the flood currents in the mangrove being oriented perpendicular to the creek banks, the ebb currents tending to be parallel to the creek banks. This results in a net downstream-oriented tidally averaged circulation in the swamp.

The net circulation is responsible for exporting fallen mangrove leaves and other mangrove detritus from the mangrove swamp to the creek. These leaves, as they float away from the creek into coastal waters, are often found to be aligned in a straight line, suggesting that they were aggregated by small scale three-dimensional currents[78] over a complex topography.

The presence of tidal currents strong enough to erode the bottom, and the stronger currents at ebb than at flood tides, result in the maintenance of a deep self-scouring tidal channel. Numerical experiments suggest that the width and depth of this channel increase with an increase in both vegetation density in the mangrove swamp and in the area covered by the swamp.[79]

Thus, mangrove land reclamation for developments such as golf courses, urban developments, construction of aquaculture ponds, or thinning of the vegetation results in siltation of the channel.[78]

B. TRAPPING DUE TO EVAPOTRANSPIRATION

When mangroves cover a large area, evapotranspiration effects become important. Mangrove trees extract freshwater from seawater and leave the salt behind. The salt increases the density of the water in the mangrove swamp at high tide, and this water, draining back into the tidal creek at ebb tide, sinks under the less saline creek water.[84] When freshwater inflow is negligible, such as in many mangrove creeks in Australia and in Asia, an inverse estuary circulation may result (Figure 14). This results in salinity increasing in the mangrove creek with increasing distance from the mouth. Such an effect is illustrated in Figure 15, showing the salinity distribution in the hot dry season in Dickson Inlet, a mangrove creek in northern Australia of similar shape and size as Coral Creek. Salinity reached 38 in the creek in the upper reaches. At high tide, the mangrove water of salinity >37.5 was pushed back in the mangrove swamp and had left the creek. This high salinity mangrove water returned to the mangrove creek at the following ebb tide.

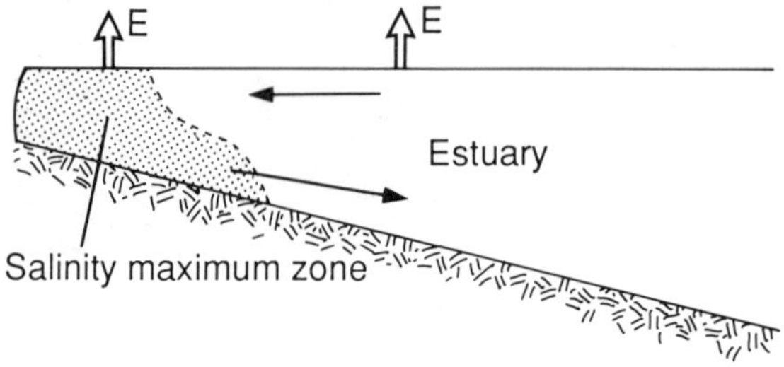

FIGURE 14. Inverse estuarine circulation generated in a mangrove creek by evapotranspiration in the mangrove swamp. (Adapted from Wolanski.[77])

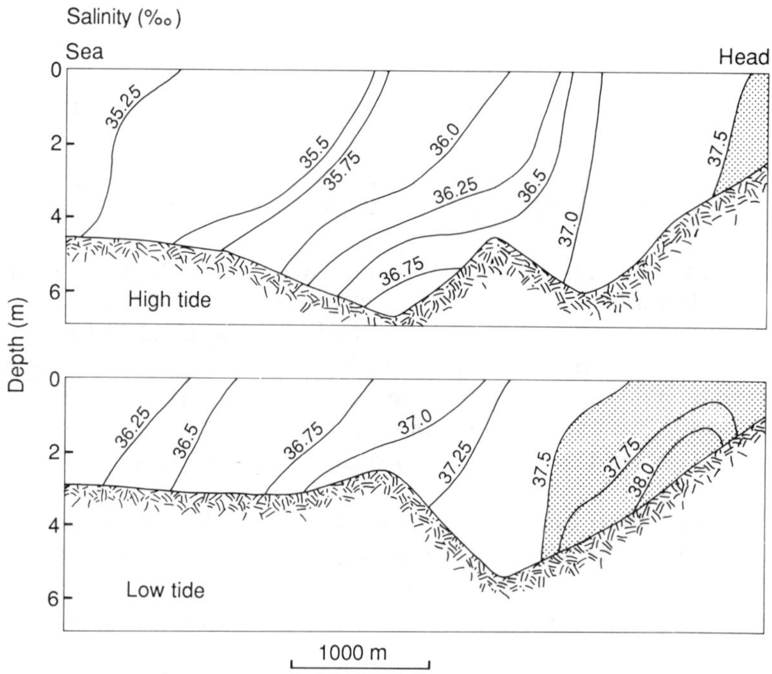

FIGURE 15. Longitudinal distribution at high and low tide of the water salinity in Dickson Inlet, a 5-km-long mangrove-fringed tidal creek, North Queensland, in the hot dry season. The shading indicates high salinity mangrove-trapped water.

This secondary circulation is ecologically important because it traps heavy (more saline) mangrove water, and mangrove detritus, on the bottom of the tidal creek. This can be seen in Figure 15, showing that at low tide the mangrove water of salinity >37.75 was trapped in the bottom layer. Since the mangrove detritus trapped in this layer decays and extracts oxygen from the water and since this bottom layer is not in contact with the atmosphere, being trapped below creek water, anoxic conditions can result near the botom in the upper reaches of mangrove creeks in the dry season, such as Coral Creek. In other mangrove creeks, anoxic conditions are not reached, but dissolved oxygen concentrations in the high salinity zone reach very low values (<1 mg l^{-1}). This is the case of Dickson Inlet in the hot dry season at low tide (Figure 16). Note the high correlation between the distribution at low tide of salinity (Figure 15) and dissolved oxygen concentration (Figure 16), implying that high-salinity, low dissolved-oxygen water was trapped in the mangrove swamp at high

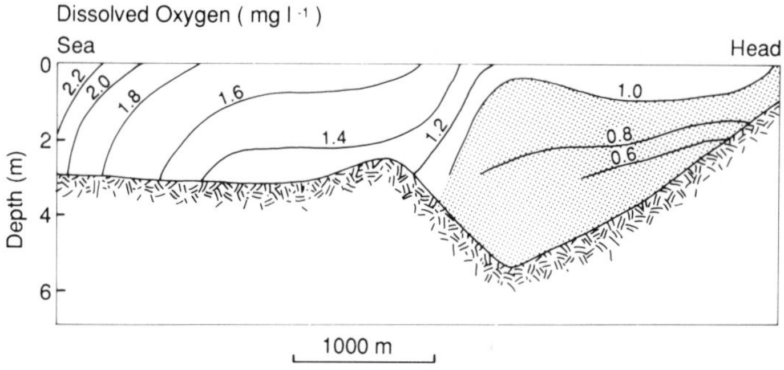

FIGURE 16. Longitudinal distribution of the dissolved oxygen concentration in Dickson Inlet at low tide in the dry season. The shading indicates low concentration values (<1 mg l^{-1}) in the upper reaches near the bottom.

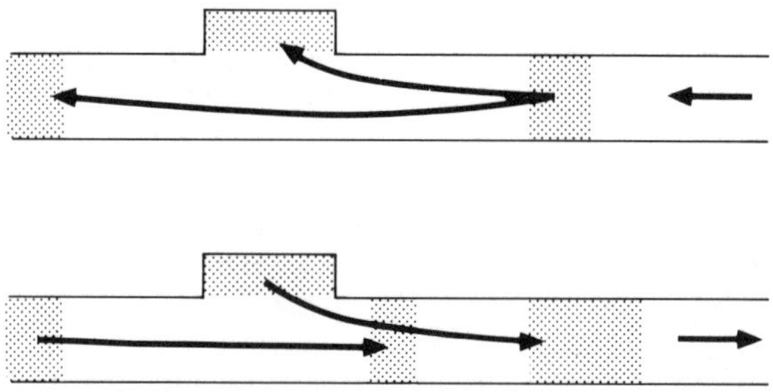

FIGURE 17. Sketch of the lateral trapping process in an embayment. Adapted from Okubo.[86]

tide and at the bottom of the creek at low tide. The ecological system must certainly be stressed during these extended periods of low dissolved-oxygen concentrations.

The Bashito-Minato mangrove-fringed tidal creek in Japan is only 500 m long and is separated from the ocean by a sandy sill. On occasion, storms build up the sill and the creek becomes ponded. In such situations, the waters become anoxic in a day, an indication of the high biological oxygen demand exerted by the mangrove detritus.[85]

The mangroves of Chwaka Bay in Zanzibar are drained through a tidal creek which is nearly ponded at low tide by the presence of a sill made of coral limestone. This inhibits mixing and permits vertical stratification. Long-term strapping results and is enhanced by the presence of a front separating creek waters from oceanic waters.[77]

These results imply that the man-induced disturbances, such as the addition of nutrients in mangrove swamps or the modification of the tidal hydrodynamics of the creek, especially in time of stress such as in the hot dry season, may very severely stress a system by making the water anoxic.

C. FLUSHING OF MANGROVE CREEKS

Tidal flushing of mangrove creeks is an exceedingly complex phenomenon to model accurately because of the role of lateral trapping and of buoyancy. As is sketched in Figure 17, lateral trapping in an embayment, a phenomenon first studied by Okubo,[86] means that water located initially in the tidal creek at low tide is, at flood tide, trapped partially in the embayment while the rest of the water moves further upstream. However, at ebb tide, embayment water returns to the creek ahead of the returning flow. The one body of water shaded in Figure 17 at the beginning of the tidal cycle is broken in two distinct diluted water

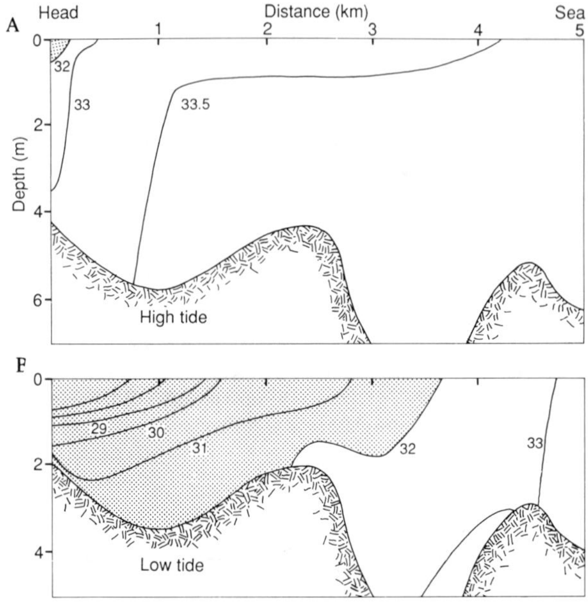

FIGURE 18. Longitudinal distribution of salinity in Dickson Inlet at high and low tide 2 weeks after cessation of runoff. The shading points to low salinity mangrove-trapped water.

masses after one tidal cycle. This reults in enhanced longitudinal diffusion. A mangrove swamp can be seen as an embayment along the whole length of the tidal creek. Hence, lateral trapping in mangrove swamps enhances longitudinal diffusion in mangrove creeks.[73,87]

Such a process enhances tidal flushing of mangrove creeks near its lower reaches. However, this process is inefficient near the upper reaches of the creek in the absence of freshwater runoff.[88] In the upper reaches at rising tide, all the creek water leaves the creek and enters the swamp, being pushed out of the creek by the incoming ocean water. Creek waters and swamp waters are then physically separated and do not mix. At the following ebb tide, the mangrove water stored in the upper regions of the swamp at high tide simply returns to the upper reaches of the creek where little mixing occurs because tidal currents in the upper reaches of the creek are extremely sluggish as, indeed, they vanish at the top of the creek. This mangrove water is not exported to the ocean by tidal currents because the tidal excursion is smaller than the length of the creek. Mathematical simulations of this phenomenon suggest a residence time in the upper regions of the order of 10 d for a typical 5-km-long tidal creek surrounded by a 500-m-wide mangrove swamp on both banks.

In the presence of buoyancy effects, this trapping effect can be even more pronounced. Such was the case, for instance, in Dickson Inlet in February, 1989. Figure 18 shows the longitudinal distribution of salinity in the tidal creek at both high and low tide. Two weeks before the study, freshwater runoff occurred, but had ceased when the study took place. In Figure 18a, at high tide, brackish water, a remnant of the runoff 2 weeks earlier, was trapped in the mangrove swamp at high tide, and only a slight amount of brackish water was present in the uppermost regions of the creek. At low tides, this mangrove-trapped brackish water returned to the creek (Figure 18b) but floated on the surface. Buoyancy effects inhibited vertical mixing and thus increased the residence time of water (and contaminants) in the system to at least 2 weeks' duration.

When freshwater inflow is small but does not vanish, a salinity maximum zone may be established near the mouth of tropical esturaries in the dry season, such as has been observed in northern Australia.[89] As sketched in Figure 19, water upstream from the salinity maximum

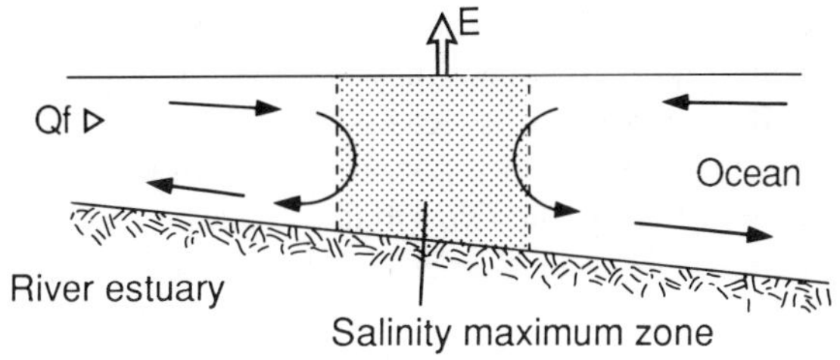

FIGURE 19. Sketch of the internal circulation generated by the salinity maximum zone (shaded). Q_f is the freshwater inflow and E is the evapotranspiration. (Adapted from Wolanski.[89])

zone may be trapped for months and unable to reach the ocean. There is, thus, no flushing at all of these estuarine waters for long periods. This salinity maximum zone can also be present in the mangrove-fringed coastal zone and has been observed in a wide range of tropical areas in the dry season, including India, Mozambique, Thailand, and tropical Australia.

D. GROUNDWATER FLOW

Another important mangrove swamp flushing mechanism is probably groundwater flow. Mazda et al.[90] studied the behavior of groundwater in the mangrove soil between a ponded mangrove lagoon and the open sea. The seepage flow had three components: first, a quasisteady flow toward the ocean; second, a reciprocating tidal flow with exponentially damped amplitude toward the lagoon; third, a residual flow toward the lagoon caused by the exponentially damped tidal flow. When lagoon waters flow underground toward the sea, they entrain with them nutrients and benthic algae. Conversely, when groundwater flows in the lagoon, bottom mixing is enhanced, and the occurrence of near-bottom anoxic conditions can be prevented.

IV. TRAPPING IN THE COASTAL ZONE

Material outwelled from mangrove swamps or discharged from rivers is often trapped along the coast. When land runoff is important, the material is trapped in a river plume, and a sharp front develops separating turbid brackish waters from clearer offshore water. Models for such plumes in shallow coastal waters have been proposed by Murray and Young.[91] These plumes can have a profound influence on the distribution and ecology of fringing and outer reefs such as those in Nicaragua.[47] These plumes can be coastally trapped, mixing slowing with offshore waters, for at least 100 km from the river mouth. The dynamics and the fate of river plumes in some tropical areas may be different from those in temperate areas. In the latter, the freshwater discharge varies slowly in time. In the former, the freshwater discharge can be highly unsteady. For instance, the Burdekin River, a major river in the Great Barrier Reef region, typically flows strongly only a few days a year. While the river flood lasts, Coriolis and buoyancy effects dictate that the river plume stays coastally trapped and moves north with the coast on its left (Southern Hemisphere) against the prevailing southward current on the shelf.[92] However, after cessation of the freshwater runoff, the plume rapidly breaks up in large patches. These patches are common in river plumes with an unsteady discharge or in areas of rugged coastal topography, strong tides, or near the equator, such as the Norwegian coastal current,[93] the Amazon River,[94] the Burdekin River,[92] and the Sepik River in Papua New Guinea.[95] This patchiness may be generated by baroclinic instabilities.[96] In the case of the Burdekin River, once the river discharge ceases, the plume-generated northward movement ceases, patches form in 1 or 2 d, and the patches are advected southward by the prevailing currents. These patches can impinge on the midshelf coral reefs where they lower salinity by typically up to 4% for typically half a day.[92] By comparison, direct rainfall under tropical cyclones also lowers salinity over coral reefs, far from runoff influences, by typically 1 to 2% as was observed during tropical cyclone Winifred on the Great Barrier Reef.

The coastline along the path of the Burdekin River plume is rugged with several capes, headlands, and embayments. As shown above, coastal trapping of river plumes does not occur there since the plume breaks up in patches. There may be a tendency, however, for coastally derived contaminants in the southeast wind season to be trapped in a wide coastal boundary layer, separating coastal waters from reefal waters on the midshelf. In Figure 5, the offshore extent of the coastal boundary layer is the zone of zero velocity separating northward flowing wind-driven water inshore from southward-flowing oceanic-driven water offshore.

By contrast, the Gulf of Carpentaria, located in tropical Australia, has a smooth coastline uninterrupted by headlands and a wide zone of shallow coastal waters. Field measurements suggest that brackish water derived from land runoff can be trapped for several months, after cessation of the freshwater inflow, in a coastal boundary layer in shallow coastal waters.[97] The existence of this barotropic coastal boundary layer is further suggested by the results of numerical experiments on the combined wind- and tide-driven circulation, by analytical modeling, by field measurements of currents, and by remote sensing. The ecological and management implications of such long-term coastal trapping have not yet been explored but may be profound. On physical grounds, one may expect similar long-term trapping phenomena also to occur in other wide and shallow gulfs, such as possibly the Eastern Yellow Sea and the Gulf of Thailand.

Coastal trapping is further enhanced by tidal motions along a shallow, mangrove-fringed coastline. Because the water is shallow, bottom friction effects are dominant and prevent the occurrence of eddies and tidal jets enhancing mixing. Instead, the currents do not flow much across the depth contours. A lateral velocity shear and a barotropic front develop between fast-flowing tidal currents in deeper waters and more slowly moving waters along the coast. Water is exchanged at tidal frequency between the tidal creek-mangrove system and the shallow coastal waters, but little cross-slope mixing occurs across the barotropic front separating offshore and coastal waters.[98] This results in coastal trapping of water and contaminants. Figure 20 shows examples of a numerical simulation, at slack high and low tides in calm weather, of the plume of aquaculture effluents discharged in a mangrove creek facing Hinchinbrook Channel in northern Australia. The channel in this area is very shallow (depth $<$ a few meters at high-tide) on its western side where it drains a vast mangrove swamp. At low tide, a plume of effluent is found in shallow coastal waters. At high tide, the bulk of this effluent returns to the tidal creek and the surrounding mangrove swamp. The effluent is, thus, coastally trapped.

V. SEDIMENT DYNAMICS

It is not possible to review here in detail the numerous studies, most of them undertaken in temperate countries, of sediment dynamics. Details of noncohesive sediment dynamics, such as sand, can be found in the proceedings of the biennial international conferences on coastal engineering. Applications of such principles to coral cays were reviewed by Gourlay.[99] The role of mangrove swamps in the sediment budget of mangrove creeks has been discussed by Wolanski et al.[79]

The principles governing the dynamics of cohesive sediment[100,101] apply also in tropical waters, but the various organic and inorganic kinetic rates determining flocculation processes may be different from those in temperate waters because of the difference in water tem-

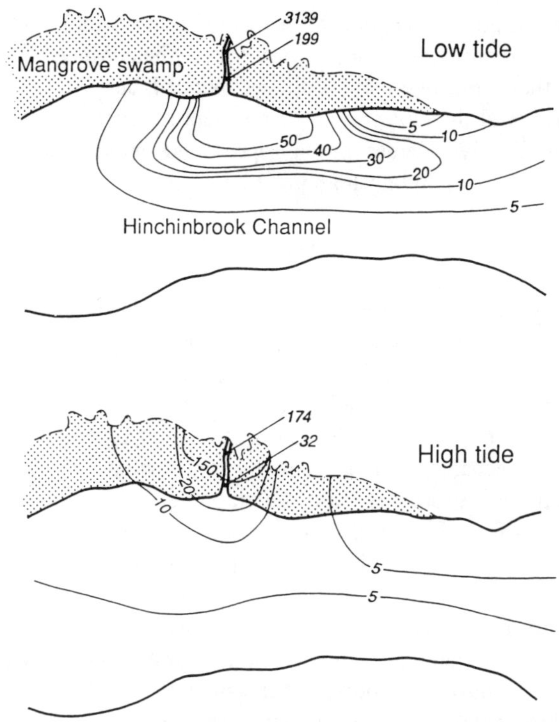

FIGURE 20. Numerically predicted distribution at both slack high and low tides of the plume from an effluent discharge in a mangrove creek in the northern region of Hinchinbrook Channel. Calm weather conditions are assumed. The western coast of the Channel is very shallow and fringed by a vast mangrove swamp. There is no surface water in the swamp at low tide. The numbers indicate concentrations in parts per million. (Adapted from Wolanski et al.[98])

perature. Mud dynamics are important in dredging, spoil dumping, mine tailing disposal, mixing of polluted mud in coastal waters, nutrient exchanges between the sediment and the water column, and water quality of mangrove creeks and turbid coastal waters. Dredging of navigation channels frequently breaks through the noncohesive surface sediment layer and penetrates, hence making it available for erosion and increasing turbidity of coastal waters, the subsurface cohesive sediment that is often found in tropical areas and that results from the previous colonization of the area by mangrove swamps. This can result in this cohesive sediment being eroded and silting previously sandy areas.

One of the most important fluid-mud processes for engineering applications, such as dredging, spoil dumping, stability of mud banks, siltation in harbors and mangrove swamps, and maintenance of a navigable channel, is the formation, under wind or tidal currents, of a two-layer fluid with a thin lutocline separating a fluidized fluid mud layer in the bottom layer from clearer upper waters, even in the absence of any temperature or salinity stratification. In the presence of a fluid-mud layer, surface waves can be attenuated sufficiently by losing energy to internal waves on the lutocline, so that they do not break on the shore.

This two-layer fluid is due both to the nonlinear dependence of the fall velocity of the suspended sediment on the concentration of the suspended sediment and to the sediment affecting both the turbulence and the buoyancy.[102-105]

Sometimes, two lutoclines can be present; one can be located in midwater, with suspended sediment concentration (SSC) above the lutocline, usually <2 g l^{-1}, and SSC below the lutocline, often >8 g l^{-1}; the second lutocline can be located less than a meter from the bottom, with SSC below the lutocline commonly >100 g l^{-1}, a thick slury.

Movements of the fluid-mud layer can result in massive siltation of dredged channels, harbors, and marinas. Trapping and release of nutrients, heavy metals, and other sediment-bound pollutants is then controlled by sediment-induced buoyancy-controlled dynamics. These aspects of fluid-mud dynamics are still not readily amenable to modeling and need to be studied further if only because of their profound engineering implications for the management of turbid areas including mangrove swamps.

VI. CONCLUSIONS

Important hydrodynamic processes of coral reefs include the modificaton of the large-scale circulation of coral reefs, the complex two- and three-dimensional flows around reefs, the process controlling the flushing of coral reef lagoons, tropical cyclones, and upwelling mechanisms near coral reefs. New techniques are needed for modeling the near-reef water circulation and the fate of buoyant particles (such as coral eggs) and pollutants (such as oil).

Important physical mechanisms in mangrove swamps include tidal dynamics, their effects on outwelling and siltation processes, and the trapping of contaminants of in mangrove swamp. Evidence exists for trapping of contaminants in tropical coastal waters, either in a river plume or in a barotropic coastal boundary layer. Coastal trapping is most pronounced in mangrove-fringed coastal waters. The formations of two-layer flows when mud is fluidized has profound engineering and management implications.

Small-scale circulation features control the fate of water and the water-borne contaminants near coral reefs, mangrove swamps, and the shallow turbid coastal zone, all areas that are targeted for development or are already under strong human pressure. Clearly, the shallow water hydrodynamic and diffusive processes in tropical waters are in great need of attention and research.

REFERENCES

1. **Hamner, W. M. and Wolanski, E.,** Hydrodynamic forcing functions and biological processes on coral reefs: a status review, in *Proc. 6th Int. Coral Reef Symp.*, Townsville, Vol. 1, 1988, 103.
2. **Wolanski, E., Ridd, P., and Inoue, M.,** Currents through Torres Strait, *J. Phys. Oceanogr.*, 18, 1535, 1988.
3. **Hamon, B. V.,** Diurnal and semi-diurnal tides on the Great Barrier Reef, Rep. No. 163, CSIRO Mar. Lab. (Australia), 1984.
4. **Flinders, M.,** *A Voyage to Terra Australis,* 2 Vols., G. W. Nicol, London, 1814.
5. **Middleton, J. H., Buchwald, V. T., and Huthnance, J. M.,** The anomalous tides near Broad Sound, *Cont. Shelf Res.*, 3, 359, 1984.
6. **Black, K. P.,** The relationship of reef hydrodynamics to variations in number of planktonic larvae on and around coral reefs, in *Proc. 6th Int. Coral Reef Symp.*, Townsville, Vol. 2, 1988, 125.
7. **Black, K. P. and Gay, S. L.,** Eddy formation in unsteady flows, *J. Geophys. Res.*, 92, 9514, 1987.
8. **Falconer, R. A., Wolanski, E., and Mardapitta-Hadjipandeli, L.,** Modeling tidal circulation in an island's wake, *J. Waterway Port Coastal Eng. ASCE,* 112, 234, 1986.
9. **Geyer, W. and Signell, R. P.,** Measurements of tidal flow around a headland with a shipboard acoustic Doppler current profiler, *J. Geophys. Res.*, in press.
10. **Pingree, R. D. and Maddock, L.,** The tidal physics of headland flows and offshore tidal bank formation, *Mar. Geol.*, 32, 269, 1979.
11. **Wolanski, E.,** Island wakes in shallow waters, *J. Geophys. Res.*, 93, 1335, 1988.
12. **Wolanski, E. and King, B.,** Flushing of Bowden Reef lagoon, Great Barrier Reef, *Estuarine Coastal Shelf Sci.*, in press.
13. **Wolanski, E., Imberger, J., and Heron, M. L.,** Island wakes in shallow coastal waters, *J. Geophys. Res.*, 89, 10553, 1984.
14. **Wolanski, E., Burrage, D., and King, B.,** Trapping and dispersion of coral eggs around Bowden reef, Great Barrier reef, following mass coral spawning, *Cont. Shelf Res.*, 9, 479, 1989.

15. **Wilson, P. R.,** Boundary layer flows past coral reefs, Rep. to the Great Barrier Reef Marine Park Authority, 1985.
16. **Wolanski, E.,** Some evidence for boundary mixing near coral reefs, *Limnol. Oceanogr.*, 32, 735, 1987.
17. **Hamner, W. M. and Hauri, I.,** Effects of island mass: water flow and plankton patterns around a reef in the Great Barrier reef lagoon, Australia, *Limnol. Oceanogr.*, 26, 1084, 1981.
18. **Wolanski, E., Pickard, G. L., and Jupp, D. L.,** River plumes, coral reefs and mixing in the Gulf of Papua and the northern Great Barrier Reef, *Estuarine Coastal Shelf Sci.*, 291, 1984.
19. **Wolanski, E., Drew, E., Abel, K. M., and O'Brien, J.,** Tidal jets, nutrient upwelling and their influence on the productivity of the alga *Halimeda* in the Ribbon Reefs, Great Barrier reef, *Estuarine Coastal Shelf Sci.*, 26, 169, 1988.
20. **Pattiaratchi, C., James, A., and Collins, M.,** Island wakes and headland eddies: a comparison between remotely sensed data and laboratory experiments, *J. Geophys. Res.*, 92, 783, 1986.
21. **Ingram, R. G. and Chu, V. H.,** Flow around islands in Rupert Bay: an investigation of the bottom friction effect, *J. Geophys. Res.*, 92, 14521, 1987.
22. **Tomczak, M.,** Island wakes in deep and shallow water, *J. Geophys. Res.*, 93, 5153, 1988.
23. **Wolanski, E.,** Water circulation in a topographically complex environment in *Physics of Shallow Estuaries and Bays,* J. van de Kreeke, Ed., Springer-Verlag, Berlin, 1986, 154.
24. **Deleersnijder, E., Wolanski, E., and Norro, A.,** Numerical simulation of the three-dimensional tidal circulation in an island's wake, in *Computer and Experiments in Fluid Flow,* Carlomagno, G. M. and Brebbia, C. A., Eds., Computational Publications, Springer-Verlag, Berlin, 1989, 355.
25. **Wolanski, E. and Hamner, W. M.,** Topographically controlled fronts in the ocean and their biological influence, *Science,* 241, 177, 1988.
26. **Alldredge, A. L. and Hamner, W. M.,** Recurring aggregation of zooplankton by a tidal current, *Estuarine Coastal Mar. Sci.*, 10, 31, 1988.
27. **Hamner, W. M. and Hauri, I.,** Fine-scale currents in the Whitsunday Islands, Queensland, Australia: effect of tide and topography, *Austr. J. Mar. Freshw. Res.*, 28, 333, 1977.
28. **Oliver, J. K. and Willis, B. L.,** Coral-spawn slicks in the Great Barrier Reef: preliminary observations, *Mar. Biol.*, 94, 521, 1987.
29. **Signell, R. P. and Geyer, W. R.,** Numerical simulation of tidal dispersion around a coastal headland, in *Residual Currents and Long-Term Transport in Estuaries and Bays,* Cheng, Ed., R. T., Springer-Verlag, Berlin, in press.
30. **Dight, I. J., James, M. K., and Bode, L.,** Models of larval dispersal within the central Great Barrier Reef: patterns of connectivity and their implications for species distributions, in *Proc. 6th Int. Coral Reef Symp.*, Townsville, Vol. 3, 1988, 217.
31. **Provis, D. G. and Lennon, G. W.,** Eddy viscosity and tidal cycles in a shallow sea, *Estuarine Coastal Shelf Sci.*, 16, 351, 1983.
32. **Wolanski, E. and Thomson, R. E.,** Wind-driven circulation on the northern Great Barrier Reef continental shelf in summer, *Estuarine Coastal Shelf Sci.*, 18, 271, 1984.
33. **Brighton, P. W. M.,** Strongly stratified flow past three-dimensional obstacles, *Q. J. R. Met. Soc.*, 104, 289, 1978.
34. **Narimousa, S. and Maxworthy, T.,** Laboratory modeling of coastal upwelling, *Eos,* 67, 2, 1985.
35. **Johannes, R. E.,** *Words of the Lagoon,* University of California Press, Berkeley, 1981.
36. **Hogg, N. G.,** Steady flow past an island with applications to Bermuda, *Geophys. Fluid Dyn.*, 4, 55, 1972.
37. **Hogg, N. G., Katz, E. J., and Sanford, T. B.,** Eddies, islands and mixing, *J. Geophys. Res.*, 83, 2921, 1978.
38. **Wolanski, E.,** Island wakes and internal tides in stratified shelf seas, *Ann. Geophys.*, 4, 425, 1986.
39. **Schumann, E. H., Perrins, L. A., and Hunter, I. T.,** Upwelling along the south coast of the Cape Province, South Africa, *S. Afr. J. Sci.*, 78, 238, 1972.
40. **von Arx, W. S.,** The circulation systems of Bikini and Rongelap lagoons, *Trans. Am. Geophys. Union,* 29, 861, 1948.
41. **Atkinson, M., Smith, S. V., and Stroup, E. D.,** Circulation in Enewetak Atoll lagoon, *Limnol. Oceanogr.*, 26, 1073, 1981.
42. **Sourina, A. and Ricard, M.,** Donnees sur l'hydrologie et la productivite du lagon d'un atoll ferme (Takapoto, iles Tuamotu), *Vie et Milieu,* 26, Ser. B, 243, 1976.
43. **Willis, B. L. and Oliver, J. K.,** Inter-reefal dispersal of coral larvae following the annual mass spawning on the Great Barrier Reef, in *Proc. 6th Int. Coral Reef Symp.*, Townsville, Vol. 2, 1988, 853.
44. **Kinsey, D. W.,** The functional role of back-reef and lagoonal Reef, in *Proc. 5th Int. Coral Reef Congr., Tahiti, Vol. 2,* 1985, 203.
45. **Kingsford, M. J., Wolanski, E., and Choat, J. H. C.,** The influence of tidally induced fronts and Langmuir circulations on the distribution and movements of presettlement fish around a coral reef, *Mar. Biol.*, in press.

46. **Hearn, C. J., Hatcher, B. G., Masini, R. J., McComb, A. J., and Simpson, C. J.,** Oceanographic processes on the Ningaloo coral reef, Western Australia, *Environ. Dynamics Rep.*, ED-86-171, The University of Western Australia, 1986.
47. **Roberts, H. H. and Suhayda, J. N.,** Wave-current interaction on a shallow reef (Nicaragua, Central America), *Coral Reefs,* 1, 209, 1983.
48. **Buddemeier, R. W.,** The geohydrology of Enewetak Atoll islands and reefs, in *Proc. 4th Int. Coral Reef Symp.*, Manila, Vol. 1, 1981, 339.
49. **Buddemeier, R. W. and Oberdorfer, J. A.,** Hydrogeology and hydrodynamics of coral reef pore waters, in *Proc. 6th Int. Coral Reef Symp.*, Townsville, Vol. 2, 1988, 485.
50. **Dill, R. F.,** The blue holes–geologically significant sink holes and caves off British Honduras and Andros, Bahama Islands, in *Proc. 3rd Int. Coral Reef Symp.*, Miami, Vol. 2, 1977, 237.
51. **Oberdofer, J. A. and Buddemeier, R. W.,** Coral reef hydrogeology, *Proc. 5th Int. Coral Reef Congr.*, Tahiti, Vol. 3, 1985, 307.
52. **Rougerie, F. and Wauthy, B.,** The concept of endo-upwelling dans le fonctionnement des atolls-oasis, *Oceanolog. Acta,* 9, 133, 1986.
53. **Rougerie, F. and Wauthy, B.,** The endo-upwelling concept: a new paradigm for solving an old paradox, in *Proc. 6th Int. Coral Reef Symp.*, Townsville, Vol. 3, 1988, 21.
54. **Kjerfve, B., Magill, K. E., Porter, J. W., and Woodley, J. D.,** Hindcasting of hurricane characteristics and observed storm damage on a fringing reef, Jamaica, West Indies, *J. Mar. Res.*, 44, 119, 1986.
55. **Done, T.,** Cyclone Winifred: an overview of geomorphic and ecological effects, in *Proc. Workshop on the Offshore Effects of Cyclone Winifred,* Great Barrier Reef Marine Park Authority, Townsville, 1986.
56. **Young, I. R.,** Wave transformation over coral reefs, *J. Geophys. Res.*, 94, 9779, 1989.
57. **Roberts, H. H., Murray, S. P., and Suhayda, J. N.,** Physical processes in a fringing reef system, *J. Mar. Res.*, 33, 233, 1975.
58. **Andrews, J. C.,** Thermal waves on the Queensland shelf, *Austr. J. Mar. Freshw. Res.*, 34, 81, 1983.
59. **Andrews, J. C. and Gentien, P.,** Upwelling as a source of nutrients for the Great Barrier Reef ecosystems: a solution to Darwin's question?, *Mar. Ecol. Progr. Ser.*, 8, 257, 1982.
60. **Furnas, M. J. and Mitchell, A. W.,** Phytoplankton dynamics in the central Great Barrier Reef. I. Seasonal changes in biomass and community structure and their relation to intrusive activity, *Cont. Shelf Res.*, 6, 363, 1986.
61. **Rougerie, F.,** Le lagon sud-ouest de Nouvelle Caledonie: specificite hydrologique, dynamique et productivite, *Coll. Etudes et Theses,* O.R.S.T.O.M., Paris, 1986.
62. **Thompson, R. O. R. Y. and Golding, T. J.,** Tidally-induced upwelling by the Great Barrier Reef, *J. Geophys. Res.*, 86, 6517, 1981.
63. **Thomson, R. E. and Wolanski, E. J.,** Tidal period upwelling within Raine island Entrance, Great Barrier reef, *J. Mar. Res.*, 42, 787, 1984.
64. **Nof, D. and Middleton, J. H.,** Geostrophic pumping, inflows and upwelling in barrier reefs, *J. Phys. Oceanogr.*, 29, 874, 1989.
65. **Simpson, J. H., Tett, P. B., Argote-Espinoza, M. L., Edwards, A., Jones, K. J., and Savidge, G.,** Mixing and Phytoplankton growth around an island in a stratified sea, *Cont. Shelf Res.*, 1, 15, 1982.
66. **Phillips, O. M., Shuy, J.-H., and Salmun, H.,** An experiment on boundary mixing: mean circulation and transport rates, *J. Fluid Mech.*, 173, 473, 1986.
67. **Sammarco, P. W. and Andrews, J. C.,** Localised dispersal and recruitment in Great Barrier Reef corals: the Helix experiment, *Science,* 239, 1422, 1988.
68. **Boto, K. G. and Bunt, J. S.,** Tidal export of particulate matter from a northern Australian mangrove system, *Estuarine Coastal Shelf Sci.*, 13, 247, 1981.
69. **Boto, K. G. and Wellington, J. T.,** Seasonal variations in concentrations and fluxes of dissolved organic and inorganic materials in a tropical tidal-dominated mangrove waterway, *Mar. Ecol. Prog. Ser.*, 50, 151, 1988.
70. **Gunther, G.,** Studies on estuarine-marine dependency, in *Oceanography: the Past,* Sears, M. and Merriman, D., Eds., Springer-Verlag, New York, 1980, 54.
71. **Macnae, W.,** Mangrove forests and fisheries, FAO/UNDP Indian Ocean Fishery Programme, Indian Ocean Fishery Commission, Publ. IOFC/Dev/74/34, 1974.
72. **Mann, K. H.,** Ecology of coastal waters: a systems approach, University of California Press, Berkeley, 1982.
73. **Ridd, P., Sandstrom, M., and Wolanski, E.,** Outwelling from tropical tidal salt flats, *Estuarine Coastal Shelf Sci.*, 26, 243, 1988.
74. **Williams, D. McB.,** Significance of coastal resources to sailfish and blackmarlin in NE Australia: an ongoing research program, in *Planning the Future of Billfish: Research and Management in the 90's and Beyond. Part 2,* Stroud, R. H., Ed., National Coalition for Marine Conservation, Savannah, Ga, in press.
75. **Somers, I. F.,** Scientific advice and fisheries management in the Australian northern prawn fishery, in Proc. Workshop on Scientific Advice for Fisheries Management: Getting the Message Across, Williams, M., Ed., Australian Dept. Primary Industry, Bureau of Rural Science, 1988.

76. **Vance, D. J., Staples, D. J., and Keer, J.,** Factors affecting year-to-year variation in the catch of banana prawns (Penaeus merguiensis) in the Gulf of Carpentaria, Australia, *J. Conseil, Conseil Int. Explor. Mer,* 42, 83, 1985.
77. **Wolanski, E.,** Measurements and modeling of the water circulation in mangrove swamps, UNESCO-COMARAF, Ser. Doc. No. 3, 1989.
78. **Wolanski, E. and Wattayakorn, G.,** Mixing, trapping and outwelling in the Klong Ngao mangrove swamp, Thailand, UNDP/UNESCO Proc. Reg. Mangroves Proj., RAS/86/120, 1989.
79. **Wolanski, E., Jones, M., and Bunt, J. S.,** Hydrodynamics of a tidal creek-mangrove swamp system, *Austr., J. Mar. Freshw. Res.,* 31, 431, 1980.
80. **Lugo, A. E., Sell, S. M., and Snedaker, S. C.,** Mangrove ecosystem analysis, in *Systems Analysis and Simulation Ecology,* Vol. 4, Patten, B. C., Ed., Academic Press, New York, 1974.
81. **Johannes, R. E.,** The ecological significance of the submarine discharge of groundwater, *Mar. Ecol. Prog. Ser.,* 3, 365, 1980.
82. **Pritchard, D. W.,** Estuarine circulation patterns, *Proc. Am. Soc. Civil Eng.,* 81, 1, 1955.
83. **Kjerfve, B., Stevenson, J. A., Proehl, T. H., Chrznowski, T. H., and Kitchens, W. H.,** Estimation of material fluxes in an estuarine cross-section: a critical analysis of spatial measurement density and errors, *Limnol. Oceanogr.,* 26, 325, 1981.
84. **Wolanski, E.,** Circulation anomalies in tropical Australian estuaries in *Hydrodynamics of Estuaries,* Vol 2, Kjerfve, B. J., Ed., CRC Press, Boca Raton, FL, 1988, 53.
85. **Mazda, Y., Sato, Y., Swamoto, S., Yokochi, H., and Wolanski, E.,** Links between physical, chemical and biological processes in Bashita-minato, a mangrove swamp in Japan, *Estuarine Coastal Shelf Sci.,* 31, 817, 1990.
86. **Okubo, A.,** Effect of shoreline irregularities on streamwise dispersion in estauries and other embayments, *Netherlands J. Sea Res.,* 6, 213, 1973.
87. **Wolanski, E. and Ridd, P.,** Tidal mixing and trapping in mangrove swamps, *Estuarine Coastal Shelf Sci.,* 23, 759, 1986.
88. **Ridd, P. V., Wolanski, E., and Mazda, Y.,** Longitudinal diffusion in mangrove fringed tidal creeks, *Estuarine Coastal Shelf Sci.,* 31, 541, 1990.
89. **Wolanski, E.,** An evaporation-driven salinity maximum zone in Australian tropical estuaries, *Estuarine Coastal Shelf Sci.,* 22, 415, 1986.
90. **Mazda, Y., Yokochi, H., and Sato, Y.,** The behaviour of groundwater in a mangrove area and the influence on the properties of water and bottom mud, *Estuarine Coastal Shelf Sci.,* 31, 621, 1990.
91. **Murray, S. P. and Young, M.,** The nearshore current along a high-rainfall, tradewind coast–Nicaragua, *Estuarine Coastal Shelf Sci.,* 21, 687, 1985.
92. **Wolanski, E. and van Senden, D.,** Mixing of Burdekin River flood waters in the Great Barrier Reef, *Austr. J. Mar. Freshw. Res.,* 34, 49, 1983.
93. **Carstens, T.,** The Norwegian coastal current: on wind mixing and on separation from the coast, in Proc. IUTAM Symp. on Mixing in Stratified Fluids, Environm. Dynamics Rep. No. 85-120, Univ. of Western Australia, 1985.
94. **Bouyne, P. and Roux, J. D.,** Remarque sur la circulation d'eaux turbides d'origine amazonienne le long de la cote des Guyanes, *C.R. Acad. Sci. Paris Ser. D,* 287, 203, 1978.
95. **Morgan, G. A.,** A survey of surface waters in the Bismarck Sea, May 1973, Weapons Res. Establishment, Adelaide, South Australia, Tech. Note-1095, 1974.
96. **Stern, M. E.,** Geostrophic fronts, bores, breaking and blocking waves, *J. Fluid Mech.,* 99, 687, 1980.
97. **Wolanski, E. and Ridd, P.,** Mixing and trapping in Australian tropical coastal waters, in *Residual Currents and Long-Term Transport in Estuaries and Bays,* Cheng, R. T., Ed., Springer-Verlag, Berlin, in press.
98. **Wolanski, E., Mazda, Y., King, B., and Gay, S.,** Dynamics, flushing and trapping in Hinchinbrook Channel, a giant mangrove swamp, Australia, *Estuarine Coastal Shelf Sci.,* in press.
99. **Gourlay, M. R.,** Coral cays: products of wave action and geological processes in a biogenic environment, in *Proc. 6th Int. Coral Reef Symp.,* Townsville, Vol. 2, 1988, 491.
100. **Dronkers, J. and van Leussen, W., Eds.,** *Physical Processes in Estuaries,* Springer-Verlag, Berlin, 1988.
101. **Mehta, A., Anally, W., Hayter, E., Teeter, A., Schoellhamer, D., Heltzel, S., and Carey, W.,** Cohesive sediment transport. I. Process description. II. Applications, *J. Hydraul. Div. ASCE,* 115, 1076, 1989.
102. **Kenney, B. C.,** Sediment resuspension and currents in Lake Manitoba, *J. Great Lakes Res.,* 11, 85, 1985.
103. **McCave, I. N. and Jones, K. P.,** Deposition of ungraded muds from high-density non-turbulent turbidity currents, *Nature (London),* 33, 250, 1988.
104. **Parker, G., Fukushima, Y., and Pantin, H. M.,** Self-accelerating turbidity currents, *J. Fluid Mech.,* 171, 145, 1986.
105. **Wolanski, E., Asaeda, T., and Imberger, J.,** Mixing across a lutocline, *Limnol. Oceanogr.,* 34, 931, 1989.
106. **King, B. and Wolanski, E.,** Coastal dynamics along a rugged coastline, *Proc. 5th Int. Conf. Physics Estuaries Coastal Seas,* 1990.

Chapter 2

THE BEHAVIOR OF NUTRIENTS IN TROPICAL AQUATIC ECOSYSTEMS

Miles J. Furnas

TABLE OF CONTENTS

I. INTRODUCTION

Nutrients are the material currency of energy flow and structural form in biological systems. All essential elements, molecular structures not synthesized endogenously, and useful biochemical energy may be treated as nutrient entities. Such a definition makes any complete description of ecosystem or organismal nutrient dynamics dauntingly complex. For simplicity, nutrient dynamics are most commonly considered on the basis of individual elements or definable classes of compounds.

Biochemical and geochemical nutrient processes in aquatic habitats have been reviewed on a number of occasions.[25,43,62,65,85,218,236] The general literature is largely based on studies carried out in mid- to high-latitude environments. Fundamental molecular and physiological processes of nutrient uptake, metabolism, and storage in tropical aquatic ecosystems are identical to those operating in terrestrial and high latitude aquatic ecosystems. Within this biochemical unity, the nutrient metabolism of specific tropical organisms and quantitative nutrient dynamics of tropical communities or ecosystems have evolved.

This chapter will emphasize aspects of nutrient processes which pertain to considerations of tropical aquatic ecosystems. As the literature on nutrient concentrations, biochemistry, and ecosystem processes is extensive, general references will be made to the above reviews and recent summary articles. Where possible, specific references are made to pertinent studies carried out in the tropics. The references cited represent the biases of the author. The reader is encouraged to use them as portals to a wide and varied literature.

A. GEOGRAPHIC CONSIDERATIONS

Geographically, the tropics lie between the Tropics of Cancer (23.5°N) and Capricorn (23.5°S), the north-south limits of the solar zenith.[223] From a climatological perspective, boundaries encompassing "tropical" conditions are modified by ocean currents, regional meteorology, and topography. The biogeography of tropical ecosystems largely follows bounds set by regional climatology or ocean circulation.

The tropics are not climatologically homogenous, ranging from the wet tropics, where rainfall occurs frequently or continuously throughout the year (e.g., the Papua New Guinea highlands), to the dry tropics, characterized by near-constant drought or low annual rainfall occurring infrequently (the Atacama Desert, Peru). Between these extremes occur regional climatologies characterized by seasonal shifts between dry and wet: "monsoonal climates", driven by large-scale movements of air between the oceans and continents (e.g., southern Asia). Climatology profoundly effects hydrology, ecosystem structure, productivity, and nutrient dynamics in both marine and freshwater habitats. No consideration of tropical nutrient dynamics can be divorced from the hydrological climate or oceanographic regime in which it occurs.

B. NUTRIENT ELEMENTS AND COMPOSITION RATIOS

The elemental composition of organisms, community biomass, and derived detritus reflect organismal and ecosystem nutrient status. The composition of individual organisms varies to greater or lesser degree in response to nutrient availability, environmental conditions, and physiological state. This is most apparent in unicellular algae[66,114] and bacteria,[120] although similar fluctuations can be found in macroalgae[125] and vascular macrophytes.[107,270] The compositition of organisms, however, is not wholly dependent on the nutrient composition of the environment. All species have mechanisms to stabilize their composition at the expense of growth or biomass.

Table 1 lists elements essential for the metabolism of aquatic organisms. The major nutrient elements C, N, P, and S are the primary structural components of biomolecules. Hydrogen and oxygen are, of course, readily obtained from water. Diatoms, sponges, and

TABLE 1
Elements Essential to Aquatic Organisms

Element	Role
Carbon	Structure of all biomolecules, energy metabolism
Nitrogen	Amino acids, nucleic acids, metabolic cofactors
Phosphorus	Nucleic acids, phospholipids, energy metabolism (ATP)
Sulfur	Amino acids (cysteine, methionine), metabolic cofactors (biotin, coenzyme A)
Silicon	Diatom cell walls, grasses
Sodium	Plasma membrane ATPase
Potassium	Catalysis (e.g., pyruvate phosphokinase), enzyme activation
Magnesium	Photosynthesis (heme center in chlorophyll), enzyme activation
Calcium	Catalysis (amylase), mitochondrial electron transport
Iron	Chlorophyll synthesis, electron transport, photosynthesis, nitrogen fixation, peroxidase, catalase, nitrification, denitrification
Boron	Carbohydrate metabolism
Manganese	Catalysis (e.g., phosphotransferases), enzyme activation, nitrification
Zinc	Catalysis (e.g., carbonic anhydrase, carboxy peptidase)
Copper	Electron transport, catalysis (e.g., tyrosinase), denitrification
Cobalt	Vitamin B_{12}
Nickel	Catalysis (e.g., methanogenesis, urease)
Molybdenum	Nitrogen fixation, nitrate reduction, nitrification
Vanadium	Nitrogen fixation (some bacteria)

certain aquatic grasses require silicon for the formation of cell walls and spicules, or lay down silica in stems and leaves. Elements in Table 1 between sodium and calcium largely act as enzyme activators/cofactors or as solutes maintaining cellular turgor and membrane charge balance. The remaining elements, most of which are metals (boron excepted), are usually only present in minute quantities, catalyzing redox reactions as bound or associated cofactors in specific enzymes. Although essential in small amounts, some of these metals (e.g., copper) are toxic at only slightly elevated free ion concentrations.[149]

Major classes of aquatic organisms or organic material entering aquatic ecosystems have characteristic elemental ratios which are useful for scaling ecosystem nutrient fluxes and pool sizes (Table 2). The earliest and perhaps most widely used of these ratios, the "Redfield Ratio"[256] broadly fits the composition of nutrient-replete marine phytoplankton.[118] Similar ratios derived for other plant types growing in or contributing organic matter to aquatic ecosystems differ significantly from the composition of phytoplankton and bacteria.[14,26,113,143] Bacteria have high N and P contents relative to vascular plants and macroalgae.[85,120] Macrophytes are carbon-rich because of their structural polymers (e.g., cellulose, lignin). Atkinson and Smith[14] calculated a mean C:N:P molar ratio of 550:30:1 for a broad range of marine macrophytes. A similar compilation has not been made for freshwater macrophytes, but composition ratios appear to be of similar order. Interestingly, fast-growing, noxious floating macrophytes such as *Salvinia molesta* and *Eichhornia crassipes* may exhibit low (10 to 20) N to P ratios more characteristic of microalgae. Fresh plant litter is usually depleted in both nitrogen and phosphorus relative to ratios in living leaves, but the relative N and P content can increase with time due to microbial colonization of the detritus[268,315] or retention of humic material.[258]

Deviations from elemental ratios characteristic of living organisms are informative in that they point to nutrient limitation,[114] the dilution of specific pools, or act as passive tracers for unbalanced relationships between processes affecting individual nutrient species.[293,298,299] The selection of appropriate ratios for scaling processes is essential.[11] Nutrient ratios in large aquatic plants can vary seasonally,[125] with the degree of colonization by periphyton,[75] or in response to growth[108] and with regeneration following grazing.[360]

TABLE 2
Elemental Ratios (by Atoms) in Aquatic Organisms or Biomass[a]

Organism	C	N	P	Source
Marine phytoplankton	106	16	1	256
Marine bacteria	47	7	1	85
	45	9	1	120
Tropical zooplankton	144	29	1	21
Aloricate ciliates	242—56	16	1	25
Oceanic seston (<200 μm)	122	15	1	187
(<3 μm)	181	20	1	187
(Total, 0—250 m)	120	16		121
(Total, surface)	132	15	1	186
Atoll lagoon seston	490—55	64—9	1	45
Sedimenting detritus	410	29	1	174
Dissolved organics	300—400	19	1	156
Turf algae (prokaryote)	432	36	1	14
Macroalgae (eukaryotic)	696	36	1	14
Seagrasses (leaves)	458	21	1	14
(Roots and rhizomes)	596	12	1	14
Mangrove leaves (live)	1133	29	1	27, 28
Mangrove litterfall	4567	24	1	26
Mangrove creek DOM	700—250	35—18	1	28
Nonmarine emergent macrophytes				
Cyperus papyrus (live)		45	1	107
(papyrus detritus)	4775	577	1	108
Typha domingensis (live)	715	24	1	165
T. domingensis (dead)	1107	4	1	165
Lepironia articulata	1920	16		103
Freshwater nonemergent macrophytes				
Utricularia flexuosa	480	16		103
Rainforest leaves (live)	516	20	1	113
	516—401	38—26	1	353
Rainforest litterfall	678	25	1	113
Floating macrophytes	480	27	1	143
Salvinia molesta (in sewage lagoon)		14—10	1	
Azolla pinnata	329	40	1	44
Paspalum repens	880—416	16		75
Eichhornia crassipes	267—166	16		75
Amazon floodplain lake seston	253	27	1	75
Macrophyte epiphytes	230—85	25—12	1	75

[a] C:N and N:P ratios were calculated relative to N = 16 and P = 1 where data for full C:N:P ratios could not be obtained.

C. NEW AND REGENERATED NUTRIENTS, CYCLES, AND SPIRALS

Dugdale and Goering[68] introduced the useful convention of differentiating "new" and "regenerated" nutrients to the formal bookkeeping of nutrient budgets. Briefly, "regenerated" nutrients are stocks available within the system of interest which are derived from local mineralization processes. "New" nutrients are nutrients added from outside the system of interest to either augment existing stocks or offset losses from the system.[78] For example, a system encompassing a vertical profile through the mixed layer of a tropical ocean, ammonia and amino acid stocks are largely "regenerated" nutrients, while most of the nitrate mixed upward from below the thermocline would be "new". Not all of the nitrate is "new", however, nor all of the ammonium "regenerated". Some nitrate is produced within the euphotic zone by ammonium-oxidizing bacteria.[342] Conversely, "new" ammonium may be added to surface waters by precipitation or to coastal waters by river discharge.

Determining which nutrients are regenerated and new depends greatly on the system under consideration, its size, degree of geochemical closure, and the time scale being considered. Where biological or geochemical transformation rates are rapid relative to system residence times, nutrient cycling (regenerated nutrients) predominate or assume a greater importance. In such systems, consideration of the specific turnover of nutrient pools gives an indication of their activity. Conversely, when water residence times are short relative to time scales of nutrient processes, local dynamics will be dominated by external inputs (new nutrients), regardless of their character. Physical dilution of pools will be an important consideration. Mathematically, dilution and specific turnover can be treated in a similar fashion,[111] but it is important to recognize the differences and consequences.

In nonhomogeneous, advection-dominated aquatic systems (streams, wetlands, coral reefs, estuaries), zones of nutrient input, production, and consumption may be physically separate[141,169,300] or water movement so rapid that products of nutrient transformation processes are displaced before they can be consumed locally: for any fixed small area, all nutrients are "new". To compare such systems, the concept of nutrient "spiraling" can be used (reviewed by Howard-Williams).[141] A spiral, the linearization of a cycle, is the mean linear distance a nutrient atom or molecule would travel downcurrent before being retransformed to its original state (e.g., free nutrient-uptake-mineralization-free nutrient). Spiral length depends on both the rates of nutrient transformation and velocity of water movement. In closed or horizontally homogenous water bodies (e.g., the oceanic surface layer or a pond), the spiral collapses to the more familiar concept of a cycle.

II. NUTRIENT POOLS AND CYCLES IN TROPICAL AQUATIC ECOSYSTEMS

A. CONCENTRATIONS OF MAJOR NUTRIENTS

The generalization is frequently made that nutrient levels in tropical waters are lower than in comparable temperate systems.[131] While this view may be broadly valid in open water marine systems, clear and quantitative demonstrations are confounded by considerable variability within and between individual ecosystems. The perception of low nutrient levels in tropical waters has been shaped by a preponderance of studies carried out in unambiguous, low-nutrient habitats (the open tropical ocean, coral reefs, large pristine tropical rivers) and a general shortage of data on transient events (floods, cyclones) which can dramatically, but briefly, increase nutrient levels in normally oligotrophic tropical systems.[91,99] Many tropical water bodies (e.g., Lake Valencia, Venezuela)[195] are highly polluted by human activities or receive high nutrient loads from surrounding watersheds.[320] Sets of data about such water bodies are difficult to obtain and are of uneven quality.

Relatively few detailed nutrient inventories have been constructed for tropical aquatic ecosystems, particularly those which attempt to resolve temporal and spatial variability of organic nutrient pools.[90,296] Table 3 summarizes dissolved and particulate nutrient concentrations in a number of tropical aquatic ecosystems. Obviously, this list is not exhaustive, but does indicate the relative size of important pools.

The most obvious difference between marine and freshwater systems, as illustrated in Table 3, are the high absolute nutrient concentrations in nonmarine relative to marine systems. These higher concentrations reflect high local loading rates per unit volume or area,[195] the local retention of detrital nutrient stocks,[108] high benthic area-to-volume ratios in shallow nonmarine systems, and reduced exchange rates with low-nutrient sinks (e.g., the open tropical ocean).

Nutrient surveys typically focus on the concentrations of the readily measured dissolved inorganic macronutrients (NH_4^+, NO_2^-, NO_3^-, $PO_4^=$, $Si[OH]_4^=$) taken up directly by macrophytes and phytoplankton. This emphasis obscures the fact that most of the C, N, and P in

TABLE 3
Dissolved and Particulate Nutrient Concentrations in Tropical Aquatic Ecosystems

	NH_4	NO_2—NO_3	DON	PON (µmol/l)	PO_4	DOP	POP	Si	Ref.
Oceanic (near surface)									
Gulf of Mexico	0.0—0.7	0.0—0.1		0.1—0.6	0.0—0.5			0—3	100
Subtropical North Pacific		0.0—0.1		0.2—0.3	0.1—0.1			2—3	79,127
Equatorial Pacific			6.2—13.8						317
Equatorial Atlantic	0.0—0.1	0.0—2.0		−0.3			0.0—0.1		187
South Pacific (Moorea)		−0.1			−0.4			−2	190
Coastal (near surface)									
Campeche Bank	0.0—2.7	0.0—0.3		0.1—5.1	0.0—0.4		0.0—0.1	0—3	100
Gulf of Papua/Torres Strait	0.0—3.2	0.0—3.1	0.9—13.7		0.3—0.6	0.0—0.2		1—30	212
Central GBR (18-20S)	0.0—0.5	0.0—0.5	2.4—14.8	1.0—3.8	0.0—0.3	0.0—0.8	0.0—0.2	0—2	102
Barbados	0.5—2.7	0.4—5.1			0.1—0.2				321
Upwelling									
Peru	0.0—3+	0.0—20+		3.0—14+	0.0—2.5+			0—25	50,231
Arabian Sea		0.0—20+			0.0—2.0+			0—16	50
Estuary (near surface)									
Cochin Backwater (India)		0.0—20.3		0.0—150.0					180
Missionary Bay (Australia)	0.1—0.4	2.0—7.0		0.1—0.4	0.2—0.6				28
Coral reefs, Oceanic and atolls									
Canton Atoll	0.1—1.3	0.0—2.4			0.0—0.5			2—3	292,294, 295
Enewetak Atoll	0.2—0.3	0.1—0.3	1.7—2.3		−0.2	−0.2			228,345, 346
Tonga Lagoon	0.1—0.7	0.1—1.0	1—23.0	3.0—10.0	0.1—0.9			17—91	358
Gilbert Islands	0.3—0.5	0.0—2.6	3.8—5.6		0.0—0.4				162
Takapoto Atoll	−0.1	−0.2			−0.1			−0	301
Moorea		−0.1			−0.5			−2	190
Coral reefs, Shelf and fringing									
Kaneohe Bay	0.4—2.4	0.1—2.6	3.4—7.5		0.2—1.0	−0.4			296
Jamaica	0.1—3.8				0.0—0.7				336
Lizard Island (GBR)	0.1—0.2	0.2—1.0	3—5.0		0.2—0.4			1—2	18
Davies and Old Reef lagoons (GBR)	−0.2	−0.4			−0.2			−1	102

Abrolhos Islands	0.1—11.0	0.8—5.2			0.2—2.9	0.0—4.9		1—7	56
Rivers									
Amazon		−8.5			−0.2			−128	73
Ganges	0.7—31.4	35.7—770.0			1.9—652.0			21—2869	180
Maroni (S. America)		5.0—9.0			−0.2			−191	326
Niger	−1.0	−7.6						−250	326
Bermejo (S. America)	0.8—35.7	5.5—52.1			0.8—3.3	0.1—0.8	1.5—230.0		242
Orinoco					−0.2			−91	326
Zaire	−0.4	6.7—7.0			0.1—0.8			165—342	210,326
Streams									
Savannah Rivers (Uganda)	<0.4—3.9	0.0—61.0			1.1—30.3			87—603	332
Wet forest streams (Uganda)	<0.4—9.7	0.0—171.0			0.1—8.2			118—825	332
Lesser Antilles	<2.8—62.8	<8.1—179.0			0.1—1.5			42—533	303
Lakes									
Lake Waigani (PNG)	−4.4	−21.6	−723.0		−42.6	−22.0			
Lake Victoria (Uganda)	−28.6	−0.4			−0.3				107
Lake Nabugado (Kenya)	−<7				−21.0				279
Lake Kyoga (Kenya)	−14.0	14.0			−16.8			−326	279
Lake Chilwa (Malawi)		2180.0—5114.0			0.0—229.0				165
Lagartijo Reservoir		−4.3			−0.3			−478	279
Lakes Robertson/Mcllwaine (Zimbabwe)	0.4—8.8	1.3—28.2			0.0—11.8				344
Lake Calado (Brazil)	0.0—1.4	0.0—3.2			0.1—0.3				280
Lake La Plata (Puerto Rico)	0.0—3.5	0.0—53.0	−46.0		1.5—6.8				320
Swamps and Wetlands									
Kawaga Swamp (Uganda)	−8.6	−0.3			−0.9				107
North Swamp (Kenya)	−3.2	−4.6	−63.0	−871.0		−1.9	−30.6		107
Tasek Bara (Malaysia)	0.0—54.7	0.7—20.7	4—109.0		0.0—3.4	1.1—25.2		−52	103
Papyrus swamps (Uganda)	−<0.4	0.0—30.3			0.2—1.3			34—558	332

Note: 0.0 = nondetectable.

both marine and freshwater aquatic systems are usually sequestered in organic forms,[108,156,312] particularly in the benthos.[296] Total dissolved organic nitrogen (DON) and dissolved organic phosphorus (DOP) concentrations are not infrequently several times dissolved inorganic concentrations,[90,210] particularly in marine waters.[157] Concentrations of free amino acids, which are taken up by both algae and bacteria, are generally in the nanomolar range in seawaters,[85,97,188] but particulate amino acid concentrations may be many times this level.[153]

Analytical methodology and sampling practices strongly color our perceptions of nutrient pool size, variability, and, through them, ecosystem dynamics. Routine colorimetric methods for inorganic nutrients[123,306] have normal working sensitivities and precisions on the order of 0.1 μM,[61] although with care,[251] sensitivity on the order of nanomoles per liter can be achieved. Concentrations of this order have become a baseline on which cycling rates in oligotrophic waters have been estimated.[79,111] Recent advances in instrumentation and extraction procedures have lowered limits for detection of a range of inorganic and organic nutrient species well into the nanomolar range,[34,105,215] but such methods are often tedious. Such techniques now clearly show that open ocean concentrations of NO_2^- and NO_3^- are in the nanomolar range.[80] Conversely, new instrumental techniques for dissolved organic carbon (DOC) and DON[310,312] suggest that organic nutrient concentrations in oceanic surface waters may be several times values determined by digestion methods, although the technique is still controversial.[341] Fluxes into and out of dissolved organic pools have emerged as an important determinant of ecosystem-scale nutrient availability.[155,156] The less precise indirect methods for analysis of total organic pool concentrations and activity[15] constrain progress in this area. The ramifications of high organic nutrient concentrations have yet to be examined in nonmarine ecosystems where dissolved organic concentrations are generally higher than reported for marine waters.[153]

B. NUTRIENT BEHAVIOR IN TROPICAL WATERS

1. Carbon

As carbon is the major chemical constituent of all organic molecules, carbon dynamics occupy the central place in models of aquatic nutrient cycling. Ecosystem fluxes of other nutrients are, to a large extent, scaled to carbon flows and transformations. Figure 1 schematically illustrates carbon pools and flows in the water column of an aquatic ecosystem. This schematization greatly simplifies the vast diversity, magnitude, and rates of organic reactions occurring within any one water body.[85] All of the processes shown also occur, sometimes preferentially, within sediment pore waters (anoxic or otherwise)[41] and microzones within particulate matter.[2,235] In sediments, zooplankton would be functionally replaced by metazoan meiofauna and protozoans, while the role of phytoplankton is taken by macrophytes and autotrophic benthic microflora.

Dissolved inorganic carbon (DIC) is the most abundant form of carbon (about 2 mM) in seawater and largely invariant in concentration. Equilibrium reactions between CO_2, HCO_3^{3-}, $CO_3^{=}$ maintain seawater pH values within a relatively narrow range (usually 7.8 to 8.2). Inorganic carbon concentrations and pH values vary over far wider ranges in nonmarine waters, varying from pH $>$10 (Lake Magadi)[314] to 4 or less (Amazon blackwater rivers).[162] While inorganic carbon is likely not a major limiting nutrient in most tropical aquatic systems, free CO_2 concentrations may be low (pH $<$5). Hydrogen ion activity (pH), which is strongly affected by the carbonate buffering system and concentrations of dissolved organic matter, directly[35,322] or indirectly[284] influences the speciation and solubility of trace metals. Relatively small diel fluctuations in seawater DIC concentrations associated with biological activity can be measured,[163,291] but the detection and interpretation of such fluctuations continue to challenge analytical technology and theoretical insight.

DOC and particulate organic carbon (POC) are arbitrarily defined size classes of organic materials on a continuum of organic structural elaboration ranging from methane (CH_4) to

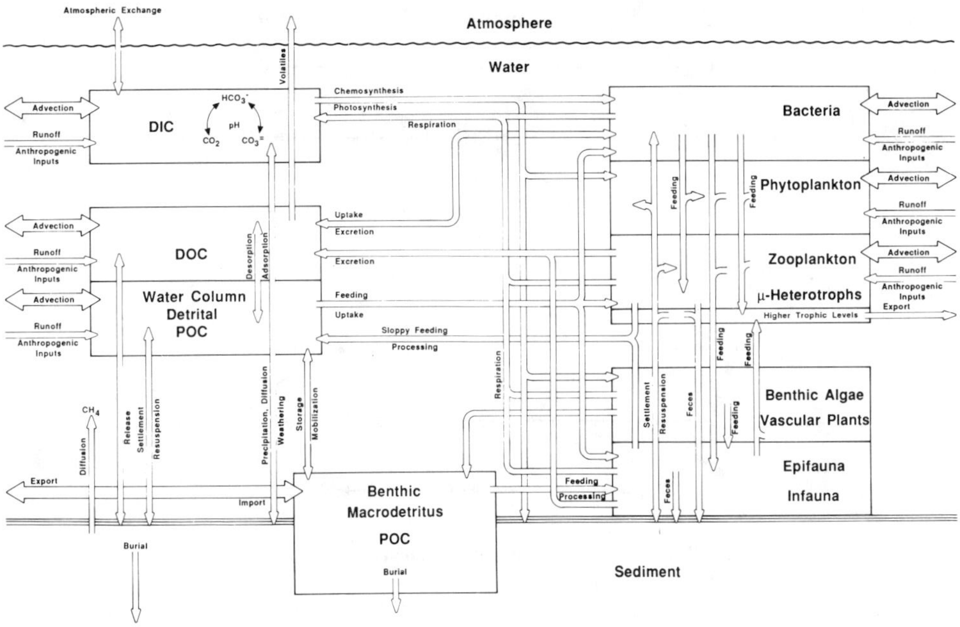

FIGURE 1. Schematic model of carbon pools and fluxes in aquatic systems.

logs. DOC concentrations exceed those of POC in almost all water bodies. There is a clear inverse relationship between particle size and "particulate" carbon concentration. Most organic carbon is present as high molecular weight dissolved or colloidal material.[282,310] Operationally, DOC is taken as the organic carbon which passes through a filter (e.g., 0.2 to 0.5-μm pore diameter) and POC as that retained. Most appropriately, DOC and POC should be differentiated as to those forms which can and cannot be directly transported or diffuse across a cellular membrane.

With the exception of lotic systems directly receiving large amounts of terrestrial detritus, photosynthesis by aquatic macrophytes, macroalgae, and phytoplankton is usually the predominant source of organic carbon in aquatic systems. Chemolithotrophic nitrifying bacteria[24,168] fix some inorganic carbon, but this is usually a minor input. Primary production within tropical aquatic ecosystems ranges from <0.2 g C m^{-2} d^{-1} in the open oligotrophic ocean (although this is open to some question) to about 25 g C m^{-2} d^{-1} in some tropical swamps[64] or floating macrophyte mats.[255] Primary production in tropical lakes (1.7 to 3.2 g C m^{-2} d^{-1}; summarized by Nixon)[226] generally matches the productivity of marine shelf ecosystems (0.5 to 6.3 g C m^{-2} d^{-1}).[48] High gross primary production rates also occur on coral reef flats (modal values 5 to 10 g C m^{-2} d^{-1})[172] and seagrass beds (1 to 18 g C m^{-2} d^{-1}).[305]

Organic materials are ultimately susceptible to degradation and mineralization; however, rates of carbon turnover in aquatic environments vary widely.[98,202,268,281,307] The breakdown of particulate materials and large dissolved organic molecules to low molecular weight products is largely carried out by bacteria and fungi.[84] The process is often accelerated through the mechanical disruption of the particulate matter by detritivores feeding on the resident microbial community.[266,315] The relatively low C to N and C to P ratios of microbes relative to detrital material (Table 2) and dissolved organic matter,[156] particularly that coming from macrophytes and higher plants[14] provide the impetus for processing the detritus. Low concentrations of readily assimilated organic carbon in oceanic waters may limit the cycling and uptake of inorganic N and P by bacteria.[173] Although vascular macrophytes and algae

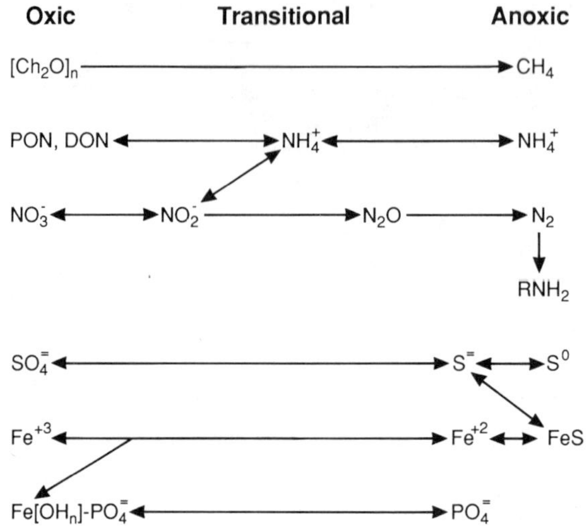

FIGURE 2. Changes in nutrient speciation occurring across oxic-anoxic gradients in aquatic habitats.

have roughly similar compositional ratios, degradation rates *in situ* and the functional role of algal and vascular plant detritus can differ greatly.[202,281,351] Mann[202] put forward the view that the precipitation and processing of dissolved (or colloidal) organic matter in aquatic ecosystems may be quantitatively more important than the processing of the more obvious particulate detritus.

Pathways and rates of carbon utilization are strongly dependent upon the presence or absence of oxygen.[41,85] In oxygenated waters and sediments, decomposition proceeds rapidly from the leaching of soluble minerals[107,142] to the complete oxidation of low molecular weight organic acids and carbohydrates in oxidative catabolic pathways.[85]

Where respiratory demand for oxygen exceeds local photosynthetic, diffusive, or advective inputs of oxygen (e.g., most sediments, lake hypolimnions, mats of detritus), anoxia develops and degradation of carbon compounds slows greatly, with fermentive and dissimilatory pathways assuming relatively greater importance,[85] leading to the production of organic acids or methane.[41] The balance between oxic-anoxic conditions not only affects carbon metabolism, but the speciation and availability of major nutrient elements (Figure 2). The development and maintenance of anoxia is facilitated by continuous waterlogging of sediments and detritus. Beds of peat are often found underlying mangrove forests[263] and continuously filled swamps.[107,152] In contrast, near-complete breakdown of detrital organic matter appears to occur in lakes and swamps subject to annual or frequent drying (e.g., Lake Chilwa,[139] Amazon floodplains).[164] In such systems, seasonal accumulations of detritus are mineralized when the detritus dries, is reoxygenated, and becomes accessible to terrestrial as well as aquatic detritivores.

Oxidative metabolism is continued within anoxic habitats by specialized bacterial assemblages using alternate electron acceptors ($SO_4^=$, NO_3^-, NO_2^-, N_2O, Fe^{+3}, Mn^{+4}, CO_2).[85] Because of the high C to N or C to S ratios of algae and vascular macrophytes, concentrations of alternative electron acceptors are rarely sufficient to complete the oxidation of the carbon present. Smith et al.[299] suggested that over sufficiently long time periods, stoichiometric imbalances between low C to N ratios in decomposers and high C to N ratios in organic matter may lead to systems-level nitrogen limitation as nitrate is consumed to sustain carbon oxidation. Although complicated by internal N-fixation processes (see below). N limitation of this nature would be expected in tropical wetland systems with high carbon loading from vascular macrophyte production.[164,334]

2. Nitrogen

Nitrogen is the second most abundant nutrient element and, like carbon, exists in a wide range of organic forms although most nitrogen-containing molecules (proteins, nucleic acids) in living things tend to be constructed from a narrow range of organic building blocks (e.g., amino acids, nucleotides). Unlike inorganic carbon, transformations between inorganic nitrogen species occur almost exclusively through biological processes, with physical equilibria assuming a minor role.

Ammonium (NH_4^+) and nitrate (NO_3^-) are the principal sources of fixed nitrogen for aquatic plants and bacteria. Nitrite (NO_2^-), urea [$CO(NH_2)_2$] and free amino acids ($R\text{-}NH_2$) can also be directly taken up, but concentrations of these nitrogen species are usually considerably lower than those of ammonium and/or nitrate in most unpolluted aquatic systems. Marine and freshwater microalgae collectively exhibit the capacity to utilize a broad range of nitrogenous compounds.[240]

Because of the energy requirements to reduce nitrate to ammonia prior to assimilation ($6\ e^-/N$), ammonium is usually the preferred nitrogen source.[207,208] Ammonia, once taken up, may be directly assimilated into organic molecules via several pathways.[82] Nitrate uptake and reduction by marine phytoplankton appears to be suppressed at ammonium concentrations $>1\ \mu M$.[208,231] Macroalgae may exhibit more diversity in this regard.[125] The ATP and NADPH required for nitrate reduction in phytoplankton are usually taken from photosynthetic pathways,[82] whereas nonphotosynthetic bacteria use energy from catabolic pathways. Nitrate uptake and assimilation by autotrophs are light dependent, with nitrate uptake and assimilation being suppressed, although not fully near the base of the euphotic zone[186,198] or at night.[197] In cases of severe nitrogen limitation, however, nitrate uptake may continue throughout the diel cycle.[201] Ammonium utilization is much less affected by ambient light intensity per se or time of day, but may be coupled to diel fluctuations in growth.[186,197]

Limitation of nitrate uptake may not be an important determinant of microalgal growth in natural aquatic ecosystems.[318] Potential uptake rates for nitrogenous substrates by N-starved plants can be an order of magnitude greater than growth rates.[117] As a result, high relative growth rates of most algae can readily be sustained at sharply attenuated nitrate uptake and reduction rates.

Urea and amino acids are also preferred N sources for marine algae. Although concentrations of dissolved free amino acids are usually $\leqslant 1\ \mu mol\ l^{-1}$ in tropical marine waters,[86,97,188] tracer experiments indicate dissolved pools may turn over on time scales of hours.[97,98] Concentrations of urea span a wide range (about $0.1\ \mu mol\ N\ l^{-1}$ to several micromoles N per liter.[79,337] Where urea is available, uptake of urea-N is equivalent to that of ammonia, but greater than nitrate.[79,129] There are fewer data on relative nitrogen preferences and the utilization of organic N species by tropical nonmarine aquatic plants.

Marine phytoplankton, macroalgae, and vascular macrophytes typically exhibit high affinities (i.e., low Michaelis-Menten K_s values) for inorganic nitrogen species in uptake experiments. Estimated K_s values range from <0.1 to several micromoles N per liter.[117,125] Some macroalgae may have multiple nitrogen uptake systems operating over a wide range of concentrations.[94] Similar kinetic estimates for nonmarine vascular macrophytes and phytoplankton are scarce. The complex geometry and anatomy of macrophytes and close associations with assemblages of periphyton and attached sediment complicate measurement and interpretation of instantaneous uptake rates. While connections between low ambient dissolved inorganic nitrogen (DIN) concentrations and low K_s values may seem obvious from an evolutionary standpoint, extrapolation of experimentally derived uptake kinetics to behavior over diurnal or diel time scales is difficult. Nitrogen (or any nutrient, for that matter) uptake kinetics are now known to be strongly dependent on the nutrient, prehistory of cells,[205] ambient conditions (temperature, irradiance), the experimental method selected, and the time period over which the uptake is measured.[119] Under most ecological scenarios, growth and measured nutrient acquisition are not likely to be tightly coupled over daily or longer time periods. Aquatic plants can take up and store sufficient nutrient for growth within short time intervals,[205,255] and half-saturation values for nutrient limited growth can be up to tenfold lower than K_s values for uptake,[116,117] allowing high relative growth rates even though uptake per se is not saturated.

Even when scaled with appropriate stoichiometric ratios (Table 2), the high productivity of many tropical acquatic ecosystems requires inputs of nitrogen in excess of that available from dissolved pools, advective inputs, and local remineralization. This demand appears to be met, in part, by the widespread fixation of atmospheric di-nitrogen (N_2) by free-living and symbiotic bacteria. Nitrogen fixation has been documented (Table 4) in all types of tropical aquatic habitats (coral reefs,[345,347] sediments,[39] wetlands,[108] seagrass beds,[40,112] floating macrophyte mats,[270] rice paddies,[65,261] the oceanic pelagial.[42].

Nitrogenase, the principal enzyme involved in nitrogen fixation, is poisoned by molecular oxygen, requiring that fixation reactions take place under anaerobic conditions. These conditions are met by isolating the nitrogenase within specialized cells (heterocysts), by the formation of low-oxygen microzones in colonies of cells,[232] by symbiotic associations,[343] through growth in low-oxygen microhabitats,[55,233] or by the temporal decoupling of nitrogen fixation from photosynthesis.[213]

N-fixation rates in tropical aquatic communities rarely constitute more than a small fraction of N inputs to particular systems or N demand scaled to primary production.[146] Direct benefits of fixation accrue solely to the fixing organisms, pure, undisturbed assemblages of which are not frequently used for *in situ* experiments. The fixed nitrogen may be exported,[345,347] accumulate in detrital pools,[106] or be denitrified elsewhere in the system.[276] While downstream benefits of N fixation provide a useful nitrogen supplement in many habitats, important organisms or constituent components in N-fixing communities may remain N limited.[220,352]

Nitrogen-fixing organisms or assemblages are not distributed uniformly in either freshwater or marine habitats. Coral reefs, seagrass beds, and freshwater swamps, in particular, have been identified as sites of high N-fixation rates (Table 4). Within the context of coral reefs, the highest benthic N-fixation rates are associated with the turf-forming cyanobacteria, *Calothrix crustacea* and *Scytonema hotmannii*[181,345] and are inversely related to the degree of disturbance of the community,[181] with the possible exception of ''farmed'' fish territories[349] (but see Larkum et al.[181]).

The oxidation of ammonium to nitrate (nitrification) and dissimilatory reduction of nitrate to nitrogen gas (denitrification) are carried out by specialized assemblages of bacteria. Comprehensive reviews of nitrification and denitrification in marine and freshwaters have recently been presented.[25,85,167,276] Despite the considerable and growing body of literature on these processes in aquatic ecosystems, relatively few direct measurements have been made in tropical ecosystems (Table 5).

Nitrification occurs aerobically as a two-step process mediated by different assemblages of bacteria. Ammonium is first oxidized to nitrite (by *Nitrosococcus, Nitrosomonas*), which is released and subsequently oxidized to nitrate (*Nitrobacter, Nitrococcus*).[85] Nitrite may also be released by phytoplankton during nitrate reduction.[171,324] Nitrification rates are affected by a range of environmental factors. Nitrifying bacteria appear to have high optimum temperatures for growth and nitrification (>30°C), suiting them well for tropical ecosystems.[10,85,133] Nitrification rates are dependent on concentrations of ammonium, nitrite, and oxygen (summarized in Kaplan).[167] Nitrifying bacteria have a high affinity for oxygen and appear able to maintain high relative nitrification rates ($v/v_{max} > 0.5$) at oxygen concentrations above 31 μmol l^{-1},[236] In contrast, measured Michaelis constants (K_s) for ammonium (71 to 714 μ*M*) and nitrite (360 to 600 μ*M*)[85] suggest that nitrifying bacteria could be moderately substrate limited in many tropical marine sediments and severely limited in almost all near-

TABLE 4
Nitrogen Fixation Rates in Tropical Aquatic Systems (μmol N m^{-2} d^{-1} unless specified)

Location	Rate	Ref.
Marine		
Open water		
Trichodesmium	0.4—3.9	146
Rhizosolenia mats	0.8—17.6	146
Sediment		
Kaneohe Bay (Hawaii)	600—2400	127
Barbados (seagrass)	30—60	239
Tampa Bay (Florida)	6—508	363
PNG mangroves	r—96	333
Coral Reefs		
Enewetak	8100	345
	391—7822	200
Aldabra Atoll	2347	249
Great Barrier Reef	258	350
Heron Island (GBR)	665—1330	181
Puerto Rico	9—86	51
Macrophyte		
Thalassia testudinum	0.2—0.6 μmol N g^{-1} d^{-1}	239
Epiphytes	9000	112
Syringodium filiforme	0.04—0.4 μmol N g^{-1} d^{-1}	239
Diplanthera wrightii	0.07—0.1 μmol N g^{-1} d^{-1}	239
Microdictyon spp. macroalgae	63—96 μmol N g^{-1} d^{-1}	40
	7—46 μmol N g^{-1} d^{-1}	40
Nonmarine		
Open water		
Lake Naivasha (Kenya)	0.6—1.7	334
Lake Sonachi (Kenya)	0.4—5.1	334
Lake George (Uganda)	860	138
Lake Valencia (Venezuala)	254	193
Benthic		
Lake Waiau (Hawaii)	2.4	183
Wetlands		
North Swamp (Kenya)	11088[a]	108
Papyrus marsh (Kenya)	2	334
Salvinia molesta mat	428	270
Rice Paddies		
S.E. Asia	70—333	343
Ivory Coast	3519—4610	261

[a] Areal rate estimated from N budget considerations.

surface marine waters (Table 2). High concentrations of oxygen and ammonium do not often co-occur in tropical ecosystems. Maximal rates of both water column and sediment nitrification, therefore, usually take place in gradients between oxic/anoxic conditions or high/low ammonium concentrations.[7,157] Because *in situ* ammonium concentrations are usually low, nitrification rates respond to pronounced chronic or episodic inputs of ammonium or organic nitrogen.[22] Coupling between N loading and nitrification has been inferred for a number of estuarine or shelf systems.[99,196,206]

Pelagic ammonium oxidizing bacteria appear to be inhibited by high light levels.[229] This is not likely a problem in sediments, but in clear oceanic waters, maximal nitrification rates

TABLE 5
Nitrification and Denitrification Rates Measured in Tropical and Subtropical Aquatic Habitats (μmol N m^{-2} d^{-1}, unless specified)

Location	Rate	Ref.
Nitrification		
Marine		
Puerto Rico reefs	6424	51
Enewetak reef flat	328	346
Dictyosphaeria spp. Kaneohe Bay	984—4900	296
Caribbean Reef sponge communities		
Chondrilla nucula	11,500	52
Anthosigmella varians	48	52
Nonmarine		
Rainforest soil (Brazil)	0.01—0.05 μmol N kg^{-1} d^{-1}	214
(Costa Rica)	10—430 μmol N kg^{-1} d^{-1}	269
Denitrification		
Marine Pelagic		
E. tropical Pacific	292—14,500	167
E. tropical Atlantic	97,800	167
Arabian Sea	301—3010	167
Marine Benthic		
Four-League Bay (LA)	48—1776	287
Ochackonee Bay (FL)	0—5040	275
Puerto Rico reef sediments	1213—2427	51
Bahamas	1200	277
Nonmarine		
Lake Okeechobee (FL)	48—240	209
Lake Naivasha (Kenya)	107—1242	334
Lake Sonachi (Kenya)	29—36	334
Rice (LA–potential)	8	254
Wetland plants (LA)	7—9	254

occur near the base of the euphotic zone,[342] and net accumulations of nitrite may occur during the day.[95]

There is good evidence that nitrification in benthic systems occurs preferentially within specialized microhabitats. Corredor et al.[52] measured net releases of nitrate/nitrite from Caribbean reef sponges. Sponge flagellates mineralized particulate organic nitrogen (PON) and DON captured from the water column which was subsequently converted to nitrate by resident populations of nitrifying bacteria in the sponge matrix. Risk and Muller[262] found elevated nitrate concentrations within the bioeroded framework of coral heads. A number of studies have demonstrated coupled nitrification/denitrification within the rhizosphere of seagrasses, rice, and other wetland plants.[254] Spatial separations between conditions suitable for nitrification and denitrification in consolidated sediments may be on the order of 100 μM.[157] The co-occurrence of both nitrification and denitrification processes in coral reef carbonate sediments[51] argues for considerable spatial heterogeneity within the sediment. Irrigation of sediments by infauna[3,124] can expand the volumetric extent of oxygenated microhabitats in which nitrification can occur.

External inputs of nitrate to tropical systems and the activities of nitrifying bacteria rarely lead to the accumulation of persistent high concentrations of nitrate in open waters

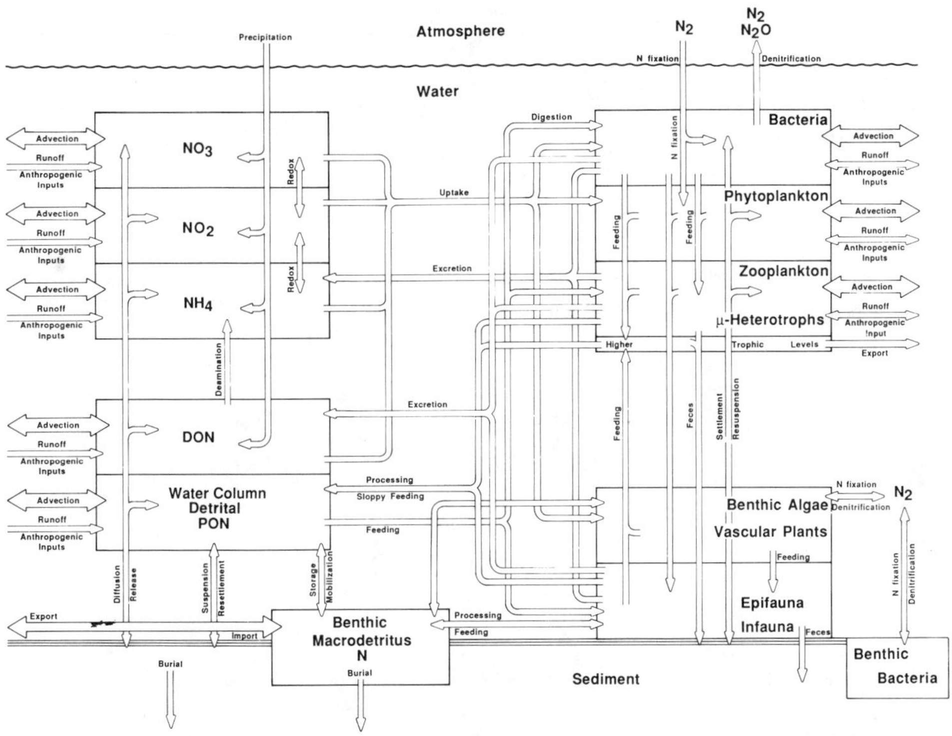

FIGURE 3. Schematic model of nitrogen pools and fluxes in aquatic systems.

or sediment porewaters, even where ammonium is relatively abundant. Nitrate not taken up directly by aquatic plants is denitrified in anaerobic sediments and microzones. Denitrification is most intense in waterlogged soils,[252,254] anaerobic sediments, and accumulations of detritus.[334] While denitrification is known or presumed to occur in most tropical aquatic ecosystems, few system-level estimates of denitrification rates have been made for nonagricultural systems.[276] The most extensive series of denitrification estimates are for the eastern tropical ocean margins, where between 0.01 and 230 $\times$ 10^6 ton N per year are estimated to be removed from the oceans.[167] Denitrification in temperate and subtropical estuarine sediments may remove up to one third of exogenous inputs.[275] Mesocosm studies show that denitrification increases with the rate of N loading,[278] but temperature dependencies are unclear.[276]

The implication for tropical systems is that denitrification will be most intense where detritus accumulates, in water bodies subject to enhanced nutrient loading from pollution,[296] water bodies with long residence times,[290] and in wetland ecosystems subject to periodic drying,[164,165,252] where oxygen inputs during drying periods stimulate coupled mineralization-nitrification-denitrification within organically rich sediments.

3. Phosphorus

The movement and storage of phosphorus in aquatic systems (Figure 4) is dominated by exchanges between dissolved pools (PO_4, DOP), biomass, and mineral or mineral-bound phases on suspended particles and in sediments. Froelich[96] provides a recent review of aquatic phosphorus geochemistry. Ortho-monophosphate and many organic phosphorus species are reasonably soluble in both oxic and anoxic fresh- and saltwaters. There is abundant evidence that the partitioning of dissolved P into inorganic and organic P pools on the basis of differences between molybdate-reactive phosphate and total P does not fully describe the

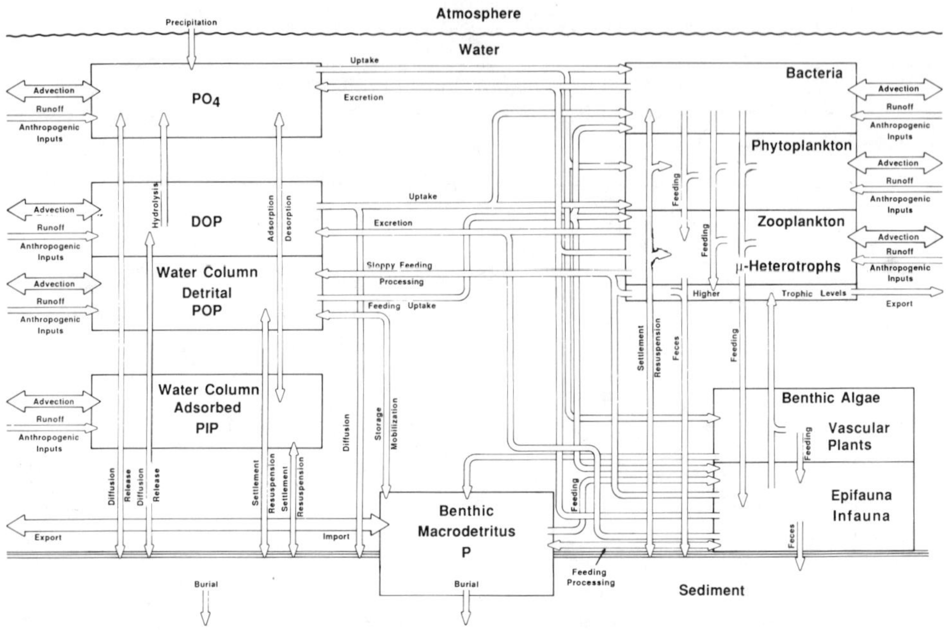

FIGURE 4. Schematic model of phosphorus pools and fluxes in aquatic systems.

speciation of P in natural waters.[260,230] Likewise, partitioning schemes for solid-phase or bound P are operational in nature and subject to overlap between species.[96]

Froelich[96] elaborated a two-stage model of mineral-P interaction whereby an initial stage of adsorption on particle surfaces is followed by incorporation of the P into the mineral structure. Phosphate ions are adsorbed onto clay or carbonate mineral particles[166] or directly incorporated into apatite (Ca, F) and ferric hydroxide (FeOH) minerals.[74,96] Iron-P associations may be reversed under anoxic conditions, leading to the release of P into sediment porewaters and the overlying water column.[178,199] Incorporation of P into some minerals, however, is largely irreversible under normal conditions. While tropical shelf sediments may contain appreciable amounts of P,[6,76] relatively little appears to be released in well-oxygenated water columns following resuspension events.[323]

Phosphorus ions undergo a complex series of desorption-adsorption reactions in estuarine systems and lake sediments.[47,93,330,331] Soluble P in estuarine waters are buffered by interactions with suspended materials and mineral sediments.[96] Processes similar to those occurring in estuaries should also be expected to operate when rivers flow into saline or high ionic-strength lakes. Total P associated with suspended matter in unpolluted tropical rivers normally ranges between 20 to 60 μmol g^{-1} (mean = about 30 to 40).[96] The proportion of this particle-bound P in an exchangeable form on particle surfaces (e.g., 15 μmol g^{-1} for Amazon River suspended particulates)[47] can vary widely. Desorption experiments demonstrate the capacity for rapid P release within estuaries ($<$1 d), with final P concentrations in the medium stabilizing between 0.5 and 2.5 μM,[47,93] concentrations of similar order to dissolved reactive P (0.5 to 1.0 μM) in the river waters.[210] Subsequent extractions with fresh seawater may yield additional P. Unambiguous estimates of P release from suspended materials are difficult to make as P fluxes from biological uptake and mineralization,[73,169,245] bouyancy driven upwelling,[326] organic sedimentation, and Fe-P interactions[96] also occur. Overall, Froelich[96] estimated that 5 μmol P g^{-1}, on average, was desorbed from fluvial material entering the oceans, approximately 15% of the total particle-bound P. In lake systems, marginal swamps or wetlands process nutrients in a manner analagous to estuaries.[165]

DOP and particulate organic P (POP) comprise a significant fraction of total water column P in many aquatic systems.[46,90,230] The same operational definition of "dissolved" applying to organic carbon and nitrogen also applies to phosphorus. Ecosystem-scale usage of DOP by water column organisms is clearly indicated by the negative relation between DOP concentrations and chlorophyll in oceanic surface waters.[156] Few direct measurements of DOP contributions to total phosphorus fluxes in tropical waters have been made.[230] Boto and Wellington[27,28] found that Australian tidal mangrove forests were net importers of DOP and exporters of dissolved inorganic P (DIP). Tropical oceanic phytoplankton, coral reef benthic algal communities, and water column microplankton express alkaline phosphatase activity, implying that local dissolved inorganic P resources are sufficient to meet demand.[11,70,243,264] While phosphorus demand is likely coupled to primary production, evidence for diel activity in DOP utilization is conflicting. Perry[243] found no diel or light-dependent differences in phosphatase activity in the subtropical North Pacific. In contrast, Rivkin and Swift[264] and Dunlap[70] found clear diel changes in phosphatase activity in an oceanic dinoflagellate and in coral reef waters, respectively.

Unlike carbon and nitrogen, there is no significant gaseous source of phosphorus. Exchanges between aquatic and atmospheric pools of phosphorus do occur,[122] but are small compared to aquatic P fluxes. Phosphorus budgets can therefore be accurately resolved from hydrological, geochemical, and biological processes.

The dynamics of phosphorus uptake by submerged and emergent plant communities are complex. Phytoplankton in oligotrophic tropical marine waters have a high affinity (low K_s) for dissolved inorganic P.[244] Affinity constants for freshwater phytoplankters are also low, but not to the degree seen for marine species,[221] although P is frequently regarded as limiting in freshwater ecosystems. To explain the high primary productivity of tropical benthic, coral reef, and wetland ecosystems in low-P waters, it has become an element of accepted wisdom that P is closely held within biomass and recycled. Phosphorus in coral reef flat communities or sediments is several times that present in water overlying the benthos.[11,76] Atkinson[12] further showed that residence times for water over reef flats was far shorter than the time necessary for the benthos to take up the waterborne P. Despite studies showing uptake and release of P by coral reef benthic communities or organisms,[11,60,248] a clear demonstration of "tight" recycling is still lacking.

4. Sulfur

Sulfur is a component to two essential amino acids (cysteine, methionine) and several enzymatic cofactors (e.g., Coenzyme A). In oxic waters, sulfate ($SO_4^=$) is the thermodynamically stable form of inorganic sulfur. Bacteria, phytoplankton, and macrophytes (Reference 57 and references therein) take up, reduce, and assimilate sulfur from sulfate ions. The biogeochemistry of sulfur, however, is determined less by the nutrient requirements for S than by the use of sulfate ($SO_4^=$) as an alternative electron acceptor under anaerobic conditions. Sulfate concentrations are sufficiently high in marine waters (about 28 m*M*) that sulfate is the primary alternative electron acceptor for certain bacteria in anaerobic sediments and microhabitats.[41] Sulfate concentrations in tropical fresh or nonmarine water bodies vary widely.[279,314] Where sulfate concentrations are low relative to oxygen demand, CO_2 becomes the primary alternative electron acceptor, leading to the formation of methane.[41]

Hydrogen sulfide (H_2S) and the sulfide ion ($S^=$) are the principal reduced-sulfur species in aquatic systems and exert a number of indirect effects on nutrient cycling. Many metals, for example, iron, form highly insoluble sulfides and, as a result, are removed from solution in anaerobic waters or porewaters. Phosphate solubilization can occur in anaerobic sediments because insoluble Fe-P minerals formed under aerobic conditions are dissolved[36] and the iron reprecipitated as pyrite (FeS).[144] Sulfide is suggested to be a metabolic poison of nitrification,[85] but this may or may not be important because sulfide concentrations are usually very low in oxic environments. Conversely, sulfide is an occasional reductant in dissimilatory nitrate reduction.[85] Interactions between nitrifying, denitrifying, and sulfate-reducing bacteria occuring within oxyclines bordering reducing habitats are likely complex, with rates governed by local nutrient transport, diffusion, and sediment-mixing rates.

5. Silicate

Diatoms, a small number of other Chrysophyta, some vascular macrophytes (e.g., papyrus) and sponges require silicon. Silicate concentrations in tropical freshwaters[279,314,332] are considerably higher than those measured in marine waters (Table 3). Dissolved silicate in freshwaters is derived directly from weathered silicate minerals, whereas in the oceans, upwelling and recycling of biologically deposited opal silica assume greater importance[73,222] relative to external inputs.

Because of the high concentrations of silicate generally found in tropical freshwaters (Table 3), it is unlikely that silicate strongly limits production or biomass. Silicate concentrations are considerably lower ($<2\ \mu M$) in most nonestuarine and nonupwelling marine waters, but with the possible exception of equatorial and coastal upwelling zones, it is also unlikely that silicate is a major limiting nutrient in tropical seas. Diatoms comprise a considerable fraction of total community biomass in upwelling zones and may deplete Si concentrations to near undetectable levels.[340] This demand is offset, in part, by high rates of Si remineralization in the euphotic zone.[222] Benthic flux measurements in tropical coastal waters indicate much of the Si flux may be abiogenic in origin.[6,323] Elsewhere, N to Si and P to Si ratios of marine diatoms[33] are sufficiently high such that N and P would be depleted before the uptake of all silicate by flagellate and picoplankton dominated phytoplankton assemblage.

6. Trace Metals

Relatively little is known regarding the quantitative role of trace metals as nutrients in tropical ecosystems, particularly in nonmarine waters. Huntsman and Sunda[149] present a general review of trace metal behavior as it affects phytoplankton. Only within the last 15 years have instrumentation, processing techniques, and, perhaps most important, clean sampling methods advanced to a plane permitting relatively unambiguous determinations of metal concentrations. Of perhaps greater importance has been the development of theory[136,217,322] and methods[274] to address the problem of metal speciation *in situ*.

Concentrations of "soluble" metal ions in natural waters and their biological availability or toxicity[16,17,311] are strongly dependent on biological processes, redox potential, pH, ionic strength, activities of organic and inorganic chelators,[217,308] and scavenging processes.[49,274] All of the above parameters vary greatly between tropical nonmarine water bodies and, to a lesser degree, in the ocean. A number of trace elements (Cu, Ni, Zn, Se) behave geochemically like nutrients, with vertical distributions related to the scavenging and mineralization of organic particulates in the water column.[274] High irradiance levels at tropical latitudes promote the formation of hydrogen peroxide,[361] superoxide ions, and free radicals in surface waters which in turn affect the redox state of metal ions or chelators.[357,359] In tropical freshwaters containing substantial amounts of humic materials (blackwaters) metal concentrations and speciation are overwhelmingly dependent on metal-organic chelation reactions. Direct measurements of total soluble (filter passing) metal concentrations give little indication as to metal speciation and availability. More realistic estimations must be indirectly obtained by solubility and complexation modeling[217,322] following some estimation of the chelation potential of the organic matter,[148] although the latter process is fraught with analytical and theoretical uncertainty.

In oxygenated marine waters and sediments, soluble iron concentrations, in particular, are determined by equilibria with insoluble iron hydroxides. Inputs and solubility may limit phytoplankton production in open oceanic waters.[204] In contrast, Gaudet[106] reported accumulations of iron oxides within mats of vegetation and detritus in a papyrus swamp. While the focus in coral reef nutrient studies has largely been directed toward the major nutrients, Entsch et al.[77] presented biochemical evidence suggesting that reef algae may, in fact, be iron-limited. This approach deserves wider application.

Molybdenum and iron are required for nitrogen fixation. Low molybdenum concentrations in particular lake systems[134] may have led to nitrogen limitation due to low rates of N fixation.[145] Extending this view, Howarth et al.[147] suggested that ecosystem-level geochemical controls on the concentration and speciation of molybdenum and/or iron may affect ecosystem rates of N fixation. High sulfate concentration (e.g., 28 m*M* in seawater) could potentially inhibit N fixation if compensatory mechanisms are not present. The high N-fixation rates in marine ecosystems, however, indicate that marine N fixers have evolved cellular mechanisms to mitigate sulfate inhibition, or more likely, that N fixation largely occurs in anaerobic habitats, where sulfate is not present or largely reduced to sulfide.

7. Vitamins

Organic growth factors such as thiamin, biotin, and cobalamin (vitamin B_{12}), their precursors, or other organic factors are required by some, although certainly not all, aquatic plants.[313] Vitamins are synthesized and excreted by bacteria and specific algae, or may be released during cell death. Early measurements of vitamin concentrations in natural waters required tedious bioassay experiments and suggested concentrations in the nanogram-per-liter range.[313] Modern high performance liquid chromatography methodologies now permit direct measurements.[69,71] These recent results indicate that benthic communities and free-living bacteria in coral reef systems release a wide range of bioactive compounds. Time series measurements show that soluble pteridine and flavin compounds are highly photoreactive with half-lives of hours in full sunlight.[72] While no comparable measurements are reported for tropical freshwaters, it is likely that bacterial populations in these organic and particle-laden systems provide sufficient vitamins in most situations.

C. MINERALIZATION PROCESSES

The high biomass-specific and absolute rates of primary production occurring in many tropical aquatic systems at low to moderate dissolved water column nutrient concentrations provides presumptive evidence for active recyling of nutrients. Estimates of nutrient demand from primary production must, of course, be properly scaled to carbon fluxes (Table 2); nonetheless, there is good reason for the view that an important fraction of nutrients are actively recycled between dissolved, biomass, and detrital pools (e.g., Figures 1, 3, and 4).

From a geochemical standpoint, mineralization rate processes are strongly weighted toward the activities of water column microorganisms within the size range 1 to 100 μm (Table 6). Detailed comparisons between individual processes in different systems are confounded by the lack of suitable data to normalize individual studies. Isotope dilution experiments carried out in both marine[98,111,128,289] and freshwaters[89,219] indicate water column nutrient demand and mineralization within microplankton communities are of similar magnitude, with inorganic N and P turnover times ranging from <1 h to many days. The low C to N and C to P ratios of bacteria (Table 2) further suggest that bacteria may be nutrient sinks, rather than conduits for N and P flux in the decomposition of organic matter with a high carbon content.[4,67] Microflagellates feeding on bacteria were found to be more efficient at N remineralization than their bacteria prey,[115] releasing up to 50% of the ingested nitrogen as ammonium.[115,283]

Despite the quantitative importance of microorganisms, animals are important agents for mineralization and nutrient cycling in aquatic systems. Ammonium[23,177] and phosphate are the principal N and P excretion products of aquatic animals. On the basis of biomass

TABLE 6
Comparisons between the Magnitude of Mineralization Rates in Several Tropical Aquatic Habitats (μmol m^{-2} d^{-1})

	N	P	Ref.
Kaneohe Bay, Hawaii inner basin, presewage diversion (10-m depth)			
Microplankton (estimated)	6360		296
(measured)	240—3900		37
Macrozooplankton	1210	452	296
Benthos	1434	280	296
Sargasso Sea (upper 100 m)			
Microplankton (measured)	4800—155,000		110
Macrozooplankton	1183		328
North Queensland Shelf (30 m depth)			
Microplankton	1224		137
Macrozooplankton	611—1161	32—65	151
Benthos	447—2550	41—142	6
Lagoon, Davies Reef, Australia, winter (10-m depth)			
Microplankton	408		137
Benthos	0.2	0.6	126
Amazon Floodplain lake			
Microplankton	60,000		192
Zooplankton	41—1850	5—199	192
Benthos	6000	1320	192
Lake George (Uganda) (depth unspecified)			
Zooplankton	108		104
Benthos	Nil	Nil	330

and specific mineralization activity, small animals make the greatest contribution to metazoan mineralization.[150,158,328] Large animals are not without their contribution, however. Viner[332] calculated that hippopotami were significant contributors to organic and nutrient loading in Lake George, Uganda. While the hippos did not feed directly on lake vegetation, they voided substantial amounts of organic material grazed from the surrounding watershed into the lake.

Nutrient mineralization processes in tropical aquatic ecosystems operate over a continuum of time scales. The immediate (<1 d) nutrient demands of phytoplankton and water column bacteria appear to be largely met by water column microbial mineralization with supplementary inputs from larger zooplankton. Water column sources of organic matter are largely of microbial (algae, bacteria, protozoa) origin, with relatively low C to N and C to P ratios that reflect low levels of structural polymers. Time scales for water column nutrient pool turnover range from minutes to hours.[89,111,128] Macroalgae, submerged and floating vascular macrophytes, periphyton, and leaves from terrestrial vegetation appear to be mineralized in large part over time scales ranging from several days to several months.[281,307,351] The decomposition of wood may take years.[268] Local increases in dissolved nutrient levels may occur when detritus is aggregated by hydrodynamic processes[130,362] or direct sedimentation.[106,271]

Although tropical regions do not undergo the pronounced seasonal fluctuations in temperature and irradiance occurring at temperate latitudes, virtually all tropical systems are affected by some type of episodic, seasonal, or interannual fluctuation in freshwater inputs,[91] water levels,[165] physical disturbance,[99] and macrophyte standing crop,[165] with the coupled release, mineralization, and uptake of substantial stocks of nutrients incorporated into sediments and biomass. Forcing events in the tropics are often of short duration, leading to short-term fluctuations in water column nutrient levels while soluble nutrients are rapidly reincorporated into plant materials or leached from sediments and detritus, closing the cycle once again.

D. SEDIMENT-WATER COLUMN EXCHANGES

Water column-benthic exchanges are an important pathway for nutrient cycling in shallow aquatic habitats.[4-6,92] Benthic nutrient fluxes vary with temperature,[224] rates of organic deposition,[225] and the composition of deposited organic matter[309] integrating both surficial[170] and subsurface mineralization,[178] denitrification,[276] inorganic exchange/solution processes occurring above and below the oxycline,[199,309] and burial.[339] In general, benthic nutrient fluxes in tropical marine systems are lower than in temperate systems (summarized by Nixon),[225] largely due to lower rates of primary production and organic deposition. One significant exception occurs in low latitude coastal upwelling systems where high primary production and deposition rates prevail.[189,271,339]

Rooted aquatic plants transfer nutrients between the sediments and water column depending on the species and conditions prevailing, with nutrients being taken up from either the sediments or water column[238] and returned to the sediments as detritus[165] or leached into the water column.[142]

Relative contributions of the benthos to total system mineralization rates vary considerably between particular systems (Table 6). With the exception of very shallow water bodies, water column microbial and zooplankton mineralization processes supply the bulk of plankton inorganic N and P demand. On an areal basis, porewater and detrital nutrient stocks within the benthos overwhelmingly dominate nutrient inventories.[108,296] Turnover rates of aggregate benthic nutrient stocks are difficult to predict as different components have different levels of lability or accessibility to decomposers. The large size of benthic detritus stocks imparts considerable hysteresis to temporal variability in benthic flux rates and water column nutrient concentrations in changing environments.[182,296] Conversely, benthic resuspension events[83,99,285,338] can lead to significant increases in water column nutrient concentrations.

Over the long term, benthic inputs, outputs, and storage must be balanced. However, for the shorter time and smaller spatial scales usually used for nutrient budgets, there is no compelling reason for fluxes of individual nutrients to be tightly coupled.[300] Nutrient sources for the benthos include sedimentation or capture of particles,[176] lateral transport of detritus,[109,130] direct uptake by benthic communities, and the decomposition of algae and macrophytes *in situ*. Outputs include benthic excretion, burial, and the lateral transport of organic detritus and minerals. In the case of carbon and nitrogen, additional losses to the atmosphere via respiration and denitrification must be included.

Discrepancies between directly measured benthic release rates and diffusive flux rates calculated from porewater concentration gradients are often noted.[192,319,323] While both methods are subject to experimental error and misinterpretation, such discrepancies are not necessarily an artifact. Bacterial and microalgal populations living at the sediment-water interface actively intercept and process organic carbon and nutrients.[29,170,175,302] Irrigation of sediments by benthic infauna also contribute significantly to benthic flux rates.[3,124]

E. TEMPORAL AND SPATIAL SCALES

Nutrient transformation processes operate over a broad range of spatial and temporal scales. Microorganisms take up nutrients and lose waste products by molecular diffusion down concentration gradients on the order of microns in thickness.[237] Time scales for the formation and dissipation of nutrient patches or gradients of that thickness are on the order of milliseconds to seconds.[154] In a nutritionally dilute medium, uptake mechanisms have evolved to operate in such short time and distance scales.[114,117,191,205] Larger microorganisms (phytoplankton, protists) use morphological structures, motility, or sinking[286] to increase shear near the cell surface and steepen diffusive gradients, but the physics of diffusion favors the small.

Bulk water motion, turbulence, and boundary layer modification at organism surfaces assume greater importance for nutrient uptake and loss from larger aquatic plants, benthic surfaces, or animals.[185,335] The structural form of aquatic plants and primary producers like corals or sponges may actively modify turbulence to physiologically useful levels.[8] Surfaces in tropical aquatic ecosystems, whether living or not, are rarely uncomplicated and never remain clear and unaltered. Most are quickly and extensively colonized by assemblages of bacteria, fungi, microalgae, protozoa, and metazoans which are, in turn, actively grazed. Prominent examples include the epilithic turf algal community of coral reefs or the epibiotic assemblages colonizing seagrasses and freshwater macrophytes.[75] Much remains to be known about the physics of and nutrient dynamics within these living boundary layers.[13] One important unresolved problem is how to estimate realistically the extent and roughness of biologically active surface areas at the size scale of individual classes of organisms (e.g., a bacterium or arthropod) within more readily comprehended areal units (square meters, hectare). Some first-order comparisons of the two-dimensional roughness of coral reefs at different length scales have been made through fractal theory,[31,32] but have not been extended as yet to more realistic three-dimensional estimates. It will be extremely interesting if fractal geometries can also be applied to other geometrically complex aquatic systems such as wetlands or submerged macrophyte beds.

Colonizing growths extend the stagnant boundary layer at surfaces, restricting diffusion of nutrients to and from the surface. While thick boundary layers may retard nutrient exchanges with the bulk fluid, they also retain nutrients released by grazers and predators living within the layer and, therefore, have nutrient dynamics considerably different from those in the bulk fluid or local community as a whole.[13,211,254,257] Epibiotic communities on hard substrata,[181,345] macroalgae, and seagrasses[38,112] are known to fix gaseous dinitrogen actively and exhibit uptake kinetics different from their substratum.[140] The accumulation of nutrients within turf communities makes them a source of food for large and small grazers.[76,304] For communities or individual organisms on this size scale, variability in nutrient concentrations may exist over centimeter distances.[288] A considerable amount of work needs to be done to resolve nutrient dynamics at this most relevant ecological scale.

For areas larger than the scale of individual organisms, relationships between water residence times and time scales for biological processes becomes a major factor behind nutrient dynamics and perceived variability.[13,56,162] The residence time of water and nutrients in proximity to biologically active boundaries is greatly facilitated by tortuous or permeable structures[9,137,211] or the formation of boundary layers.[355] For enclosed or semienclosed water bodies, residence times depend on volume, flushing rates, and internal frictional forces.[354]

III. SOME SPECIAL SITUATIONS

A. FLOATING FRESHWATER MACROPHYTES

Floating macrophytes, in particular, the water hyacinth, *Eichhornia crassipes* (Mart.) Solm., water fern, *Salvinia molesta* (Mitchell), and water fern, *Azolla* spp. have significant effects on nutrient dynamics and water chemistry in tropical freshwater systems.[30] *Eichhornia* and *Salvinia*, originally from South America, are now pan-tropical weeds.[20] The impact of *Azolla* is perhaps more benign. *Azolla* strains support a symbiotic association with the N-fixing cyanobacterium *Anabena azollae* and in some locations are cultivated in rice paddies as living fertilizer.[216] Extensive beds of floating macrophytes or rooted macrophytes with floating leaves occur widely in tropical wetlands, seasonally covering vast areas in floodplain lakes.[162]

With adequate nutrient supplies, high temperatures, and high irradiance levels, both *Eichhornia* and *Salvinia* are capable of extremely high primary production rates and explosive growth,[253,273] forming dense mats of floating vegetation which can completely cover water surfaces. Relative to other macrophytes (Table 2), *Eichhornia* and *Salvinia* can have high relative N and P contents. Internal N and P contents of *Eichhornia* are directly related to areal nutrient loading rates,[255] while growth rates are influenced by temperature, irradiance, and plant nutrient content.[270]

Direct impacts of *Eichhornia* and *Salvinia* mats on nutrient processes include the fixation of atmospheric, not dissolved CO_2 into organic matter; occlusion of the water surface, limiting growth of phytoplankton and submerged macrophytes; direct uptake of inorganic nutrients from the water column;[253] and large-scale introductions of detritus from sinking and decaying vegetation. Decomposition of this detritus can lead to reductions in pH and formation of anoxic conditions which can alter the balance between oxic and anoxic microbial pathways and the speciation of a range of nutrients (Figure 2). Because of the high growth rate and nutrient demand of floating macrophytes, they have repeatedly been considered for and used as nutrient traps in wastewater treatment plants.[253] de Busk et al.[59] compared nutrient removal rates in continually harvested and unharvested ponds of *Eichhornia*. While the uptake of N and P from the water into biomass (363 and 115 mg m^{-2} d^{-1}) was greater in harvested ponds, net removal of nitrogen on an areal basis was greater in the unharvested ponds (875 vs. 360 mg N m^{-2} d^{-1}) due to enhanced rates of detritus-fueled denitrification. Uptake of phosphate into unharvested biomass continued until ponds reach their carrying capacity. Thereafter, little additional net P removal occurred.

B. RICE PADDIES

Rice paddies are a special type of tropical aquatic ecosystem. Land planted to rice covers extensive areas throughout the tropics. Rice paddies are highly productive, receive variable, usually managed, nutrient inputs in the form of organic and inorganic fertilizers, and paddy soils (sediments) are highly disturbed (bioturbated) by cultivation practices. There is an extensive literature on nutrient relations in rice plants, paddies, and soils.[1,65] Rice paddy nutrient dynamics lie on an interface between physical chemistry, wetland ecology, and benthic microbiology. The composition and structure of paddy soils and interstitial waters exert a strong effect on nutrient relationships through the binding of ammonia and phosphorus. Nitrification, denitrification, and N fixation[261] all occur in paddy soils. N fixation is encouraged in paddy soils as a natural fertilization process, while nitrification and denitrification are discouraged through cultivation and fertilization practices to minimize leaching or gaseous losses. In high pH soils, direct gaseous evasion of ammonia can lead to significant N losses from paddies.[87] Rice plants directly modify nutrient cycles within anaerobic paddy soils through transport of oxygen into the rhizosphere[252,254] and transpiration of gaseous ammonia.[58] Similar processes would also occur in highly loaded tropical wetlands.

C. GROUNDWATER INPUTS

Groundwaters may contribute significantly to nutrient inputs in specific aquatic ecosystems.[159] Groundwaters and elevated nutrient concentrations associated with them have been detected in a number of locations where they impinge on coral reef systems[63,194,203] and coastal embayments.[358] Measured inputs to nonmarine systems have ranged from nonsignificant[108] to dramatic.[183] The measurement or estimation of groundwater nutrient fluxes is fraught with considerable uncertainty;[325] nonetheless, where it might be suspected, some attention to groundwater measurements may be rewarding.

Ridd et al.[259] reported another form of terrestrial input to tropical coastal waters. Where intermittently flooded marine tidal flats are exposed to high insolation, extensive amounts of dried salt may accumulate. Dissolution of these salts during episodic or seasonal high tide events may deliver substantial loads of salts and associated nutrients to coastal waterways.

IV. NUTRIENT LIMITATION

The question of whether nutrients are limiting in a particular aquatic ecosystem and if so, which one(s), continues to remain a vexing problem. Both perspective and scaling are important. On one hand, local nutrient concentrations or supply rates may be sufficiently low that specific growth is slowed by a biomass-scaled nutrient acquisition rate below that potentially obtainable under the ambient physical conditions. Alternatively, nutrient acquisition rates may not be greatly restrained by ambient concentration, but the total biomass-scaled nutrient pool(s) accessible to an individual or community is smaller than that of other nutrient elements.

From culture and field experiments with oceanic phytoplankton[184,205] we know that individual species and communities have the potential to achieve high relative (u/u_{max}) growth rates at extremely low dissolved inorganic nutrient concentrations. In conjunction with the observation that uptake and *in situ* remineralization rates of N and P are of similar magnitude, this suggests that specific instantaneous growth rates of marine phytoplankton may not be strongly nutrient limited. As dissolved N and P concentrations in tropical freshwater bodies tend to be higher than in tropical marine waters, nutrient limitation of growth rates may also not be likely.

Ryther and Dunstan[272] put forward the physiologically derived view, based on ratios between dissolved inorganic nutrient concentrations and the composition of phytoplankton, that nitrogen is the principal nutrient limiting phytoplankton growth in coastal marine and, by some extrapolation, oceanic waters. An alternative geochemical view, framed by Smith,[290] is that as additional nitrogen can be biologically fixed from gaseous sources, nutrient elements without gaseous or external sources, for example, P,[293] Fe,[204] or Mo[134] must ultimately be limiting. Although care must be applied to the interpretation of bioassay experiments, there is some suggestion that nitrogen may more frequently limit phytoplankton growth in tropical freshwater systems than in temperate lakes.[320,329] (See also Reference 280.)

Reconciliation between the two views largely depends on both perspective and scaling. Either N[280, 352] or P[290, 293] limitation may occur in discrete water bodies or parcels, depending on the nature of unbalanced relationships between N inputs, fixation, losses, and denitrification and those of other nutrients. Mineralization processes, not always apparent in sampling of the bulk fluid, must be considered. Atkinson[13] discussed scenarios whereby apparent nutrient limitation in benthic algal populations of coral reefs are dependent on the physics and chemistry of the microbial boundary layer rather than on bulk water properties.

V. SOME CONCLUDING THOUGHTS

Any exploration of the literature about tropical aquatic nutrient processes repeatedly highlights four central factors: the linkage between hydrodynamic and nutrient processes, the central role of microbial transformations, the role of detritus as a depot for nutrients, and the behavior of nutrients at surfaces.

All tropical aquatic ecosystems are geometrically complex on at least the microbial scale. Spatial complexity on larger scales occurs in coral reefs, submerged macrophyte beds, mangroves, and wetlands. Transfer processes between surfaces, organisms, and the bulk fluid, whether over large or small areas, must occur through a hierarchy of boundary layers ranging from microns to kilometers (Chapter 1) in thickness. In many cases, much of the biological and geochemical activity takes place through the activities of microbial populations operating within the thinnest boundary layers. Innovative techniques[31] are needed to quantify the structure and surface area of geometrically complex habitats and microhabitats across a broad range of spatial scales. Such quantifications are needed to provide a basis for scaling the results of experimental studies carried out with arbitrarily sized samples of substratum to micro- or macro-scales of interest.

While microbial rate processes generally increase in magnitude with temperature, temperature dependencies in individual processes, the requirements of particular species or groups, and interactions between temperature and other environmental parameters have not been widely quantified for tropical aquatic ecosystems or communities, particularly at very high temperatures which may occur in exposed, very shallow, or isolated habitats. Much of what we know about the microbially mediated behavior of nutrients in aquatic ecosystems is based on temperate systems and the extrapolated behavior of temperate species. Extension of these findings to models of nutrient behavior in the tropics will require an extended and better resolved parameterization of temperature dependencies in tropical communities and verification of rates determined in temperate organisms.

In any ecosystem, detritus comes, goes, is mineralized, but is always there. While nutrient studies focus on the frenzied turnover of biomass and inorganic nutrients, the bulk of nutrients are often stored in detrital or dissolved organic pools which cycle at vastly different and usually unquantified rates. Clearly, greater attention needs to be placed on quantifying nutrient fluxes into and out of detrital and dissolved organic pools.

ACKNOWLEDGMENT

This is AIMS contribution no. 546.

REFERENCES

1. Academia Sinica, *Proceedings of Symposium on Paddy Soil,* Springer Verlag, New York, 1981.
2. **Alldredge, A. L.,** The significance of suspended detrital aggregates of marine snow as microhabitats in the pelagic zone of the ocean, in *Recent Advances in Microbial Ecology,* Hattori, T., Ishida, Y., Maruyama, Y., Morita, R. Y., and Uchida, A., Eds., Japan Sci. Soc. Press, Tokyo, 1989, 108.
3. **Aller, R. C.,** Experimental studies of changes produced by deposit feeders on pore water, sediment and overlying water chemistry, *Am. J. Sci.,* 278, 1185, 1978.
4. **Alongi, D. M.,** Detritus in coral reef ecosystems: fluxes and fates, *Proc. 6th Int. Coral Reef Symp.,* Townsville, Australia, 1, 1989, 29.
5. **Alongi, D. M.,** The role of soft-bottom benthic communities in tropical mangrove and coral reef ecosystems, *Crit. Rev. Mar. Sci.,* 1, 243, 1989.
6. **Alongi, D. M.,** Benthic processes across mixed terrigenous-carbonate sedimentary facies on the central Great Barrier Reef continental shelf, *Cont. Shelf Res.,* 9, 629, 1989.
7. **Anderson, J. J.,** The nitrite-oxygen interface at the top of the oxygen minimum zone in the eastern tropical North Pacific, *Deep Sea Res.,* 29, 1193, 1982.
8. **Anderson, S. M. and Charters, A. C.,** A fluid dynamics study of seawater flow through *Gelidium nudifrons, Limnol. Oceanogr.,* 27, 399, 1982.
9. **Andrews, J. C. and Muller, H.,** Space-time variability of nutrients in a lagoonal patch reef, *Limnol. Oceanogr.,* 28, 215, 1983.
10. **Antoniou, P., Hamilton, J., Koopman, B., Jain, R., Holloway, B., Lyberatos, G., and Svoronos, S. A.,** Effect of temperature and pH on the effective maximum specific growth rate of nitrifying bacteria, *Water Res.,* 24, 97, 1990.
11. **Atkinson, M. J.,** Rates of phosphate uptake by coral reef flat communities, *Limnol. Oceanogr.,* 32, 426, 1987.
12. **Atkinson, M. J.,** Slow uptake of ^{32}P over a barrier reef flat, *Limnol. Oceanogr.,* 32, 436, 1987.
13. **Atkinson, M. J.,** Are coral reefs nutrient limited?, *Proc. 6th Int. Coral Reef Symp.,* Townsville, Australia, Vol. 1, 1989, 157.
14. **Atkinson, M. J. and Smith, S. V.,** C:N:P ratios of benthic marine plants, *Limnol. Oceanogr.,* 28, 568, 1983.
15. **Axler, R. P. and Reuter, J. E.,** A simple method for estimating the ^{15}N content of dissolved organic matter ($DO^{15}N$) in N-cycling studies, *Can J. Fish, Aquat. Sci.,* 43, 130, 1986.
16. **Barber, R. T. and Ryther, J. H.,** Organic chelators: factors affecting primary production in the Cromwell Current upwelling, *J. Exp. Mar. Biol. Ecol.,* 3, 191, 1969.

17. **Barber, R. T., Dugdale, R. C., MacIsaac, J. J., and Smith, R. L.,** Variations in phytoplankton growth associated with the source and conditioning of upwelled water, *Inv. Pesq.*, 35, 171, 1971.
18. **Barnes, D. J. and Crossland, C. J.,** Physical, chemical and biological properties of waters over coral reefs at Lizard Island, Great Barrier Reef (LIMER II Expedition, September, 1977), Australian Inst. of Marine Science, Data Rept. AIMS-RS-83-1, 1983.
19. **Barnes, D. J.,** Seasonality in community productivity and calcification at Davies Reef, central Great Barrier Reef, *Proc. 6th Int. Coral Reef Symp.*, 2, 521, 1989.
20. **Barrett, S. C. H.,** Waterweed invasions, *Sci. Am.*, 261, 66, 1989.
21. **Beers, J. R.,** Studies on the chemical composition of the major zooplankton groups in the Sargasso Sea off Bermuda, *Limnol. Oceanogr.*, 11, 520, 1966.
22. **Berounsky, V. M. and Nixon, S. W.,** Eutrophication and the rate of net nitrification in a coastal marine ecosystem, *Est. Coast. Shelf Sci.*, 20, 775, 1985.
23. **Bidigare, R. R.,** Nitrogen excretion by marine zooplankton, in *Nitrogen in the Marine Environment,* Carpenter, E. J. and Capone, D. G., Eds., Academic Press, New York, 1983, 357.
24. **Billen, G.,** Evaluation of nitrifying activity in sediments by dark ^{14}C-bicarbonate incorporation, *Water Res.*, 10, 51, 1976.
25. **Blackburn, T. H. and Sorensen, J., Eds.,** *Nitrogen Cycling in Coastal Marine Environments,* Blackburn, London, 1988.
26. **Boto, K. G. and Bunt, J. S.,** Carbon export from mangroves, in *The Cycling of Carbon, Nitrogen, Sulfer and Phosphorus in Terrestrial and Aquatic Ecosystems,* Galbally, I. E. and Freney, J. R., Eds., Australian Academy of Science, Canberra, 1982, 105.
27. **Boto, K. G. and Wellington, J. T.,** Phosphorus and nitrogen nutrient status of a northern Australian mangrove forest, *Mar. Ecol. Prog. Ser.*, 11, 63, 1983.
28. **Boto, K. G. and Wellington, J. T.,** Seasonal variations in concentrations and fluxes of dissolved and inorganic materials in a tropical, tidally-dominated, mangrove waterway, *Mar. Ecol. Prog. Ser.*, 50, 151, 1988.
29. **Boto, K. G., Alongi, D. M., and Nott, A. L. J.,** Dissolved organic carbon-bacteria interactions at sediment-water interface in a tropical mangrove system, *Mar. Ecol. Prog. Ser.*, 51, 243, 1989.
30. **Bowmaker, A. P.,** The physico-chemical limnology of the Mwenda River mouth, Lake Kariba, *Arch. Hydrobiol.*, 77, 66, 1976.
31. **Bradbury, R. H. and Reichelt, R. E.,** Fractal dimensions of a coral reef at ecological scales, *Mar. Ecol. Prog. Ser.*, 10, 169,
32. **Bradbury, R. H., Reichelt, R. E., and Green, D. D.,** Fractals in ecology: methods and interpretation, *Mar. Ecol. Prog. Ser.*, 14, 295, 1985.
33. **Brzezinski, M. A.,** The Si:C:N ratio of marine diatoms: interspecific variability and the effect of some environmental variables, *J. Phycol.*, 21, 347, 1985.
34. **Brzezinski, M. A.,** Colorimetric determination of nanomolar concentrations of ammonium in seawater using solvent extraction, *Mar. Chem.*, 20, 277, 1987.
35. **Byrne, R. H., Kump, L. R., and Cantrell, K. J.,** The influence of temperature and pH on trace metal speciation in seawater, *Mar. Chem.*, 25, 163, 1988.
36. **Callender, E.,** Benthic phosphorus regeneration in the Patomac River, *Hydrobiologia,* 92, 431, 1982.
37. **Caperon, J., Schell, D., Hirota, J., and Laws, E.,** Ammonium excretion rates in Kaneohe Bay, Hawaii, measured by a ^{15}N isotope dilution technique., *Mar. Biol.*, 54, 33, 1979.
38. **Capone, D. G.,** $N_2(C_2H_2)$ fixation by macroalgal epiphytes, *Proc. 3rd Int. Coral Reef Symp.*, 1, 337, 1977.
39. **Capone, D. G.,** Benthic nitrogen fixation, in *Nitrogen in the Marine Environment,* Carpenter, E. J. and Capone, D. G., Eds., Academic Press, New York, 1983, 105.
40. **Capone, D. G. and Taylor, B. F.,** Nitrogen fixation (acetylene reduction) in the phyllosphere of *Thalassia testudinum, Mar. Biol.*, 40, 19, 1977.
41. **Capone, D. G. and Kiene, R. P.,** Comparison of microbial dynamics in marine and freshwater sediments: contrasts in anaerobic carbon catabolism, *Limnol. Oceanogr.*, 33, 725, 1988.
42. **Carpenter, E. J.,** Physiology and ecology of the marine planktonic *Oscillatoria* (*Trichodesmium*), *Mar. Biol. Lett.*, 4, 69, 1983.
43. **Carpenter, E. J. and Capone, D. G., Eds.,** *Nitrogen in the Marine Environment,* Academic Press, New York, 1983.
44. **Chakrabotty, M. and Kushari, D.,** Influence of domestic sewage on growth and nitrogen fixation of *Azolla pinnata* R. Br., *Aquat. Bot.*, 24, 61, 1986.
45. **Charpy, L.,** Distribution and composition of particulate organic matter in the lagoon of Tikehau (Tuamotu Archipelago, French Polynesia), *Proc. 5th Int. Coral Reef Congress,* Tahiti, Vol. 3, 1985, 353.
46. **Charpy, L. and Charpy-Boubaud, C. J.,** Phosphorus cycle and budget in an atoll lagoon, *Proc. 6th Int. Coral Reef Symp.*, Townsville, Australia, Vol. 2, 1989, 547.
47. **Chase, E. M. and Sayles, F. L.,** Phosphorus in suspended sediments of the Amazon River, *Est. Coast. Shelf Sci.*, 11, 383, 1980.

48. **Chavez, F. P. and Barber, R. T.,** An estimate of new production in the equatorial Pacific, *Deep Sea Res.,* 34, 1229, 1987.
49. **Clegg, S. L. and Sarmiento, J. L.,** The hydrolytic scavenging of metal ions by marine particulate matter, *Prog. Oceanogr.,* 23, 1, 1989.
50. **Codispoti, L. A.,** Temporal variability in three different upwelling systems, in *Coastal Upwelling,* Richards, F. A., Ed., Vol. 1, AGU Coastal and Estuarine Series, 1980, 209.
51. **Corredor, J. E. and Capone, D. G.,** Studies on nitrogen diagenesis in coral reef sands, *Proc. 5th Int. Coral Reef Congress,* Tahiti, Vol. 3, 1985, 395.
52. **Corredor, J. E., Wilkinson, C. R., Vicente, V. P., and Morell, J. M.,** Nitrate release by Caribbean reef sponges, *Limnol. Oceanogr.,* 33, 114, 1988.
53. **Corredor, J. E. and Morell, J. M.,** Assessment of inorganic nitrogen fluxes across the sediment-water interface in a tropical lagoon, *Est. Coast. Shelf Sci.,* 28, 339, 1989.
54. **Crossland, C. J.,** Dissolved nutrients in coral reef waters, in *Perspectives on Coral Reefs,* Barnes, D. J., Ed., Cluston, Manuka, 1982.
55. **Crossland, C. J. and Barnes, D. J.,** Acetylene reduction by coral skeletons, *Limnol. Oceanogr.,* 21, 153, 1976.
56. **Crossland, C. J., Hatcher, B. G., Atkinson, M. J., and Smith, S. V.,** Dissolved nutrients of a high latitude coral reef, Houtman Abrolhos Islands, Western Australia, *Mar. Ecol. Prog. Ser.,* 14, 159, 1984.
57. **Cuhel, R. H., Jannasch, H. W., Taylor, C. D., and Lean, D. R. S.,** Microbial growth and macromolecular synthesis in the northwestern Atlantic Ocean, *Limnol. Oceanogr.,* 28, 1, 1983.
58. **da Silva, P. R. F. and Stutte, C. A.,** Nitrogen loss in conjunction with transpiration from rice leaves as influenced by growth stage, leaf position and N supply, *Agron. J.,* 73, 38, 1981.
59. **deBusk, T. A., Williams, L. D., and Ryther, J. H.,** Removal of nitrogen and phosphorus from waste water in a water hyacinth-based treatment system, *J. Environ. Qual.,* 12, 257, 1983.
60. **D'Elia, C. F.,** The uptake and release of phosphorus by reef coral, *Limnol.Oceanogr.,* 22, 301, 1977.
61. **D'Elia, C. F.,** Nitrogen determination in seawater, in *Nitrogen in the Marine Environment,* Carpenter, E. J. and Capone, D. G., Eds., Academic Press, New York, 1983, 731.
62. **D'Elia, C. F.,** The cycling of essential elements in coral reefs, in *Concepts of Ecosystem Ecology. A Comparative View,* Pomeroy, L. R. and Alberts, J. J., Eds., Springer-Verlag, New York, 1988, 195.
63. **D'Elia, C. F., Webb, K. L., and Porter, J. W.,** Nitrate-rich groundwater inputs to Discovery Bay, Jamaica: A significant source of N to local coral reefs, *Bull. Mar. Sci.,* 31, 903, 1981.
64. **Denny, P.,** Mineral cycling by wetland plants — a review, *Arch. Hydrobiol.,* 27, 1, 1987.
65. **deSatta, S. K. and Patrick, W. H., Jr.,** *The Nitrogen Economy of Flooded Rice Soils,* Martinus Nijhoff, The Hague, 1986.
66. **Droop, M. R.,** The nutrient status of algal cells in continuous culture, *J. Mar. Biol. Assoc. U.K.,* 54, 825, 1974.
67. **Ducklow, H. W., Purdie, D. A., Williams, P. J. LeB., and Davies, J. M.,** Bacterioplankton, a sink for carbon in a coastal marine plankton community, *Science,* 232, 865, 1986.
68. **Dugdale, R. C. and Goering, J. J.,** Uptake of new and regenerated forms of nitrogen in primary productivity, *Limnol. Oceanogr.,* 12, 196, 1967.
69. **Dunlap, W.,** Distribution of vitamin B_{12} in reef waters of the central Great Barrier Reef, *Proc. Inaugural Great Barrier Reef Conf.,* Baker, J. T., Carter, R. M., Sammarco, P. W., and Stark, K. P., Eds., James Cook University, Townsville, Australia, 1983, 305.
70. **Dunlap, W.,** Measurement of extracellular phosphatase enzyme activity in reef waters of the central Great Barrier Reef, *Proc. 5th Int. Coral Reef Congr.,* Tahiti, Vol. 3, 1985, 451.
71. **Dunlap, W. C. and Susic, M.,** Determination of pteridines and flavines in seawater by reverse-phase, high-performance liquid chromatography with fluorometric detection, *Mar. Chem.,* 17, 185, 1985.
72. **Dunlap, W. C. and Susic, M.,** Photochemical decomposition rates of pteridines and flavins in seawater exposed to surface solar radiation, *Mar. Chem.,* 19, 99, 1986.
73. **Edmond, J. M., Boyle, E. A., Grant, B., and Stallard, R. F.,** The chemical mass balance in the Amazon plume. I. The nutrients, *Deep-Sea Res.,* 28, 1339, 1981.
74. **Emerson, S. and Widmer, G.,** Early diagenesis in anaerobic lake sediments. II. Thermodynamic and kinetic factors controlling the formation of iron phosphate, *Geochim. Cosmochim. Acta,* 42, 1307, 1978.
75. **Engle, D. A. and Maleck, J. M.,** Floating meadow epiphyton: biological and chemical features of epiphytic material in an Amazon floodplain lake, *Freshwater Biol.,* 23, 479, 1989.
76. **Entsch, B., Boto, K. G., Sim, R. G., and Wellington, J. T.,** Phosphorus and nitrogen in coral reef sediments, *Limnol. Oceanogr.,* 28, 465, 1983.
77. **Entsch, B., Sim, R. G., and Hatcher, B. G.,** Indications from photosynthetic components that iron is a limiting nutrient in primary production on coral reefs, *Mar. Biol.,* 73, 17, 1983.
78. **Eppley, R. W. and Peterson, B. H.,** Particulate matter flux and planktonic new production in the deep ocean, *Nature (London),* 282, 677, 1979.
79. **Eppley, R. W., Renger, E. H., Venrick, E. L., and Mullin, M. M.,** A study of plankton dynamics and nutrient cycling in the central gyre of the North Pacific Ocean, *Limnol. Oceanogr.,* 18, 534, 1973.

80. **Eppley, R. W., Swift, E., Redalje, R. D., Landry, M. R., and Haas, H. W.,** Subsurface chlorophyll maximum in August-September 1985 in the CLIMAX area of the North Pacific, *Mar. Ecol. Prog. Ser.,* 42, 289, 1988.
81. **Falkowski, P. G.,** The enzymology of nitrogen assimilation, in *Nitrogen in the Marine Environment,* Carpenter, E. J. and Capone, D. G., Eds., Academic Press, New York, 1983, 839.
82. **Falkowski, P. G. and Stone, D. P.,** Nitrate uptake in marine phytoplankton: energy sources and the interaction with carbon fixation, *Mar. Biol.,* 32, 77, 1975.
83. **Fanning, K. A., Garder, K. L., and Betzer, P. R.,** Sediment resuspension by coastal waters: a potential mechanism for nutrient recycling on the ocean's margin, *Deep Sea Res.,* 29, 953, 1982.
84. **Fell, J. W. and Master, I. M.,** Fungi associated with the degradation of mangrove (*Rhizophora mangle* L.) leaves in South Florida, in *Estuarine Microbiol Ecology,* Stevenson, C. L. H. and Colwell, R. R., Eds., University of South Carolina Press, Columbia, 1973, 455.
85. **Fenchel, T. and Blackburn, T. H.,** *Bacteria and Mineral Cycling,* Academic Press, New York, 1979.
86. **Ferguson, R. L. and Sunda, W. G.,** Utilization of amino acids by planktonic marine bacteria: importance of clean technique and low substrate additions, *Limnol. Oceanogr.,* 29, 258, 1984.
87. **Fillery, I. R. P. and Vlek, P. L. G.,** Reappraisal of the significance of ammonia volatilization as an N loss mechanism in flooded rice fields, in *Nitrogen Economy of Flooded Rice Soils,* deDatta, S. K. and Patrick, W. H., Jr., Eds., Martinpus Nijhoff, The Hague, 1986, 79.
88. **Fisher, T. R. and Parsley, P. E.,** Amazon lakes: water storage and nutrient stripping by algae, *Limnol. Oceanogr.,* 24, 547, 1979.
89. **Fisher, T. R., Doyle, R. D., and Peele, E. R.,** Size-fractionated uptake of ammonium and phosphate in a tropical lake, *Verh. Int. Verein, Limnol.,* 23, 637, 1988.
90. **Forsberg, B. R.,** Nutrient processing in Amazon floodplain lakes, *Verh. Int. Verein. Limnol.,* 22, 1294, 1984.
91. **Forsberg, B. R., Devol, A. H., Richey, J. E., Martinelli, L. A., and dos Santos, H.,** Factors controlling nutrient concentrations in Amazon floodplain lakes, *Limnol. Oceanogr.,* 33, 41, 1988.
92. **Forsberg, C.,** Importance of sediments in understanding nutrient cycles in lakes, *Hydrobiologia,* 176, 262, 1989.
93. **Fox, L. E., Sager, S. L., and Wofsy, S. C.,** The chemical control of soluble phosphorus in the Amazon estuary, *Geochim. Cosmochim. Acta,* 50, 783, 1986.
94. **Friedlander, M. and Dawes, C. J.,** In situ uptake kinetics of ammonium and phosphate and chemical composition of the red seaweed *Gracillaria tikvahiae, J. Phycol.,* 21, 448, 1985.
95. **French, D. P., Furnas, M. J., and Smayda, T. J.,** Diel changes in nitrite concentrations in the chlorophyll maximum in the Gulf of Mexico, *Deep Sea Res.,* 30, 707, 1984.
96. **Froelich, P. N.,** Kinetic controls of dissolved phosphate in natural rivers and estuaries: a primer on the phosphate buffer mechanism, *Limnol. Oceanogr.,* 33(4), 649, 1988.
97. **Fuhrman, J. A. and Ferguson, R. L.,** Nanomolar concentrations and rapid turnover of dissolved free amino acids in seawater: agreement between chemical and microbiological studies, *Mar. Ecol. Prog. Ser.,* 33, 237, 1986.
98. **Fuhrman, J. A.,** Close coupling between release and uptake of dissolved free amino acids in seawater studied by an isotope dilution approach, *Mar. Ecol. Prog. Ser.,* 37, 45, 1987.
99. **Furnas, M. J.,** Cyclonic disturbance and a phytoplankton bloom in a tropical shelf ecosystem, in *Red Tides: Biology, Environmental Science and Toxicology,* Okaichi, T., Anderson, D. M., and Nemoto, T., Eds., Elsevier, New York, 1988, 271.
100. **Furnas, M. J. and Smayda, T. J.,** Inputs of subthermocline waters and nitrate onto the Campeche Bank, *Cont. Shelf Res.,* 7, 161, 1984.
101. **Furnas, M. J., Mitchell, A. W., Wellington, J., and Brady, B.,** Dissolved and particulate nutrients in waters of the Whitsunday Island group, February, 1988, Rep. to the Great Barrier Reef Marine Park Authority, Townsville, 1988.
102. **Furnas, M. J., Mitchell, A. W., Gilmartin, M., and Revelante, N.,** Phytoplankton biomass and primary production in the semi-enclosed reef lagoons of the central Great Barrier Reef, *Coral Reefs,* 9, 1, 1990.
103. **Furtado, J. I. and Mori, S.,** Biogeochemical cycling, in *Tasek Bera, The Ecology of a Freshwater Swamp,* Monogr. Biol., 1982, 359.
104. **Ganf, G. G. and Blazka, P.,** Oxygen uptake ammonia and phosphate excretion by zooplankton of a shallow equatorial lake (Lake George, Uganda), *Limnol. Oceanogr.,* 19, 313, 1974.
105. **Garside, C.,** A chemiluminescent technique for the determination of nanomolar concentrations of nitrate and nitrite in seawater, *Mar. Chem.,* 11, 159, 1982.
106. **Gaudet, J. L.,** Nutrient relationships in the detritus of a tropical swamp, *Arch. Hydrobiol.,* 78, 213, 1976.
107. **Gaudet, J. L.,** Uptake, accumulation, and loss of nutrients by papyrus in tropical swamps, *Ecology,* 58, 415, 1977.
108. **Gaudet, J. L.,** Seasonal changes in nutrients in a tropical swamp: North Swamp, Lake Naivasha, Kenya, *J. Ecol.,* 67, 953, 1979.

109. **Gerber, R. P. and Marshall, N.,** Ingestion of detritus by the lagoon pelagic community at Enewetak Atoll, *Limnol. Oceanogr.*, 19, 815, 1974.
110. **Glibert, P. M.,** Regional studies of daily, seasonal and size fraction variability in ammonium mineralization, *Mar. Biol.*, 70, 209, 1982.
111. **Glibert, P. M., Lipschultz, F., McCarthy, J. J., and Altabet, M. M.,** Isotope dilution models of uptake and remineralization of ammonium by marine plankton, *Limnol. Oceanogr.*, 27, 639, 1982.
112. **Goering, J. J. and Parker, P. L.,** Nitrogen fixation by epiphytes on sea grasses, *Limnol. Oceanogr.*, 17, 320, 1972.
113. **Golley, F. P.,** Nutrient cycling and nutrient conservation, in *Ecosystems of the World, 14A, Tropical Rainforests,* Golley, F. P., Ed., Elsevier, Amsterdam, 1983, 137.
114. **Goldman, J. C.,** Physiological processes, nutrient availability and the concept of relative growth rate in marine phytoplankton ecology, *Brookhaven Symp. Biol.*, Vol. 31, Plenum Press, New York, 1980, 179.
115. **Goldman, J. C. and Caron, D. A.,** Experimental studies on an omnivorous microflagellate: implications for grazing and nutrient regeneration in the marine microbial food chain, *Deep Sea Res.*, 32, 899, 1985.
116. **Goldman, J. C. and McCarthy, J. J.,** Steady state growth and ammonium uptake of a fast growing marine diatom, *Limnol. Oceanogr.*, 23, 695, 1978.
117. **Goldman, J. C. and Glibert, P. M.,** Kinetics of inorganic nitrate uptake by phytoplankton, in *Nitrogen in the Marine Environment,* Carpenter, E. J. and Capone, D. G., Eds., Academic Press, New York, 1983, 233.
118. **Goldman, J. C., McCarthy, J. J., and Peavey, D. G.,** Growth rate influence on the chemical composition of phytoplankton in oceanic waters, *Nature (London),* 279, 210, 1979.
119. **Goldman, J. C., Taylor, C. D., and Glibert, P. M.,** Nonlinear time-course uptake of carbon and ammonium by marine phytoplankton, *Mar. Ecol. Prog. Ser.*, 6, 137, 1981.
120. **Goldman, J. C., Caron, D. A., and Dennett, M. R.,** Regulation of gross growth efficiency and ammonium regeneration bacteria by substrate C:N ratio, *Limnol. Oceanogr.*, 32, 1239, 1987.
121. **Gordon, D. G.,** Distributions of particulate organic carbon and nitrogen at an oceanic station in the central Pacific, *Deep Sea Res.*, 18, 1127, 1134.
122. **Graham, W. F. and Duce, R. A.,** Atmospheric inputs of phosphorus to remote tropical islands, *Pac. Sci.*, 35, 241, 1981.
123. **Grasshoff, K.,** *Methods of Sea Water Analysis,* Verlag Chemie, 1976.
124. **Grundmanis, V. and Murray, J. W.,** Nitrification and denitrification in marine sediments from Puget Sound, *Limnol. Oceanogr.*, 22, 804, 1977.
125. **Hanisak, M. D.,** Nitrogen relationships of marine macroalgae, in *Nitrogen in the Marine Environment,* Carpenter, E. J. and Capone, D. G., Eds., Academic Press, New York, 1983, 699.
126. **Hansen, J. A., Alongi, D. M., Moriarity, D. J. W., and Pollard, P. C.,** The dynamics of benthic microbial communities at Davies Reef, central Great Barrier Reef, *Coral Reefs,* 6, 63, 1987.
127. **Hanson, R. B. and Gunderson, K.,** Influence of sewage discharge on nitrogen fixation and nitrogen flux from coral reefs in Kaneohe Bay, Hawaii, *Appl. Environ. Microbiol.*, 31, 942, 1976.
128. **Harrison, W. G. and Harris, L. R.,** Isotope dilution and its effects on measurements of nitrogen and phosphorus uptake by oceanic microplankton, *Mar. Ecol. Prog. Ser.*, 27, 253, 1986.
129. **Harvey, W. A. and Caperon, J.,** The rate of utilization of urea, ammonium and nitrate by natural populations of marine phytoplankton in a eutrophic environment, *Pac. Sci.*, 30, 329, 1976.
130. **Hatcher, A. I. and Frith, C. A.,** The control of nitrate and ammonium concentrations in a coral reef lagoon, *Coral Reefs,* 4, 101, 1985.
131. **Hatcher, B. G., Johannes, R. E., and Robertson, A. I.,** Review of research relevant to the conservation of shallow tropical marine ecosystems, *Oceanogr. Mar. Biol. Annu. Rev.*, 27, 337, 1989.
132. **Hattori, A.,** Detritification and dissimilatory nitrate reduction, in *Nitrogen in the Marine Environment,* Carpenter, E. J. and Capone, D. G., Eds., Academic Press, New York, 1983, 191.
133. **Helder, W. and deVries, R. T. P.,** Estuarine maxima and nitrifying bacteria (Ems Dollard estuary), *Neth. J. Sea Res.*, 17, 1, 1983.
134. **Henry, R. and Tundisi, J. G.,** Evidence of limitation by molybdenum and nitrogen on the growth of the phytoplankton community of the Lobo Reservoir (Sao Paulo, Brazil), *Rev. Hydrobiol. Trop.*, 15, 201, 1982.
135. **Henry, R., Tundisi, J. G., and Curi, P. R.,** Effects of phosphorus and nitrogen enrichment on the phytoplankton in a tropical reservoir (Lobo Reservoir, Brazil), *Hydrobiologia,* 118, 177, 1984.
136. **Hirose, K.,** Metal-organic ligand interactions in seawater: multi-metal complexation model for metal titration, *Mar. Chem.*, 25, 39, 1988.
137. **Hopkinson, C. S., Sherr, B. F., and Ducklow, H. W.,** Microbial regeneration of ammonium in the water column of Davies Reef, Australia, *Mar. Ecol. Prog. Ser.*, 41, 147, 1987.
138. **Horne, A. J. and Viner, A. B.,** Nitrogen fixation and its significance in tropical Lake George, Uganda, *Nature (London),* 232, 417, 1971.
139. **Howard-Williams, C.,** Vegetation changes in a shallow African lake: response of the vegetation to a recent dry period, *Hydrobiologia,* 47, 381, 1975.

140. **Howard-Williams, C.**, Studies on the ability of a *Potamogeton pectinatus* community to remove dissolved nitrogen and phosphorus compounds from lake water, *J. Appl. Ecol.*, 18, 619, 1981.
141. **Howard-Williams, C.**, Cycling and retention of nitrogen and phosphorus in wetlands: a theoretical and applied perspective, *Freshwater Biol.*, 15, 391, 1985.
142. **Howard-Williams, C. and Howard-Williams, W.**, Nutrient leaching from the swamp vegetation of Lake Chilwa, a shallow African lake, *Aquat. Bot.*, 4, 257, 1978.
143. **Howard-Williams, C. and Junk, W. J.**, The chemical composition of central Amazonian aquatic macrophytes with special reference to their role in the ecosystem, *Arch. Hydrobiol.*, 79, 446, 1977.
144. **Howarth, R. W. and Merkel, S.**, Pyrite formation and the measurement of sulfate reduction in salt marsh sediments, *Limnol. Oceanogr.*, 29, 598, 1984.
145. **Howarth, R. W. and Cole, J. J.**, Molybdenum availability, nitrogen limitation, and phytoplankton growth in natural waters, *Science*, 229, 653, 1985.
146. **Howarth, R. W., Mariano, R., Lane, J., and Cole, J. J.**, Nitrogen fixation in freshwater, estuarine and marine ecosystems. I. Rates and importance, *Limnol. Oceanogr.*, 33(4), 669, 1988.
147. **Howarth, R. W., Mariano, R., and Cole, J. J.**, Nitrogen fixation in freshwater, estuarine and marine ecosystems. II. Biogeochemical controls, *Limnol. Oceanogr.*, 33(4), 688, 1988.
148. **Huizenga, D. L. and Kester, D. R.**, Protonation equilibria of marine dissolved organic matter, *Limnol. Oceanogr.*, 24, 145, 1979.
149. **Huntsman, S. A. and Sunda, W. G.**, The role of trace metals in regulating phytoplankton growth with emphasis on Fe, Mn and Cu, in *The Physiological Ecology of Phytoplankton*, Morris, I., Ed., University of California Press, Berkeley, 1980, 285.
150. **Ikeda, T., Hing Fay, E., Hutchinson, S. A., and Boto, G. M.**, Ammonia and inorganic phosphate excretion by zooplankton from inshore waters of the Great Barrier Reef, Qld. I. Relationship between excretion rate and body size, *Austr. J. Mar. Freshwater Res.*, 33, 55, 1982.
151. **Ikeda, T., Carlton, J. H., Mitchell, A. W., and Dixon, P.**, Ammonia and phosphate excretion by zooplankton from the inshore waters of the Great Barrier Reef. II. Their *in situ* contributions to nutrient regeneration, *Austr. J. Mar. Freshwater Res.*, 33, 683, 1982.
152. **Ikusima, I.**, Primary production and population ecology of the aquatic sedge *Leptopironia articulata* in a tropical swamp, Tasek Bera, Malaysia, *Aquat. Bot.*, 4, 269, 1978.
153. **Ittekkot, V., Martins, O., and Seifert, R.**, Nitrogenous organic matter transported by the major world rivers, *Mitt. Geol. Palaont. Inst. Univ. Hamburg*, 55, 119, 1983.
154. **Jackson, G. A.**, Phytoplankton growth and zooplankton grazing in oligotrophic oceans, *Nature (London)*, 284, 439, 1980.
155. **Jackson, G. A.**, Implications for high dissolved organic matter concentrations oceanic properties and processes, *Oceanography*, 2, 28, 1988.
156. **Jackson, G. A. and Williams, P. M.**, Importance of dissolved organic nitrogen and phosphorus to biological nutrient cycling, *Deep Sea Res.*, 32, 223, 1985.
157. **Jenkins, M. C. and Kemp, W. M.**, The coupling of nitrification and denitrification in two estuarine sediments, *Limnol. Oceanogr.*, 29, 609, 1984.
158. **Johannes, R. E.**, Phosphorus excretion and body size in marine animals: microzooplankton and nutrient regeneration, *Science*, 146, 923, 1964.
159. **Johannes, R. E.**, The ecological significance of the submarine discharge of groundwater, *Mar. Ecol. Prog. Ser.*, 3, 65, 1980.
160. **Johannes, R. E. et al.**, The metabolism of some coral reef communities: a team study of nutrient and energy flux at Eniwetok, *Bioscience*, 22, 541, 1972.
161. **Johannes, R. E., Kimmerer, W., Kinzie, R., Shiroma, E., and Walsh, T.**, The impacts of human activities on Tarawa Lagoon, Rep. to Government, Gilbert Islands, 1979.
162. **Johannes, R. E., Wiebe, W. J., and Crossland, C. J.**, Three patterns of nutrient flux in a coral reef community, *Mar. Ecol. Prog. Ser.*, 12, 131, 1983.
163. **Johnson, K. M., Burney, C. M., and Sieburth, J. McN.**, Precise and accurate determination by infrared photometry of CO_2 dynamics in marine ecosystems, *Mar. Ecol. Prog. Ser.*, 10, 251, 1983.
164. **Junk, W. J.**, Amazonian floodplains: their ecology, present and potential use, *Rev. Hydrobiol. Trop.*, 15, 285, 1982.
165. **Kalk, M., McLachlan, A. J., and Howard-William, C.**, *Lake Chilwa. Studies of Change in a Tropical Ecosystem*, Junk, The Hague, 1979.
166. **Kanel, J. and Morse, J. W.**, The chemistry of ortho-phosphate uptake from seawater on to calcite and aragonite, *Geochim. Cosmochim. Acta*, 42, 1335, 1978.
167. **Kaplan, W. A.**, Nitrification, in *Nitrogen in the Marine Environment*, Carpenter, E. J. and Capone, D. G., Eds., Academic Press, New York, 1983, 139.
168. **Karl, D. M., Knauer, G. A., Martin, J. H., and Ward, B. B.**, Bacterial chemolithotrophy in the oceans is associated with sinking particles, *Nature (London)*, 309, 54, 1984.
169. **Kaul, L. W. and Froelich, P. N.**, Modelling estuarine nutrient geochemistry in a simple system, *Geochim. Cosmochim. Acta*, 48, 1417, 1984.

170. **Kelly, J. R. and Nixon, S. W.,** Experimental studies of the effect of organic deposition on the metabolism of a coastal marine bottom community, *Mar. Ecol. Prog. Ser.,* 17, 57, 1984.
171. **Kiefer, D. A., Olson, R. J., and Holm Hansen, O.,** Another look at the nitrite and chlorophyll maxima in the central north Pacific, *Deep Sea Res.,* 1199, 1976.
172. **Kinsey, D. W.,** Metabolism, calcification and carbon production, *Proc. 5th Int. Coral Reef Congr.,* Tahiti, Vol. 4, 1985, 504.
173. **Kirchman, D. L.,** Regulation of ammonium uptake by heterotrophic bacteria in the oceanic subarctic Pacific, in *Recent Advances in Microbial Ecology,* Hattori, T. et al., Eds., Japan Sci. Soc. Press, Tokyo, 1989, 383.
174. **Knauer, G. A., Martin, J. H., and Bruland, K. W.,** Fluxes of particulate carbon, nitrogen and phosphorus in the upper water column of the northeast Pacific, *Deep Sea Res.,* 26, 97, 1979.
175. **Koike, I. and Sumi, T.,** Nitrogen cycling in coastal sediments with special reference to ammonium metabolism, in *Recent Advances in Microbial Ecology,* Hattori, T., et al., Eds., Japan, Sci. Soc. Press, Tokyo, 1989, 365.
176. **Koop, K. and Larkum, A. W. D.,** Deposition of organic matter in a coral reef lagoon, One Tree Island, Great Barrier Reef, *Est. Coast. Shelf, Sci.,* 25, 1, 1987.
177. **Kormanic, G. A. and Cameron, J. N.,** Ammonia excretion in animals that breathe water: a review, *Mar. Biol. Lett.,* 2, 11, 1981.
178. **Krom, M. D. and Berner, R. A.,** The diagenesis of phosphorus in a nearshore marine sediment, *Geochim. Cosmochim. Acta,* 45, 207, 1981.
179. **Lakshmanan, P. T., Shynamma, C. S., Balchand, A. N., and Nambisan, P. N. K.,** Distribution and variability of nutrients in Cochin Backwaters, southwest coast of India, *Ind. J. Mar. Sci.,* 16, 99, 1987.
180. **Lakshimarayana, S. S.,** Studies on phytoplankton of the river Ganges, Varanasi, India, I. The physico-chemical characteristics of the river Ganges, *Hydrobiologica,* 15, 119, 1965.
181. **Larkum, A. W. D., Kennedy, I. R., and Muller, W. J.,** Nitrogen fixation on a coral reef, *Mar. Biol.,* 98, 143, 1988.
182. **Laws, E. A. and Redalje, D. G.,** Sewage diversion effects on the water column of a subtropical estuary, *Mar. Environ. Res.,* 6, 265, 1982.
183. **Laws, E. A. and Woodcock, A. H.,** Hypereutrophication of an Hawaiian alpine lake, *Pac. Sci.,* 35, 257, 1981.
184. **Laws, E. A., DiTullio, G. R., and Redalje, D. G.,** High phytoplankton growth and production rates in the North Pacific subtropical gyre, *Limnol. Oceanogr.,* 32, 905, 1987.
185. **Lasier, J. R. N. and Mann, K. H.,** Turbulence and the diffusive layers around small organisms, *Deep Sea Res.,* 36, 1721, 1989.
186. **LeBouteiller, A.,** Environmental control of nitrate and ammonium uptake by phytoplankton in the equatorial Atlantic Ocean, *Mar. Ecol. Prog. Ser.,* 30, 167, 1986.
187. **LeBouteiller, A. and Herbland, A.,** Carbon fixation and productivity in relation to chlorophyll and light in the equatorial Atlantic Ocean, *Oceanogr. Trop.,* 19, 161, 1984.
188. **Lee, C. and Bada, J. L.,** Dissolved amino acids in the equatorial Pacific, the Sargasso Sea and Biscayne Bay, *Limnol. Oceanogr.,* 22, 502, 1977.
189. **Lee, C. and Cronin, C.,** The vertical flux of particulate organic nitrogen in the sea: decomposition of amino acids in the Peru upwelling area, *J. Mar. Res.,* 48, 227, 1982.
190. **Legendre, L., Demers, S., Delesalle, B., and Harnois, C.,** Biomass and photosynthetic activity of phototrophic picoplankton in coral reef waters (Moorea Island, French Polynesia), *Mar. Ecol. Prog. Ser.,* 47, 153, 1988.
191. **Lehman, J. T. and Scavia, D.,** Microscale nutrient patches produced by zooplankton, *Proc. Natl. Acad. Sci. U.S.A., 79, 5001, 1982.*
192. **Lenz, P. H., Melack, J. M., Robertson, B., and Hardy, E. A.,** Ammonium and phosphate regeneration by the zooplankton of an Amazon floodplain lake, *Freshwater Biol.,* 16, 821, 1986.
193. **Levine, S. N. and Lewis, W. M., Jr.,** A numerical model of nitrogen fixation and its application to Lake Valencia, Venezuela, *Freshwater Biol.,* 17, 265, 1987.
194. **Lewis, J. B.,** Measurement of groundwater seepage flux into a coral reef: spatial and temporal variations, *Limnol. Oceanogr.,* 32, 1165, 1987.
195. **Lewis, W. M. and Weibezahn, F. H.,** Phosphorus and nitrogen loading of Lake Valencia, *Acta Cient. Venez.,* 34, 345, 1983.
196. **Lipschultz, F., Wofsy, S. C., and Fox, L. E.,** Nitrogen metabolism of the eutrophic Delaware River ecosystem, *Limnol. Oceanogr.,* 31, 701, 1986.
197. **MacIsaac, J. J.,** Diel cycles of inorganic nitrogen uptake in a natural phytoplankton population dominated by *Gonyaulax polyedra, Limnol. Oceanorg.,* 23, 1, 1978.
198. **MacIsaac, J. J. and Dugdale, R. C.,** Interactions of light and inorganic nitrogen in controlling nitrogen uptake in the sea, *Deep Sea Res.,* 19, 209, 1972.

199. **Mackin, J. E., Aller, R. C., and Ullman, W. J.,** The effect of iron reduction and nonsteady-state diagenesis on iodine, ammonium and boron distributions in sediments from the Amazon continental shelf, *Cont. Shelf Res.,* 8, 363, 1988.
200. **Mague, T. H. and Holm-Hansen, O.,** Nitrogen fixation on a coral reef, *Phycologia,* 14, 87, 1975.
201. **Malone, T. C., Garside, C., Haines, K. C., and Roels, O. A.,** Nitrate uptake and growth of *Chaetoceros* sp. in large outdoor continuous culture, *Limnol. Oceanogr.,* 20, 9, 1975.
202. **Mann, K. H.,** Production and use of detritus in various freshwater, esturarine and coastal marine ecosystems, *Limnol. Oceanogr.,* 33(4), 910, 1988.
203. **Marsh, J. A.,** Terrestrial inputs of nitrogen and phosphorus on fringing reefs of Guam, *Proc. 3rd Int. Coral Reef Symp.,* Vol. 1, 1977, 331.
204. **Martin, J. H. and Fitzwater, S. E.,** Iron deficiency limits phytoplankton growth in the northeast Pacific subarctic, *Nature (London),* 331, 341, 1988.
205. **McCarthy, J. J. and Goldman, J. C.,** Nitrogenous nutrition of marine phytoplankton in nutrient depleted waters, *Science,* 203, 670, 1979.
206. **McCarthy, J. J., Kaplan, W., and Nevins, J. L.,** Chesapeake Bay nutrient and plankton dynamics. II. Sources and sinks of nitrite, *Limnol. Oceanogr.,* 29, 84, 1984.
207. **McCarthy, J. J., Taylor, W. R., and Taft, J. L.,** The dynamics of nitrogen and phosphorus cycling in the open waters of Chesapeake Bay, in *Marine Chemistry in the Coastal Environment,* Church, T. M., Ed., ACS Symp. Ser. 18, Washington, D.C., 1975, 664.
208. **McCarthy, J. J., Taylor, W. R., and Taft, J. L.,** Nitrogenous nutrition of the plankton in the Chesapeake Bay. I. Nutrient availability and phytoplankton preferences, *Limnol. Oceanogr.,* 22, 996, 1977.
209. **Messer, J. and Brezonik, P. L.,** Comparison of denitrification rate estimation techniques in a large shallow lake, *Water Res.,* 631, 1983.
210. **Meybeck, M.,** Carbon, nitrogen and phosphorus transport by world rivers, *Am. J. Sci.,* 282, 401, 1982.
211. **Meyer, J. L., Schultz, E. T., and Helfman, G. S.,** Fish schools: an asset to corals, *Science,* 220, 1047, 1983.
212. **Mitchell, A. W., Ed.,** Oceanogrphic observations in the Gulf of Papua and Torres Strait, Australian Inst. of Marine Science Data Re., AIMS-OS-82-1.
213. **Mitsui, A., Kumazawa, S., Takahashi, A., Ikemoto, H., Cao, S., and Ari, T.,** Strategy by which nitrogen-fixing unicellular cyanobacteria grow photoautotrophically, *Nature (London),* 323, 720, 1986.
214. **Montagnini, F. and Buschbacher, R.,** Nitrification rates in two undisturbed tropical rain forests and three slash-and-burn sites of the Venezuelan Amazon, *Biotropica,* 21, 9, 1989.
215. **Mopper, K. and Lindroth, P.,** Diel and depth variations in dissolved free amino acids and ammonium in the Baltic Sea determined by shipboard HPLC analysis, *Limnol. Oceanogr.,* 27, 336, 1982.
216. **Moore, A. W.,** *Azolla,* Biology and agronomic significance, *Bot. Rev.,* 35, 17, 1969.
217. **Morel, F. M. M. and Morgan, J.,** A numerical model for computing equilibria in aqueous chemical systems, *Environ. Sci. Tech.,* 6, 58, 1972.
218. **Morris, I.,** *The Physiological Ecology of Phytoplankton,* University of California Press, Berkeley, 1980.
219. **Morrissey, K. M. and Fisher, T. R.,** Regeneration and uptake of ammonium by plankton in an Amazon floodplain lake, *J. Plank. Res.,* 10, 31, 1988.
220. **Muscatine, L., Falkowski, P. G., Dubinsky, Z., Cook, P. A., and McCloskey, L. R.,** The effect of external nutrient resources on the population dynamics of zooxanthellae in a reef coral, *Proc. R. Soc. London,* B236, 311, 1989.
221. **Nalewajko, C. and Lean, D. R. S.,** Phosphorus, in *The Physiological Ecology of Phytoplankton,* Morris, I., Ed., University of California Press, Berkeley, 1980, 235.
222. **Nelson, D. M. and Goering, J. J.,** Near-surface silica dissolution in the upwelling region off northwest Africa, *Deep Sea Res.,* 24, 65, 1977.
223. **Niewolt, S.,** *Tropical Climatology,* J. Wiley & Sons, New York, 1977.
224. **Nixon, S. W., Oviatt, C. A., and Hale, S. S.,** Nitrogen regeneration and the metabolism of marine bottom communities, in *The Role of Terrestrial and Aquatic Organisms in Decomposition Processes,* Anderson, J. M. and MacFadyed, A., Eds., Blackwell Scientific, London, 1976, 269.
225. **Nixon, S. W.,** Remineralization and nutrient cycling in coastal marine ecosystems, in *Estuaries and Nutrients,* Neilsen, B. J. and Cronin, L. E., Eds., Humana, Clifton, NJ, 1981, 111.
226. **Nixon, S. W., Ed.,** Comparative ecology of freshwater and marine ecosystems, *Limnol. Oceanogr.,* 33(part 2), 1988.
227. **Nowicki, B. L. and Nixon, S. W.,** Benthic nutrient remineralization in a coastal lagoon ecosystem, *Estuaries,* 8, 182, 1985.
228. **Odum, H. T. and Odum, E. P.,** Trophic structure and productivity of a windward coral reef community on Eniwetok Atoll, *Ecol. Monogr.,* 25, 291, 1955.
229. **Olson, R. J.,** Differential photoinhibition of marine nitrifying bacteria: a possible mechanism for the formation of the primary nitrite maximum, *J. Mar. Res.,* 39, 227, 1981.

230. **Orrett, K. and Karl, D. M.,** Dissolved organic phosphorus production in surface seawaters, *Limnol. Oceanogr.*, 32, 383, 1987.
231. **Packard, T. T., Blasco, D., MacIsaac, J. J., and Dugdale, D. G.,** Variations of nitrate reductase activity in marine phytoplankton, *Inv. Pesq.*, 35, 209, 1971.
232. **Paerl, H. W. and Bebout, B. M.,** Direct measurements of O_2-depleted microzones in marine *Oscillatoria*: relation to N_2 fixation, *Science*, 241, 442, 1988.
233. **Paerl, H. W. and Carlton, R. G.,** Control of nitrogen fixation by oxygen depletion in surface associated microzones, *Nature (London)*, 332, 260, 1988.
234. **Paerl, H. W., Crocker, K. M., and Prufert, L. E.,** Limitation of N_2 fixation in coastal marine waters: relative importance of molybdenum, iron, phosphorus, and organic matter availability, *Limnol. Oceanogr.*, 32, 525, 1987.
235. **Paerl, H. W., Rebout, B. M., Prufet, L. E., and Currin, C. A.,** Mediation of aquatic N_2 fixation potentials by O_2-depleted microzones, in *Recent Advances in Microbial Ecology*, Hattori, T., Ishida, Y., Maruyama, Y., Morita, R. Y., and Uchida, A., Eds., Japan Sci. Soc. Press, Tokyo, 1989, 108.
236. **Painter, H. A.,** A review of literature of inorganic nitrogen metabolism in microorganisms, *Water Res.*, 4, 393, 1970.
237. **Pasciak, W. J. and Gavis, J.,** Transport limitation of nutrient uptake in phytoplankton, *Limnol. Oceanogr.*, 19, 881, 1974.
238. **Patriquin, D. G.,** The origin of nitrogen and phosphorus for growth of the marine angiosperm *Thalassia testudinum*, *Mar. Biol.*, 15, 35, 1972.
239. **Patriquin, D. G. and Knowles, R.,** Nitrogen fixation in the rhizosphere of marine angiosperms, *Mar. Biol.*, 16, 49, 1972.
240. **Paul, J. H.,** Uptake of organic nitrogen, in *Nitrogen in the Marine Environment*, Carpenter, E. J. and Capone, D. G., Eds., Academic Press, New York, 1983, 275.
241. **Pawly, A. L. and Alfaro, R.,** Zooplankton in the Lake Titicaca ecosystem: the importance of regenerated nitrogen to phytoplankton production, *Verh. Int. Verein. Limnol.*, 22, 1258, 1984.
242. **Pedrozo, F. and Bonetto, C.,** Nitrogen and phosphorus transport in the Bermejo River (South America), *Rev. Hydrobiol. Trop.*, 20, 91, 1987.
243. **Perry, M. J.,** Alkaline phosphatase activity in subtropical central North Pacific waters using sensitive fluorometric method, *Mar. Biol.*, 15, 113, 1972.
244. **Perry, M. J.,** Phosphate utilization by an oceanic diatom in phosphorus limited chemostat culture and in the oligotrophic waters of the central North Pacific Ocean, *Limnol. Oceanogr.*, 21, 88, 1976.
245. **Pilson, M. E. Q.,** Annual cycles of nutrients and chlorophyll in Narragansett Bay, R.I., *J. Mar. Res.*, 43, 849, 1985.
246. **Pocklington, R. and MacKinnon, M. D.,** Organic matter in upwelling off Senegal and The Gambia, *Rapp. P.-V. Reun. Cons. Int. Expl. Mer.*, 180, 254, 1982.
247. **Polunin, N. C.,** The decomposition of emergent macrophytes in freshwater, *Adv. Ecol. Res.*, 14, 115, 1984.
248. **Pomeroy, L. R., Pilson, M. E. Q., and Wiebe, W. J.,** Tracer studies of exchange of phosphorus between reef waters and organisms on the windward reef of Eniwetok Atoll, *Proc. 2nd Int. Coral Reef Symp.*, 1, 87, 1974.
249. **Potts, M. and Whitton, B. A.,** Nitrogen fixation by blue-green algal communities in the intertidal zone of the lagoon of Aldabra Atoll, *Oecologia*, 27, 275, 1977.
250. **Putt, M. and Stoecker, D. K.,** An experimental determined carbon: volume ratio for marine "oligotrichous" ciliates from estuarine and coastal waters, *Limnol. Oceanogr.*, 34, 1097, 1989.
251. **Raimault, P., Slawyk, G., Coste, B., and Fry, J.,** Feasibility of using an automated colorimetric procedure for the determination of seawater nitrate in the 0 to 100 nM range: examples from field and culture, *Mar. Biol.*, 104, 347, 1990.
252. **Reddy, K. R. and Patrick, W. H., Jr.,** Denitrification losses in flooded rice fields, in *Nitrogen Economy of Flooded Rice Soils*, deDatta, S. K. and Patrick, W. H., Jr., Eds., Martinus Nijhoff, The Hague, 1986, 99.
253. **Reddy, K. R. and Smith, W. H., Eds.,** *Aquatic Plants for Water Treatment and Resource Recovery*, Magnolia Publ., Orlando, 1987.
254. **Reddy, K. R., Patrick, W. H., Jr., and Lindau, C. W.,** Nitrification-denitrification at the plant root-sediment interface in wetlands, *Limnol. Oceanogr.*, 34, 1004, 1989.
255. **Reddy, K. R., Agami, M., and Tucker, J. C.,** Influence of nitrogen supply rates on growth and nutrient storage by water hyacinth (*Eichhornia crassipes*) plants, *Aquat. Bot.*, 36, 33, 1989.
256. **Redfield, A. C., Ketchum, B. H., and Richards, F. A.,** The influence of organisms on the composition of seawater, in *The Sea*, Hill, M. N., Ed., John Wiley & Sons, New York, 1963, 26.
257. **Riber, H. H. and Wetzel, R. G.,** Boundary layer and internal diffusion effects on phosphorus fluxes in lake periphyton, *Limnol. Oceanogr.*, 32, 1181, 1987.

258. **Rice, D. L.,** The detritus nitrogen problem: new observations and perspectives from organic geochemistry, *Mar. Ecol. Prog. Ser.*, 9, 153, 1982.
259. **Ridd, P., Sandstrom, M. W., and Wolanski, E.,** Outwelling from tropical tidal salt flats, *Est. Coast. Shelf Sci.*, 26, 243, 1988.
260. **Rigler, F. H.,** Further observations inconsistent with the hypothesis that the molybdenum blue method measures orthophosphate in lake waters, *Limnol. Oceanogr.*, 9, 511, 1968.
261. **Rinaudo, G., Balandreau, J., and Dommergues, Y.,** Algal and bacterial non-symbiotic nitrogen fixation in paddy soils, in *Biological Nitrogen Fixation in Natural and Agricultural Habitats,* Lie, T. A. and Multer, E. G., Eds., Plant and Soil Spec. Vol., 1971, 471.
262. **Risk, M. J. and Muller, H. R.,** Porewater in coral heads: evidence for nutrient regeneration, *Limnol. Oceanogr.*, 28, 1004, 1983.
263. **Risk, M. J. and Rhodes, E. G.,** From mangroves to petroleum precursors: an example from tropical northeast Australia, *Bull Am. Assoc. Pet. Geol.*, 69, 1230, 1981.
264. **Rivkin, R. B. and Swift, E.,** Diel and vertical patterns of alkaline phosphatase activity in the oceanic dinoflagellate *Pyrocystis noctiluca, Limnol. Oceanogr.*, 24, 107, 1979.
265. **Rivkin, R. B. and Swift, E.,** Characterization of alkaline phosphatase and organic phosphorus utilization in the oceanic dinoflagellate *Pyrocystis noctiluca, Mar. Biol.*, 61, 1, 1980.
266. **Robertson, A. I.,** Leaf-burying crabs: their influence on energy flow and export from mixed mangrove forests (Rhizophora spp.) in northeastern Australia, *J. Exp. Mar. Biol. Ecol.*, 102, 237, 1986.
267. **Robertson, A. I.,** Decomposition of mangrove leaf litter in tropical Australia, *J. Exp. Mar. Biol. Ecol.*, 116, 235, 1988.
268. **Robertson, A. I. and Daniels, P. A.,** Decomposition and the annual flux of detritus from fallen timber in tropical mangrove forests, *Limnol. Oceanogr.*, 34, 640, 1989.
269. **Robertson, G. P.,** Nitrification and nitrogen mineralization in a lowland rainforest succession in Costa Rica, Central America, *Oecologia,* 61, 99, 1984.
270. **Room, P. M.,** Equations relating growth and uptake of nitrogen by *Salvinia molesta* to temperature and the availability of nitrogen, *Aquat. Bot.*, 24, 43, 1986.
271. **Rowe, G. T.,** Benthic processes in coastal upwelling ecosystems, in *Coastal Upelling,* Richards, F. A., Ed., American Geophys. Union, Washington, D.C., 1981, 464.
272. **Ryther, J. H. and Dunstan, W. H.,** Nitrogen, phosphorus and eutrophication in the coastal marine environment, *Science,* 171, 1008, 1971.
273. **Sale, P. J. M., Orr, P. T., Shell, G. S., and Erskine, D. J. C.,** Photosynthesis and growth rates in *Salvinia molesta* and *Eichhornia crassipes, J. Appl. Ecol.*, 22, 125, 1985.
274. **Santschi, P. H.,** Factors controlling the biogeochemical cycles of trace elements in fresh and coastal marine waters as revealed by artifical radioisotopes, *Limnol. Oceanogr.*, 33(4), 848, 1988.
275. **Seitzinger, S. P.,** Nitrogen biogeochemistry in an unpolluted estuary: the importance of benthic denitrification, *Mar. Ecol. Prog. Ser.*, 41, 177, 1987.
276. **Seitzinger, S. P.,** Denitrification in freshwater and coastal marine ecosystems: ecological and geochemical significance, *Limnol. Oceanogr.*, 33(4), 702, 1988.
277. **Seitzinger, S. P. and D'Elia, C. D.,** Preliminary studies of dinitrification on a coral reef, *Symp. Ser. Undersea Res., NOAA Undersea Res. Prog.*, 3, 199, 1985.
278. **Seitzinger, S. P. and Nixon, S. W.,** Eutrophication and the rate of denitrification and N_2O production in coastal marine sediments, *Limnol. Oceanogr.*, 30, 1332, 1985.
279. **Serruya, C. and Pollingher, U.,** *Lakes of the Warm Belt,* Cambridge University Press, London, 1983.
280. **Setaro, F. V. and Melack, J. M.,** Responses of phytoplankton to experimental nutrient enrichment in an Amazon floodplain lake, *Limnol. Oceanogr.*, 29, 972, 1984.
281. **Sharma, K. P. and Goel, P. K.,** Studies on decomposition of two species of *Salvinia, Hydrobiologia,* 131, 57, 1986.
282. **Sharp, J. H.,** Size classes of organic carbon in seawater, *Limnol. Oceanogr.*, 18, 441, 1973.
283. **Sherr, B. F., Sherr, E. B., and Berman, T.,** Grazing, growth, and ammonium excretion rates of a heterotrophic microflagellate fed with four species of bacteria, *Appl. Environ. Microbiol.*, 45, 1196, 1983.
284. **Sholkovitz, E. R. and Copland, D.,** The coagulation, solubility and adsorption properties of Fe, Mn, Cu, Ni, Cd, Co and humic acids in a river water, *Geochim. Cosmochim. Acta,* 45, 181, 1981.
285. **Simon, N. S.,** Nitrogen cycling between sediment and the shallow-water column in the transition zone of the Potomac River and estuary. II. The role of wind-driven resuspension and adsorbed ammonium, *Est. Coast. Shelf Sci.*, 28, 551, 1989.
286. **Smayda, T. J.,** The suspension and sinking of phytoplankton in the sea, *Oceanogr. Mar. Biol. Annu. Rev.*, 8, 353, 1970.
287. **Smith, C. J., DeLaune, R. D., and Patrick, W. H., Jr.,** Fate of riverine nitrate entering an estuary. I. Denitrification and nitrogen burial, *Estuaries,* 8, 15, 1985.
288. **Smith, D. F.,** Small-scale spatial heterogeneity in dissolved nutrient concentrations, *Limnol. Oceanogr.*, 31, 167, 1986.

289. **Smith, R. E. H., Harrison, W. G., and Harris, L.,** Phosphorus exchange in marine microplankton communities near Hawaii, *Mar. Ecol. Prog. Ser.,* 86, 75, 1985.
290. **Smith, S. V.,** Phosphorus versus nitrogen limitation in the marine environment, *Limnol. Oceanogr.,* 29, 1149, 1984.
291. **Smith, S. V. and Kinsey, D. W.,** Calcification and organic carbon metabolism as indicated by carbon dioxide, in *Coral Reefs: Research Methods,* Stoddart, D. R. and Johannes, R. E., Eds., UNESCO, Paris, 1978, 469.
292. **Smith, S. V. and Henderson, R. S.,** An environmental survey of Canton Island lagoon, 1973, Rep. Naval Undersea Res. Dev. Center, San Diego, NUC-TP-395, 1973.
293. **Smith, S. V. and Atkinson, M. J.,** Mass balance of carbon and phosphorus in Shark Bay, Western Australia, *Limnol. Oceanogr.,* 28, 625, 1983.
294. **Smith S. V. and Jokiel, P. L.,** Water composition and biogeochemical gradients in the Canton Atoll lagoon. I. Lagoon description; design of system analysis; salt and water budget, *Mar. Sci. Commun.,* 1, 75, 1975.
295. **Smith, S. V. and Jokiel, P. L.,** Water composition and biogeochemical gradients in the Canton Atoll lagoon. II. Budgets of phosphorus, nitrogen, carbon dioxide, and particulate materials, *Mar. Sci. Commun.,* 1, 165, 1975.
296. **Smith, S. V., Kimmerer, W. J., Laws, E. A., and Walsh, T. W.,** Kaneohe Bay sewage diversion experiment: perspectives on ecosystem responses to nutritional perturbation, *Pac. Sci.,* 35, 279, 1981.
297. **Smith, S. V., Kimmerer, W. J., and Walsh, T. W.,** Vertical flux and biogeochemical turnover regulate nutrient limitation of net organic production in the North Pacific gyre, *Limnol. Oceanogr.,* 31, 161, 1986.
298. **Smith, S. V. and Veeh, H. H.,** Mass balance of biogeochemically active materials (C,N,P) in a hypersaline gulf, *Est. Coast. Shelf Sci.,* 29, 195, 1989.
299. **Smith, S. V., Hollibaugh, J. T., Dollar, S. J., and Vink, S.,** Tomales Bay, California: a case study for carbon-controlled nitrogen cycling, *Limnol. Oceanogr.,* 34, 37, 1989.
300. **Smith, S. V. and Kinsey, D. W.,** Why don't budgets of energy, nutrients and carbonates always balance at the level of organisms, reefs and tropical oceans?, *Proc. 6th Int. Coral Reef Symp.,* Townsville, Vol. 1, 1989, 115.
301. **Sournia, A. and Ricard, M.,** Donnees sur l'hydrologie et la productivite du lagon d'un atoll ferme (Takapoto, Ilses Tuamotu), *Vie Mileau,* 26, 243, 1976.
302. **Stanley, S. O., Boto, K. G., Alongi, D. M., and Gillan, F. T.,** Composition and bacterial utilization of free amino acids in tropical mangrove sediments, *Mar. Chem.,* 22, 13, 1987.
303. **Starmuhlner, F. and Therezien, Y.,** Resultats de la mission hydrobiologique austrofrancaise de 1979 aux iles de la Guadeloupe, de la Dominique et de la Martinique (Petites Antilles), *Rev. Hydrobiol. Trop.,* 15, 131, 1982.
304. **Steneck, R. S.,** Herbivory on coral reefs: a synthesis, *Proc. 6th Int. Coral Reef Symp.,* Townsville, Vol. 1, 1989, 37.
305. **Stevenson, C. J.,** Comparative ecology of submerged grass beds in freshwater, estuarine and marine environments, *Limnol. Oceanogr.,* 33, 867, 1988.
306. **Strickland, J. D. H. and Parsons, T. R.,** *A Practical Handbook of Seawater Analysis,* 2nd ed., Bull. Fish. Res. Bd. Can., 167, 1972.
307. **Stout, J.,** Leaf decomposition rates in Costa Rican lowland tropical rainforest streams, *Biotropica,* 12, 264, 1980.
308. **Stumm, W. and Morgan, J. J.,** *Aquatic Chemistry: An Introduction Emphasizing Chemical Equilibria in Natural Waters,* Wiley Interscience, New York, 1970.
309. **Suess, E.,** Phosphate regeneration from sediments of the Peru continental margin by dissolution of fish debris, *Geochim. Cosmochim. Acta,* 45, 577, 1981.
310. **Sugimura, Y. and Suzuki, Y.,** A high-temperature catalytic oxidation method for nonvolatile dissolved organic carbon in seawater by direct injection of liquid samples, *Mar. Chem.,* 24, 105, 1988.
311. **Sunda, W. G. and Guillard, R. R. L.,** The relation between cupric ion activity and the toxicity of copper to phytoplankton, *J. Mar. Res.,* 34, 511, 1976.
312. **Suzuki, Y., Sugimura, Y., and Ito, T.,** A catalytic oxidation method for the determination of total nitrogen dissolved in seawater, *Mar. Chem.,* 16, 83, 1985.
313. **Swift, D. G.,** Vitamins and phytoplankton growth, in *The Physiological Ecology of Phytoplankton,* Morris, I., Ed., University of California Press, Berkeley, 1980, 329.
314. **Talling, J. F. and Talling, I. B.,** The chemical composition of African lake waters, *Int. Rev. Ges. Hydrobiol.,* 50, 421, 1965.
315. **Tenore, K. R.,** Nitrogen in benthic food chains, in *Nitrogen Cycling in Coastal Marine Environments,* Blackburn, T. H. and Sorensen, J., Eds., John Wiley & Sons, New York, 1988, 171.
316. **Thomas, P. A. and Room, P. A.,** Taxonomy and control of *Salvinia molesta, Nature (London),* 320, 581, 1986.

317. **Thomas, W. H., Renger, E. H., and Dodson, A. N.,** New-surface organic nitrogen in the eastern tropical Pacific Ocean, *Deep Sea Res.,* 18, 65, 1971.
318. **Thompson, P. A., Levasseur, M. E., and Harrison, P. J.,** Light-limited growth on ammonium vs. nitrate: what is the advantage for marine phytoplankton?, *Limnol. Oceanogr.,* 34, 1014, 1989.
319. **Thornton, J. A., Nduku, W. K., and Thornton, A. J.,** Sediment chemistry, in *Lake McIlwaine, The Eutrophication and Recovery of a Tropical African Man-made Lake,* Thornton, J. A. and Nduku, W. K., Eds., Junk, Amsterdam, 1982, 59.
320. **Tilly, L. J. and Garcia-Sais, J. R.,** Limnological features of a Puerto Rico reservoir, *Arch. Hydrobiol. Suppl.,* 76, 145, 1987.
321. **Tomascik, T. and Sander, F.,** Effects of eutrophication on reef-building corals. I. Growth rate of the reef-building coral *Montastrea annularis, Mar. Biol.,* 87, 143, 1985.
322. **Turner, D. R., Whitfield, M., and Dickson, A. G.,** The equilibrium speciation of dissolved components in freshwater and seawater at 25 C and 1 atm pressure, *Geochim. Cosmochim. Acta,* 45, 855, 1981.
323. **Ullman, W. J. and Sandstrom, M. W.,** Dissolved nutrient fluxes from the nearshore sediments of Bowling Green Bay, central Great Barrier Reef lagoon (Australia), *Est. Coast. Shelf Sci.,* 24, 289, 1987.
324. **Vacarro, R. and Ryther, J. H.,** Marine phytoplankton and the distribution of nitrite in the sea, *J. Cons.,* 27, 3, 1960.
325. **Valiela, I., Teal, J. M., Volkmann, S., Shafer, D., and Carpenter, E. J.,** Nutrient and particulate fluxes in a salt marsh ecosystem: tidal exchanges and inputs by precipitation and groundwater, *Limnol. Oceanogr.,* 23, 798, 1978.
326. **van Bennekom, A. J., Berger, G. W., Helder, W., and de Vries, R. T. P.,** Nutrient distribution in the Zaire estuary and river plume, *Neth. J. Sea Res.,* 12, 296, 1978.
327. **Venugopalan, V. K. and Rajendran, A.,** Dissolved and particulate nitrogen in Vellar estuary, *Bull Dept. Mar. Sci. Univ. Cochin,* 7, 885, 1975.
328. **Verity, P. G.,** Ammonia excretion rates of oceanic copepods and implications for estimates of primary production in the Sargasso Sea, *Biol. Oceanogr.,* 3, 249, 1985.
329. **Vincent, W. F., Wurstbaugh, W., Vincent, C. L., and Richerson, P. J.,** Seasonal dynamics of nutrient limitation in a tropical high-altitude lake (Lake Titicaca, Peru-Bolivia): application of physiological bioassays, *Limnol. Oceanogr.,* 29, 540, 1984.
330. **Viner, A. B.,** The sediments of Lake George (Uganda). II. Release of ammonia and phosphate from an undisturbed mud surface, *Arch. Hydrobiol.,* 76, 368, 1975.
331. **Viner, A. B.,** The sediments of Lake George (Uganda). III. The uptake of phosphate, *Arch. Hydrobiol.,* 76, 393, 1975.
332. **Viner, A. B.,** The supply of minerals to tropical rivers and lakes (Uganda), in *Coupling of Land and Water Systems,* Hasler, A., Ed., Ecological Studies 10, Springer-Verlag, Heidelberg, 227, 1975.
333. **Viner, A. B.,** The status and transport of nutrients through the Purari River (Papua New Guinea); Purari River Hydroelectric Scheme Environ. Study, Vol. 9, 1979.
334. **Viner, A. B.,** Nitrogen fixation and denitrification in sediments of two Kenyan lakes, *Biotropica,* 14, 91, 1982.
335. **Vogel, S.,** *Life in Moving Fluids: the Physical Biology of Flow,* Willard Grant Press, Boston, 1981.
336. **Wade, B. A.,** The pollution ecology of Kingston Harbor, Jamaica, Sci. Rept. U.W.I.O.D.M., Kingston Harbor Res. Prog., Zool. Dept., U.W.I., Kingston, 1976.
337. **Wafar, M. V. M., Wafar, S., and Devassy, V. P.,** Nitrogenous nutrients and primary production in a tropical oceanic environment, *Bull. Mar. Sci.,* 38, 273, 1986.
338. **Walker, T. A. and O'Donnell, G.,** Observations on nitrate, phosphate and silicate in Cleveland Bay, Northern Queensland, Australia, *Austr. J. Mar. Freshwater Res.,* 32, 877, 1981.
339. **Walsh, J. J.,** *On the Nature of Continental Shelves,* Academic Press, New York, 1988.
340. **Walsh, J. J., Whitledge, T. E., Kelley, J. C., Huntsman, S. A., and Pillsbury, R. D.,** Further transition states of the Baja California upwelling ecosystem, *Limnol. Oceanogr.,* 22, 264, 1977.
341. **Walsh, T. W.,** Total dissolved nitrogen in seawater: a new high-temperature combustion method and a comparison with photo-oxidation, *Mar. Chem.,* 26, 295, 1989.
342. **Ward, B. B., Kilpatrick, D. A., Renger, E. H., and Eppley, R. W.,** Biological nitrogen cycling in the nitricline, *Limnol. Oceanogr.,* 34, 492, 1989.
343. **Watanabe, I.,** Biological nitrogen fixation associated with wetland rice, in *Proceedings of Symposium on Paddy Soil,* Springer-Verlag, Heidelberg, 1981, 348.
344. **Watts, C. J.,** An examination of phytoplankton nutrient limitation in Lake MacIlwaine and the Hunyani River system, in *Lake McIlwaine, The Eutrophication and Recovery of a Tropical African Man-Made Lake,* Thorton, J. L. and Nduku, D., Eds., Junk, Amsterdam, 1982, 117.
345. **Webb, K. L., DuPaul, W. D., Weibe, W., Sottile, W., and Johannes, R. E.,** Enewetak (Eniwetok) Atoll: aspects of the nitrogen cycle on a coral reef, *Limnol. Oceanogr.,* 20, 198, 1975.
346. **Webb, K. L. and Wiebe, W. J.,** Nitrification on a coral reef, *Can. J. Microbiol.,* 21, 1427, 1975.

347. **Wiebe, W. J., Johannes, R. E., and Webb, K. L.,** Nitrogen fixation in coral reef communities, *Science,* 188, 257, 1975.

348. **Wilkinson, C. R. and Fay, P.,** Nitrogen fixation in coral reef sponges with symbiotic cyanobacteria, *Nature (London),* 279, 527, 1979.

349. **Wilkinson, C. R. and Sammarco, P. W.,** Effects of fish grazing and damselfish territoriality. II. Nitrogen fixation, *Mar. Ecol. Prog. Ser.,* 13, 15, 1983.

350. **Wilkinson, C. R., Williams, D.McB., Sammarco, P. W., Hogg, R. W., and Trott, L. A.,** Rates of nitrogen fixation on coral reefs across the continental shelf on the central Great Barrier Reef, *Mar. Biol.,* 80, 25, 1984.

351. **Williams, S. L.,** Decomposition of the tropical macroalga *Caulerpa cupressoides* (West) C. Agardh. Field and laboratory studies, *J. Exp. Mar. Biol. Ecol.,* 80, 109, 1984.

352. **Williams, S. L. and Carpenter, R. C.,** Nitrogen-limited primary productivity of coral reef algal turfs: potential contribution of ammonium excreted by *Diadema antillarum, Mar. Ecol. Prog. Ser.,* 47, 145, 1988.

353. **Williams-Linera, G.,** Biomass and nutrient content in two successional stages of tropical wet forest in Uxpanapa, Mexico, *Biotropica,* 15, 275, 1983.

354. **Wolanski, E. and Ridd, P.,** Tidal mixing and trapping in mangrove swamps, *Est. Coast. Shelf Sci.,* 23, 759, 1986.

355. **Wolanski, E.,** Some evidence for boundary mixing near coral reefs, *Limnol. Oceanogr.,* 32, 735, 1987.

356. **Wright, R.,** Seasonal variations in water quality of a west African river (R. Jong in Sierra Leone), *Rev. Hydrobiol. Trop.,* 15, 193, 1982.

357. **Zafirou, O. C., Joussat-Dubien, P. F., Zepp, R. G., and Zika, R. G.,** Photochemistry of natural waters: a critical review, *Environ. Sci. Technol.,* 18, 358A, 1984.

358. **Zann, L.,** A case history from Tonga: the degradation of Fanga'uta lagoon, Tongata, Great Barrier Reef Marine Park Authority, Workshop on Nutrients in the Great Barrier Reef Region, GBRMPA, Townsville, Australia, 1988, 132.

359. **Zepp, R. G., Schlotzauer, P. F., and Sink, R. M.,** Photosensitized transformations involving electronic energy transfer in natural waters: role of humic substances, *Environ. Sci. Technol.,* 19, 74, 1985.

360. **Zieman, J. C., Iverson, R. L., and Ogden, J. C.,** Herbivory effects on *Thalassia testudinum* leaf growth and nitrogen content, *Mar. Ecol. Prog. Ser.,* 15, 151, 1984.

361. **Zika, R. G., Saltzman, E. S., and Cooper, W. J.,** Hydrogen peroxide concentrations in the Peru upwelling area, *Mar. Chem.,* 17, 265, 1985.

362. **Zimmerman, C. F. and Montgomery, J. R.,** Effects of decomposing drift algal mat on sediment pore water nutrient concentrations in a Florida seagrass bed, *Mar. Ecol. Prog. Ser.,* 19, 299, 1984.

363. **Zuberer, D. A. and Silver, W. S.,** Biological dinitrogen fixation (acetylene reduction) associated with Florida mangroves, *Appl. Environ. Microbiol.,* 35, 567, 1978.

Section II: Comparison of Tropical and Temperate Aquatic Ecosystems

Chapter 3

PHYSIOLOGICAL, TEMPERATURE TOLERANCE, AND BEHAVIORAL DIFFERENCES BETWEEN TROPICAL AND TEMPERATE ORGANISMS

P. Saenger and N. Holmes

TABLE OF CONTENTS

I. INTRODUCTION

In reviewing the differences between temperate and tropical organisms and their implications for aquatic pollution management, a useful starting point is to describe the biophysical conditions of tropical and temperate zones. By definition, these zones are based on temperature, the most significant factor regulating the distribution and abundance of aquatic organisms. At one extreme, ice crystals disrupt cells and terminate metabolic activity, while at the other, high temperatures impair physiological integration, inactivate enzymes, alter the cytoplasm, and lead to behavioral changes. In turn, latitudinal gradients in temperature lead to biogeographic zonation.[1] However, it would be simplistic to consider only temperature, for a number of other important biophysical differences exist between temperate and tropical zones.

II. THE BIOPHYSICAL SETTING

A. ENVIRONMENTAL CONDITIONS

Latitudinal differences in annual thermal regimes are readily apparent and well documented.[2-5] The high water temperatures of the tropics are also to be found in transitional regimes of the subtropical zone (Figure 1). Both in the tropics and in high latitudes there is relatively little annual thermal variation, which contrasts with a marked difference[4] in annual temperature variation in the subtropical and midtemperate zone between 20 and 50°N or S (Figure 1). Long-term offshore data from the Australian coastal region (Figure 2) show an annual seawater temperature range of 6 to 8°C from Macquarie Island (latitude 54.5°S) via Port Hacking (34°S) to Cooktown (14.5°S). Two coastal stations (Point Lonsdale, 38°S and Port Curtis, 25°S) show a wider annual range than offshore stations at similar latitude. At higher latitudes, the minimal temperatures occur in August to September whereas at low latitudes the coolest month occurs earlier, in June or July.

Although shallow inshore waters are thermally more variable than oceanic waters (Figure 2), tropical inshore waters are less variable than temperate shallow waters. The relatively greater thermal stability of tropical estuaries, for example, has been invoked to explain the higher diversity of cyclopoid copepods, a group largely confined to estuarine waters.[6]

Rainfall and humidity in the tropics are also generally higher with much of the rainfall resulting from frequent thunderstorms or tropical showers which reduce oceanic salinity (Figure 1). Combined with the higher air and water temperatures, the resultant warm, moist conditions are favorable for increased terrestrial and freshwater primary productivity. In addition, such conditions provide the evolutionary potential for the development of novel life-styles. For instance, Taylor and Innes[7] have suggested that with high humidity and temperatures above 25°C, the physiological availability of atmospheric oxygen is sufficiently increased to enable the functioning and development of lung-breathing in gilled organisms. Thus, the very existence of land hermit crabs, for example, is only possible in the tropics where such a combination of physical conditions prevails.[7]

An additional effect of higher rainfall and the consequent leaching of soluble materials from soils is that many tropical soils tend to have relatively how nutrient concentrations. In consequence, many aquatic systems in the tropics may be nutrient limited, and novel means of nutrient cycling have been described. For example, insectivory in aquatic angiosperms (e.g., *Nepenthes, Utricularia*) appears to be an adaptation largely confined to tropical oligotrophic waters. Similarly, for coral reef systems, tight nutrient coupling is widespread and nutrient enrichment has adverse effects.[8] For tropical oceanic waters, low nutrient concentrations coupled with the vertical stability of the water column results in relatively low primary productivity.[9]

There is a striking uniformity of the light regimen in the tropics,[2,3,10,11] where throughout

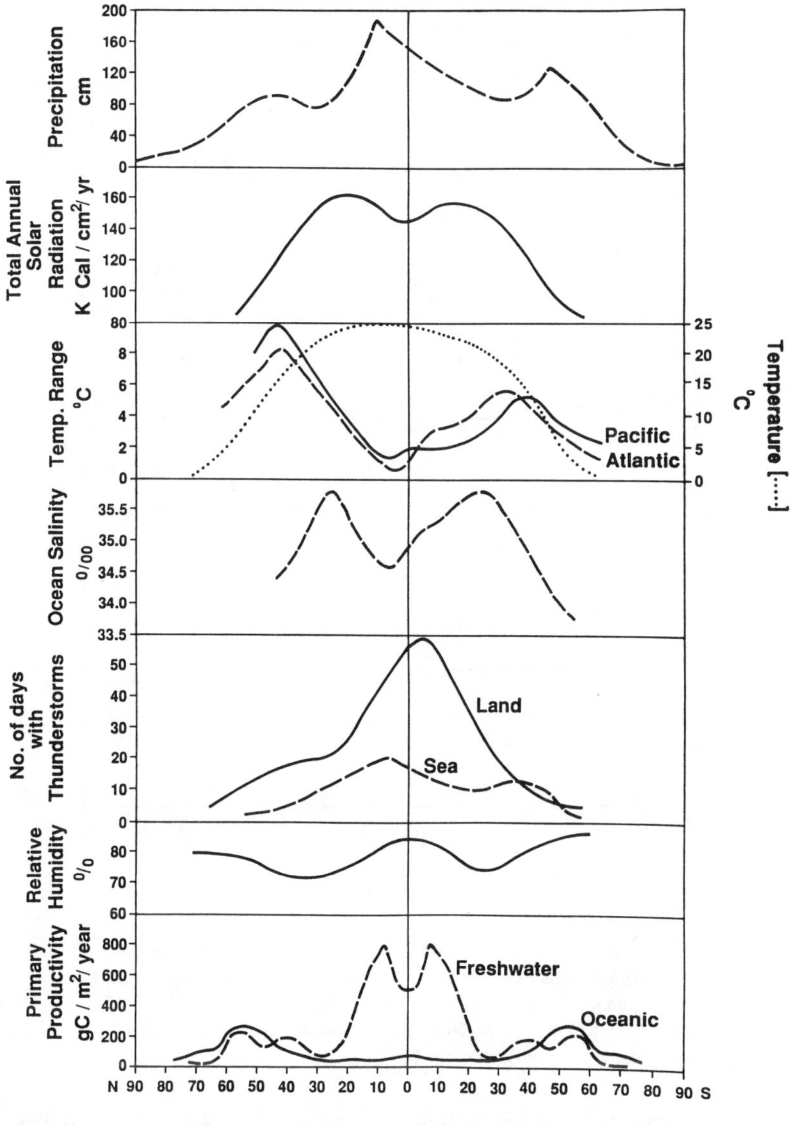

FIGURE 1. Latitudinal changes in various biophysical parameters. (Compiled from numerous sources.)

the year, the daily hours of light differ by only 1.5 h.[12] In temperate or polar regions the hours of daylight vary greatly with the season, although the total annual incident radiation is not widely different in tropical or temperate regions.[4] Such an abundance of solar energy throughout the year in the tropics provides optimal visible light for photosynthesis and radiant energy for evapotranspiration and environmental heating. In turn, these may result in high evaporation rates, high soil or water temperatures, photoinhibition of some photosynthetic pathways or photooxidation of photosynthetic pigments, or excessive ultraviolet (UV) radiation levels leading, for example, to photodisinfection. Thus, in a comparison of temperate and tropical heterotrophic nitrogen-fixing bacteria from sewage, high light intensities in clear tropical waters quickly killed the predominant Enterobacteriaceae (mainly *Klebsiella*) whereas in temperate waters, although similar bacterial numbers were found, bacteria survived for considerably longer due to low light levels and higher turbidities.[13]

Similarly, the depth of maximum biomass of submerged angiosperms in 164 tropical to

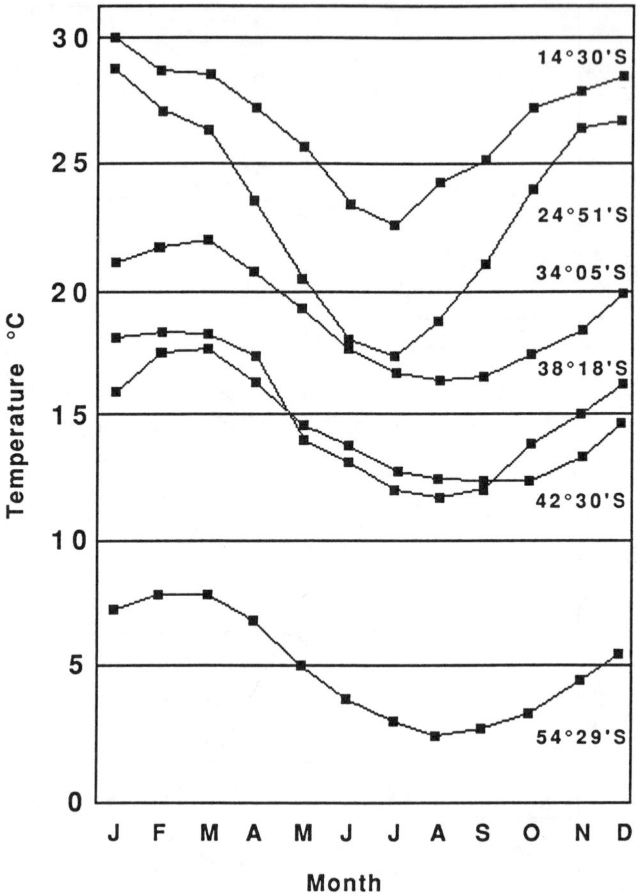

FIGURE 2. Long-term patterns of annual changes in water temperatures from offshore (Macquarie Island, 54°29′S; southeast Tasmania, 42°30′S; Port Hacking, 34°05′S; Cooktown, 14°30′S) and nearshore (Point Lonsdale, 38°18′S; Port Curtis, 24°51′S) sites from the Australian coastal region. (Compiled from numerous sources.)

temperate lakes[14] was best related to latitude as an integrating measure for length of growing season, incoming irradiance, and water temperature. A higher significant relationship between the depth of maximum biomass (Z_b), transparency (S_D measured as Secchi disc depth), and latitude was obtained by multiple regression:

$$Z_b\ (m) = 1.09 - 0.02(S_D)^2 + 19.5\ (S_D/\text{latitude}) \quad (1)$$

This equation indicates that the depth of maximum biomass increases with decreasing latitude; latitude also explained a larger fraction of the variance in Z_b than did water transparency.[14]

Apart from the direct effects of particular light conditions, light also interacts with temperature.[15] For example, growth in temperate seagrasses in saturating light intensities increased with temperature,[16] but in light intensities near the compensation point, growth decreased as temperature increased.

B. ENVIRONMENTAL PREDICTABILITY

Whereas in the wet tropics, rainfall and solar radiation are generally uniformly high with little seasonal variation,[2] subtropical areas or those described as "wet-dry" tropics have a pronounced rainfall seasonality.[3,17] In addition, the wet-dry tropics have a high interyear variability in annual rainfall when compared with the humid tropics, temperate regions, or both. This is particularly so during the transition between the annual wet and dry seasons. However, summary statistics of rainfall, such as mean monthly means, underestimate the variability of tropical growing conditions because much of the rain falls in occasional intense storms.[17]

Other sources of variability include the storm waves resulting from intense tropical storms, particularly nonseasonal ones. Such short-term events may be the most important regulators of marine benthic algal productivity in the tropics.[18]

Regular seasonality of production increases toward the poles, and this is generally accompanied by a highly seasonal but short reproductive phase. Because of the highly seasonal food availability, high-latitude organisms need an abundant supply of food reserves. Thus, midwater crustaceans of polar regions contain much more lipid than the same or related species from lower latitudes.[19] Nevertheless, regular seasonality may occur in tropical and subtropical areas[20] although it reflects rainfall periodicity rather than regular variation in light intensity or temperature.

Although aquatic environments differ latitudinally in temperature regime, illumination, and seasonality, differences also exist between habitats or life-zones at particular latitudes. There is some variation with depth. Thus, coastal habitats in temperate zones are characterized by wide seasonal changes in temperature (Figure 2) and light, as well as diurnal changes in temperature in some habitats, such as the intertidal. Similarly, estuaries are characterized at most latitudes by marked tidal, diurnal, and seasonal changes in temperature and salinity. In contrast, midlatitude habitats with a large thermal inertia, such as the pelagic zones of the ocean, may show only relatively slight diurnal and restricted seasonal changes in temperature. Tropical and polar environments, as well as the deep sea, tend to show reduced seasonal and diurnal changes in temperature. Polar regions exhibit extreme changes in day length and in the maximum values of illumination, whereas the deep sea is a constantly dark environment. Tropical environments may experience only slight seasonal changes in day length and illumination intensity, although seasonal changes in cloudiness (e.g., in the wet-dry tropics) may affect surface irradiation.

Southwood[21] has analyzed the biophysical constraints on communities and organisms that are imposed by environmental heterogeneity and concluded that two types of heterogeneity could be discerned. The first is the heterogeneity resulting from the durational stability of the particular habitat, with *r*-species associated with rapidly changing habitats and *K*-species predominating in those with long durational stability.[22] The second type of heterogeneity was based on the favorableness or adversity of the habitat, where habitats with continuously harsh conditions are relatively free from interspecific competition.[22,23] These environmental permutations have been summarized (Table 1) for open ocean waters from tropical, temperate, and polar latitudes.

What this habitat templet suggests is that organisms in polar regions must have strategies to cope with prolonged harsh conditions and be adapted to exploit the short periods of favorableness both for growth and reproduction. In the tropics, where stability and favorableness are generally high, growth and reproduction are not so constrained, and other challenges, such as interspecific competition, become the more important constraints on the organisms. For example, tropical intertidal gastropods were found to be ecologically constrained by temperature stress, on the one hand, which could be overcome by the selection of suitable microhabitats, but they were constrained in both space and time, on the other hand, by predatory fish during high tides.[24]

In contrast, the life-history strategies of temperate organisms must be able to effect their survival during short periods of physically or chemically harsh conditions and then to "make hay while the sun shines".

TABLE 1
Habitat Heterogeneity Permutations for Open Ocean Waters

	Tropical	Temperate	Polar
Durational stability			
Scale of changes	Small	Large	Medium
Intensity of change	Small	High	Medium
Frequency of changes	High	Medium	Low
Favorableness			
Temperature	High & uniform	Medium & seasonal	Low & constant
Rainfall	High & uniform	Medium & seasonal	Low & constant
Light intensity	High & uniform	Medium & seasonal	Medium & seasonal
Nutrients	Low & variable	Medium & seasonal	High & constant

III. FUNCTIONAL RESPONSES TO PREVAILING BIOPHYSICAL CONDITIONS

The importance of light and temperature in controlling primary production and respiratory metabolism suggests that latitudinal variations in these factors are likely to be matched by an evolutionary adaptation of some organisms to life in a particular latitudinal environment, although other forms may be sufficiently labile to allow a cosmopolitan distribution. Given the wide range of factors to which organisms must respond, not all of which change consistently with latitude (Figure 1), the study of latitudinal adaptations is not straightforward. Individual organisms do not respond to external factors in isolation, but to a suite of factors operating simultaneously. Laboratory study of responses to environmental factors must endeavor to simplify this complex reality and thus, can consider only one or two variables at a time. Such studies, therefore, cannot show directly the bases of adaptation to a polar, temperate, or tropical existence, but must address the physiological and other mechanisms of adaptation.

Adaptation to the environment is an ill-defined concept and to discuss it in depth is beyond the scope of this chapter. For the present it is sufficient to note that organisms can adapt at a physiological level, on short time scales, or genetically over longer or evolutionary time scales. The nature of the adaptations required may be linked to the time scales of change in the factors to which adjustment is required. Individuals in absolutely constant environmental conditions will have little need to undertake adaptive adjustments to external factors. In contrast, organisms in rapidly and continuously fluctuating environments will be able to survive only by tolerance of the changed conditions or adjustment to them.

The capacity for making short-term physiological or metabolic adjustments to environmental conditions may itself be genetically determined.[25-27] Stenoplethous organisms are unable to adapt significantly to a changed factor (e.g., temperature), while euryplethous organisms are more labile and can cope with short- or long-term fluctuations.

Any functional differences in the organisms of polar, temperate, and tropical zones ultimately are an outcome of the various organisms' adaptability to their physical environments of which the major aspects are discussed below.

A. METABOLIC RESPONSES TO TEMPERATURE

There has been considerable work on physiological responses to changed conditions, especially on short-term and seasonal temperature changes on plants and ectothermic animals. Oxygen consumption has frequently been used as a convenient and relevant marker of metabolic responses.

The most common method of quantifying rate changes in physiological processes with temperature is by the temperature coefficient (Q_{10}), which is the factor by which a process

TABLE 2
Some Indicative Q_{10} Determinations for Various Physiological Processes in a Range of Taxa

Metabolic process	Q_{10}	Temperature range of determination	Species	Ref.
Oxygen uptake	1.3—2.2	5.5—27.5	*Nereis* spp.	31
	2.98	30	*Palaemon pacificus*	32
	4.69	15	*P. pacificus*	32
	1.5	9—16	*Carcinus maenas*	33
	1.4—1.6	na	Intertidal pagurids	34
	2.1—2.4	na	Subtidal pagurids	34
	2.6—2.7	na	Supratidal pagurids	34
	1.96—2.40	2—15	*Mytilus edulis*	35
	1.44	20—29	*Macrognathus aculeatum*	36
	1.27	20—29	*Anabas testudineus*	36
	1.30	20—29	*Channa punctatus*	36
	2.91	29—32	*C. punctatus*	36
	3.05	29—32	*Macrognathus aculeatum*	36
Gill clearance	1.11	10—20	*Branchionus plicatilis*	37
Food ingestion	1.75	10—20	*B. plicatilis*	37
Ventilation/heart beat	1.9—3.6	na	Decapod crustaceans	33
Total sediment oxygen	6.5	na	Benthic community	38
Zinc accumulation	1.8	10—20	*Palaemon elegans*	39

(or reaction) rate increases over a rise of 10°C. A Q_{10} of 2 signifies a doubling of reaction velocity over the stated 10° temperature interval. Thus,

$$Q_{10} = k_{t+10}/k_t \tag{2}$$

or, expressed for any temperature differential,

$$Q_{10} = (k_2/k_1)\exp[10/t_2 - t_1] \tag{3}$$

This form is the most widely used,[28] even though it has long been established that Q_{10} is not a true constant but varies considerably at different temperatures, generally declining with increasing temperature. However, for most biological reactions Q_{10} lies between 2 and 3, and is of descriptive (and comparative) value despite the availability of a range of more sophisticated temperature formulas.[28] Typical Q_{10} determinations for several marine organisms are presented in Table 2. However, tropical, temperate, and polar organisms generally have similar oxygen consumption rates when determined at their respective habitat temperatures.[29, 30]

Temperature affects rates of physiological activity at a molecular level, but many forms show higher-level control mechanisms that may modulate the molecular, cellular, or tissue responses to changing temperature. Most ectotherms show well-developed responses to temperature change[40,41] and the nature of the responses change with the length of exposure to a new temperature. Although not all the metabolic processes are well understood, descriptive accounts of acclimation have been available in a variety of texts since the 1960s. The sequence of events is best shown in the responses to a single upward or downward temperature change, where there is an initial ''immediate phase'', followed by a ''stabilization phase'', and the attainment of a new ''steady state''.

The immediate phase occupies the first few seconds or minutes and involves changes to cellular metabolism, often accompanied by a modification of whole-organism activity. These responses may be seen not only in changes to instantaneous oxygen consumption rates, but by changes in membrane permeability,[42] phospholipid metabolism,[43] amino acid transport,[44] heart and ventilation rates,[45] and swimming speed.[46] While physiological changes seen in the immediate phase may involve adaptive responses in nervous or hormonal control of tissue or organ function, there seems to be little compensation during this phase at a cellular or molecular level, where changed reaction rates appear to result mainly from kinetic responses to temperature changes.

The length of the stabilization phase varies with the species concerned, the size of the temperature change, and the nature of the initial conditions. It may range from a period of hours to one of many days, during which time oxygen consumption may change only slowly. This phase is described by Kinne[41] as being the period during which the organism reaches a new steady state after the initial temperature change. It is during this phase that adaptation (compensation) occurs at a tissue or cellular level, with shifts in rate-temperature relationships often resulting from changes in enzyme kinetics or metabolic pathways as discussed briefly below.

Thermal acclimation may have a marked effect on thermal survival boundaries. Acclimation to a higher temperature usually results in an upward movement of both upper and lower boundaries,[47-49] with a reverse movement during cold acclimation. The size of temperature change and the rate of acclimation may affect the size of this boundary shift.[40]

The acclimation process appears to involve responses at a variety of organizational levels. At a subcellular level there is often a modulation of enzyme activity during acclimation, although there seems to be no consistent correlation between temperature and the activation energy of enzymes.[50,51] Enzyme kinetics may change during acclimation, with the induction of thermal isoenzymes with different activation energies and substrate affinities.[52,53]

Marine phytoplankton adapt to temperature, with photosynthetic or respiration rates showing temperature compensation. In one view[54] this adaptation is evident in photosynthesis over most irradiance levels. Another model suggests that temperature adaptation affects principally the maximum rate of photosynthesis in given conditions. In this view adaptation to suboptimal conditions is achieved at the cost of performance in optimal conditions.[55] As Li and Morris[56] indicate, these two models may be compatible in that they emphasize different aspects of the process of thermal acclimation.

Thermal acclimation also occurs in attached algae. For example, winter plants of the algae *Lobophora variegata* and *Zonaria tournefortii* had lower oxygen consumption at experimental temperatures than did summer plants.[57] Both species also showed some accommodation of photosynthetic activity. Similarly, production in the red alga *Gracilaria* was partially acclimated to temperature, although the quality of agar from the plant was adversely affected by temperature extremes.[58]

In trout, cold-acclimation involves an increase in the rate of glycolysis and higher levels of lipogenesis, protein metabolism, and glycogen synthesis.[59] These findings indicate that tissue and organ functions are implicated in acclimation, so that the process is not merely an integrated result of changes at a cellular level. It appears that most organisms show some sort of supracellular control of acclimation processes that may modulate the cellular responses to changing temperature. There may be some metabolic reorganization, with some pathways being activated and others inhibited.[60,61]

The freshwater fish *Channa punctatus* maintained at three different temperatures (23, 30, and 35°C), compared with a control group maintained at 14°C, showed obvious signs of stress at 30 and 35°C, as indicated by loss of weight and increase in mortality rate.[62] On the other hand, those at 14 and 25°C gained weight steadily and showed normal growth. Biochemical studies confirm the stress symptoms, as there was a steady, statistically significant fall in blood glucose level and depletion of glycogen reserves in liver and muscle within the first week at 35°C and by the fourth week at 30°C.[62]

Isolated, demembranated muscle fibers have been used to study evolutionary temperature adaptation in Antarctic fish muscle when compared with that from tropical and temperate species.[63] It was found that (1) the maximal isometric tension at 0°C in fibers from Antarctic species is consistently higher than that of fibers from temperate and tropical species at their preferred body temperatures; (2) ATP turnover rates during isometric contraction are similar at all temperatures for representative Antarctic, temperate, and tropical species; (3) at the preferred body temperature of each species, the economy of contraction is in the ratio 4:2:1 for Antarctic, temperate, and tropical species, respectively; (4) maximal contraction velocity is highly temperature dependent in all species; (5) adaptation to low temperature results in contractile proteins which are unstable at high temperatures.

Furthermore, unlike temperate and tropical fishes which freeze at −0.7°C in the presence of ice and die, polar fishes live at a relatively constant temperature of about −1.8°C, in an oxygen-rich environment. In comparison with fishes that live in temperate or tropical waters, their blood contains fewer erythrocytes and less hemoglobin,[64,65] and their survival appears to be linked to the presence of biological antifreeze compounds in their blood and in most of the other body fluids. In almost all Antarctic notothenoid fishes and Arctic gadoid fishes, the antifreeze compounds are a series of glycopeptides, the serum levels of which are not affected by warm acclimation.[66]

Clearly, acclimation must be regarded as a whole-organism phenomenon, integrating a variety of processes and adjustments at cellular and higher levels. The results of physiological acclimation are not the same in all organisms, with acclimation resulting in either a rotation of the curve of rate functions vs. temperature, or a translation or both together.[40,67]

Not all ectotherms show thermal acclimation but are able to tolerate some level of changed temperature. For instance, the asteroid *Luidia clathrata* did not acclimate to temperature in the laboratory[68] while the whelk *Bullia digitalis* did not show seasonal acclimation of oxygen consumption, unlike the closely-related *B. rhodostomata.*[69]

Overall, laboratory-derived models indicate some of the processes involved in acclimation, but do not define all of the problems that an organism must "solve" in order to survive in a naturally varying environment. In most habitats in the euphotic zone from the temperate zones to the tropics, organisms must withstand constantly changing temperatures, although the rate and scope of the changes will vary. There has been relatively little work on adaptation to changing temperatures, although the realization that heated discharges may lead to increases in temperature fluctuations has resulted in some recent interest.[70,71] In a variable temperature environment, acclimation cannot be a finite process, and organisms must be constantly adjusting at all levels of metabolic organization to changes in environmental temperature.

Biochemical acclimation in an evolutionary sense can be illustrated by the pathways of photosynthetic carbon fixation. The key enzyme in this cycle, ribulose diphosphate carboxylase-oxygenase, has two alternate roles: either it can catalyze the carboxylation of ribulose diphosphate to 3-phosphoglyceric acid or it catalyzes the oxygenation of ribulose diphosphate to form glycolic acid in the presence of atmospheric oxygen. Hence this key enzyme, which captures atmospheric CO_2 for conversion to sugar, has a considerable built-in disadvantage associated with it because, at high temperatures, its oxidative function predominates, leading to a reduction of the overall efficiency of photosynthesis.[72] In hot, humid climates this constraint has been overcome in relatively recent times in C_4 plants, where a second carbon pathway captures the external CO_2 and pumps it into special bundle sheath cells where the C_3 cycle can then operate without the inefficiency of the enzyme's oxidative function. In this sense, the C_4 pathway is a tropical phenomenon and of particular relevance to swamp communities. For example, in freshwater swamp communities in Australia, the following number of grasses show the C_4 pathways: tropical Australia, 31 of 38 genera; subtropical Australia, 9 of 15 genera; and temperate Australia, 0 of 4 genera.[73]

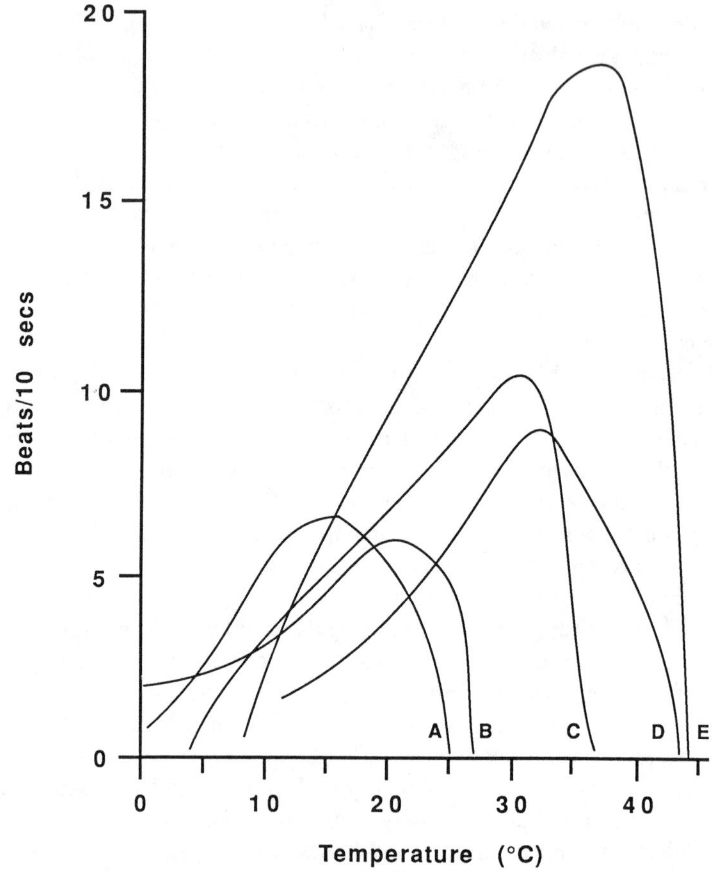

FIGURE 3. Cirral activity as a function of temperature in selected species of barnacles: (A) *Balanus crenatus* (small, sublittoral, arctic); (B) *Balanus balanus* (large, sublittoral, arctic); (C) *Chthamalus stellatus* (large, midtide, temperate); (D) *Tetraclita squamosa* (large, midtide, tropical); (E) *Balanus amphitrite* (small, sublittoral, tropical). (Modified from Reference 45.)

From a functional viewpoint, the C_4 pathway clearly provides an advantage under tropical conditions. Thus, a comparison of monotypic stands of the C_4 sedges *Cyperus papyrus* and *C. latifolius* with the C_3 *Typha domingensis* in a Kenyan swamp[74] showed maximal rates of photosynthesis for all three species to be 8 $\mu mol/m^2/s$. However, this maximal net photosynthesis occurred at 26°C for *C. papyrus*, 21°C for *C. latifolius,* and 17°C for *T. domingensis*. In addition, the quantum yield from incident light was higher in *C. papyrus* than in the other two species while both the photosynthetic water-use efficiency and nitrogen-use efficiency in the C_4 plants were more than double those of the C_3 plant.

B. WHOLE ORGANISM EFFECTS

1. Temperature Tolerance

For any organism, no matter how precise its acclimation mechanisms, there are eventual limits, so that all organisms are constrained by environmental temperature in some way (Figure 3). To measure the temperature tolerance of organisms, the concept of critical thermal maximum and minimum (CTM) was introduced,[75] although subsequently modified.[76,77] In essence, the critical thermal minimum or maximum is that temperature, respectively, below or above which orientation and locomotion of an organisms is affected to the extent where

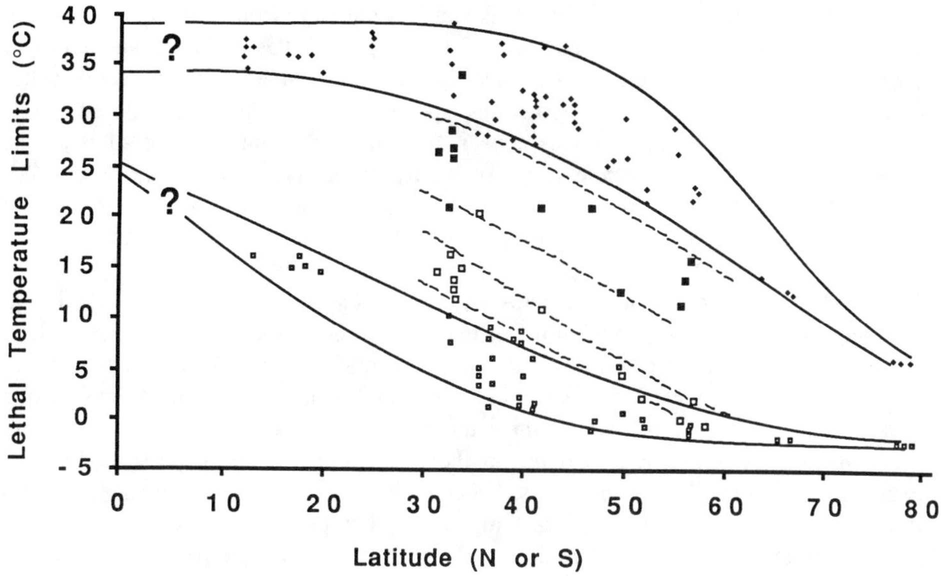

FIGURE 4. Envelopes of upper and lower lethal limits for embryonic (dashed line) and postembryonic (continuous line) fish. (Based on Reference 82, with later data inserted.)

it becomes unable to escape situations that may lead to death, either by prolonged exposure or by an increased vulnerability to predators.

While several other measures have been proposed, CTMs appear to be particularly useful, despite their methodological difficulties, for assessing the relative temperature tolerance of organisms acclimated to different temperatures or coming from a wide latitudinal range.[78]

Because of its simplicity in measurement, the concept of upper (or lower) lethal temperature (ULT) has also been used. Upper and lower lethal temperature limits have been determined for a wide variety of organisms and show a gradual narrowing both at polar and tropical latitudes (Figure 4). Lethal temperature limits depend on the previous thermal history of the individual, with warm-acclimation raising the upper lethal limit while cold-acclimation lowers it. Although this shift in thermal limits is not present in all organisms, the actual value for ULT generally is higher for short exposures and for warm-acclimated animals.[79] The shift in thermal tolerance limits following acclimation may be due to a variety of factors, associated with cellular metabolism, as well as higher level metabolic control systems. The patterns of change in enzyme kinetics seen in some organisms during acclimation, mentioned above, may be associated with a greater ability to withstand extreme temperatures, but this aspect of thermal tolerance seems to have attracted relatively little attention. The shift in upper thermal limit with acclimation has been shown in many species of fish. For example, the avoidance temperature increased with warm acclimation in several fish species from the Gulf of Thailand, as did the critical thermal maximum and incipient ULT,[80] whereas cold shock in menhaden resulted in increased mortality when temperatures were lowered rapidly.[81]

Temperature tolerance also may differ in different stages in the life cycle of the particular organism. In general, as Figure 4 shows for fish, the adults usually show a greater range of thermal tolerance than do larval or juvenile stages.[83] The temperature tolerance of post-larval and juvenile spot (fish) was linearly related to acclimation temperature up to 25°C, but increased very little between acclimation temperatures of 25 and 35°C.[84] An increase in acclimation salinity resulted in an increased resistance time but decreased upper limits.

The thermal tolerance of invertebrate larvae often can be correlated with time of their release into the water column, so that larvae normally released during the winter or spring months are more tolerant of low temperatures and less tolerant of higher temperatures than larvae released during the summer months.[85] For example, studies on larvae and adults of fiddler crabs, *Uca,* show that larvae of tropical species can survive low temperatures better than adults,[86] whereas larvae of temperate zone species show the opposite response.[87]

For some species, the larval stages appear to be more tolerant than the adults, e.g., free-living, field-acclimated juvenile stages of the freshwater crayfish *Orconectes rusticus* consistently showed CTMaxima 0.9 to 2.6°C above those of the adults.[88] Other environmental conditions such as water hardness also may alter thermal tolerance of particular species or life stages.[89]

Temperature tolerances in invertebrates also are related to thermal history, whether in the laboratory or in the field. Populations of mussels, *Mytilus edulis,* from Newfoundland were apparently less able to perform well at higher temperatures than those from New York, although the differences were less marked in the winter.[90] High shore mollusks from Hong Kong and Tanzania had higher ULT limits than those lower on the shore,[91] presumably reflecting the higher temperatures found on the high shore, confirming the findings of Broekhuysen[92] for South African mollusks. Other studies showed that the thermal tolerances of high and low shore gastropods from temperate and tropical localities were related to position on the shore, but that subtropical gastropods had higher thermal tolerances than temperate species.[93,94]

Similar patterns of temperature tolerance can be seen in bivalve mollusks. Potter and Hill[79] found that intertidal oysters showed excess mortality when high temperatures occurred at low tide. Though these oysters had median lethal temperatures of more than 40°C, the exact value depended on time of exposure. Insolation rapidly increased the tissue temperatures of tidally exposed oysters. A relationship between habitat and thermal tolerance was also found for *Tellina* and *Donax*; deeper-water *T. fabula* had a lower thermal tolerance than the intertidal *T. tenuis.*[95] The previous acclimation history of an individual affected burrowing behavior and thermal tolerance. Mediterranean *Tellina* also had a higher thermal limit than those from the North Atlantic.[95] Similar results were obtained for three species of *Donax.*[96]

Similar trends between high and low shore species or populations have been found for most other groups of invertebrates (e.g., anemones[67]).

The precise value of the upper thermal limit of many species seems to depend not only on the acclimation history but also on the rate of heating. It is possible that, during the immediate phase of adjustment to temperature change, overshoot reactions in various metabolic processes may increase mortality. However, not all reports show that more rapid heating reduces thermal limits. For instance, in the mussel *Perna perna,* slow temperature increases (0.2°C/min) gave lower thermal tolerances than faster ones (0.8°C/min).[97] In addition, small mussels had lower limits than larger ones. In this study, where heating rates were very rapid, it is possible that thermal inertia of the mussels could have resulted in tissue temperatures lagging behind those in the test aquarium. The authors noted that even the slowest temperature change was too fast to allow appreciable acclimation.

The responses to thermal shock can themselves be modified by thermal history. After a thermal shock of 10°C (in the form of a single temperature change) amphipods acclimated to 15°C or less could survive for 5 d or more.[98] Acclimation to 20 or 25°C allowed survival for only 2 d or 2 h, respectively. For gentler temperature changes, upper thermal tolerance increased up to 25°C and decreased in the range of 25 to 38°C, showing that the rate of change of temperature greatly affected measured heat tolerance.[98] For corals, upper absolute temperatures appeared to be more critical in producing coral damage than short-term temperature fluctuations near the upper lethal limits.[99]

Cold tolerance in many invertebrates seems not to have been so well studied as heat tolerance. In four mollusk species median lower lethal temperatures were well below zero.[100] There was also significant seasonal variation in cold tolerance, which could be partly explained by seasonal variations in salinity.

Studies of the relationship between cold tolerance and other habitat variables in the barnacle *Semibalanus balanoides* showed that cold-resistance was controlled by the short photoperiod, reduced food supply, and lowered temperatures present in winter.[101] Changes to any one of those factors could result in changed lower thermal limits. Breeding and cold tolerance were associated but not mutually dependent.

Cold-temperate species, especially in freshwaters or the intertidal zone, may experience subzero temperatures for extended periods. For a range of invertebrates, freezing resistant forms are able to tolerate tissue ice, possibly assisted by a switch from aerobic to anaerobic metabolism.[102] Extracellular ice is the main cause of freezing injury. Freezing tolerance is restricted to forms in the supra- or eulittoral zone of temperate and arctic regions and, again, involves a tolerance of tissue ice formation.[103] An inverse relationship between cold-hardiness and acclimation temperature has also been demonstrated in a variety of invertebrates.[103] Brackish water forms were usually less ice-resistant than saline forms, and there was some evidence in some species for an endogenous control of seasonal fluctuations in freezing tolerance.

Aquatic plants show similar characteristics and responses to those shown by animals. Thus, in cold-temperate latitudes algae may possess adaptations to avoid intracellular freezing or to tolerate intercellular ice.[104] Many species can withstand low temperatures for considerable periods, especially when emersed or under conditions of partial desiccation, as in *Porphyra*.[105] Survival under these conditions appears to be more a matter of tolerating ice within the tissues than adaptation of metabolic pathways and rates.[106] Little work has been done on the mechanisms affecting cold tolerance at temperatures above freezing.

The upper thermal limits of some marine plants vary with habitat. Algae on an exposed littoral in summer may experience periods of high temperatures, followed by rapid cooling as the tide comes in. Sublittoral forms, in contrast, experience more moderate temperature conditions, especially at depths where diurnal insolation effects are minor. Algae from the deeper sublittoral are less heat tolerant than those from shallower waters,[107] and the heat tolerance of *Bangia* was increased when the plant was desiccated during emersion.[108] Tropical algae may not have greater heat tolerance than temperate algae,[109] although the relationship between ecological conditions and thermal performance in algae has not been effectively studied.

In four *Laminaria* species in culture, differences were found between the species in thermal tolerances but no within-species differences in populations, i.e., the performance of a species did not vary with location from which the parent culture stock was collected.[110] Since *Laminaria* is a successful over a wide area, these findings illustrate the great phenotypic plasticity of the genus.

Mangroves, intertidal plants largely confined to tropical rather than temperate shorelines, generally have high temperature optima for leaf production and cease leaf production between 12 and 16°C.[111] Cold and heat tolerances were measured by the decline in induced chlorophyll fluorescence following application of a cold or heat stress to the leaf tissue, a technique initially developed for crop plants.[112] For mangroves, a wide range of cold tolerance was found with greater cold tolerance in those species occurring at higher latitudes. The decline of chlorophyll fluorescence after heating in water to 49°C for 10 min showed that all species had a very high degree of heat tolerance and that, as a group, these plants were at the extreme high end of the heat tolerance range for nonarid tropical plants.[113]

2. Acclimation at the Organismal Level

Generally, organisms from colder regions are less heat tolerant and more cold tolerant than organisms from warmer regions (Figure 4). In addition, the thermal range to which an organism is naturally exposed seems to determine its ability to acclimate. For example, because of the relatively constant thermal regime to which they are exposed, tropical and polar organisms have a lesser ability to alter their metabolic response to different temperatures in an adaptive or compensatory manner than temperate zone organisms. Many temperate zone organisms do acclimate and, rather than passively responding to thermal changes, have evolved homeostatic mechanisms, both physiological and behavioral, which allow them to adapt to temperature fluctuation.[5] Similarly, organisms from the upper regions of the intertidal zone are usually more thermally resistant than organisms inhabiting deeper waters.[85]

While some aspects of this ability to acclimate appear to be genetic, Winkler[61] has shown that in *Gambusia affinis* in warm springs, environmental induction of irreversible physiological changes is more likely than genetic adaptation. The finding that tropical species generally withstand low salinity best at high temperatures while cold-water species tolerate low salinities best at low temperatures[85] seems to suggest that some environmental cue triggers particular physiological pathways present in these organisms.

3. Stress Tolerance and "Scope for Growth"

The concept of "scope for activity" in ectothermal animals was elaborated by Fry[114] for fish and was defined as the difference between the standard (maintenance) and active (maximal) rates of oxygen consumption of acclimated organisms. The normal or routine rate, resulting from the normal spontaneous activities of the animal, lies between the standard and active rates. The difference between the standard and active rates represents the respiration available for activity. The concept has been extended as "scope for growth" or "metabolic scope" to refer to the energy available for activity, growth, and reproduction and has been used by many workers[27] to explore the effects of environmental and other stresses on organisms.

The standard metabolic rate is determined partly by the mechanisms of acclimation discussed above and partly by genetic factors.[27] In addition, the efficiency of respiratory surfaces, blood transport, food conversion, etc. may also regulate the input of energy to the organism, so affecting the active rate under given conditions. External factors such as availability of oxygen and food will also influence the active rate.

Evolutionary adaptation to latitudinal factors is evident in some forms. The cold stenotherm fish *Salvelinus fontinalis,* with an upper thermal limit of 20°C or so, shows a curve of standard rate that is fairly flat over the range of 5 to 20°C. In contrast, the eurythermal goldfish (*Carassius auratus*) shows a curve that is steep at low temperatures, flattening at acclimation temperatures of 25°C or more.[115] At 5°C *S. fontinalis* has a weight-specific oxygen consumption that is an order of magnitude greater than that of *C. auratus* at the same temperature. Several studies[116] have shown that standard metabolism may be increased by the effects of stress.

Clearly, elevation of the standard rate or depression of the active rate will lessen the amount of available energy above that needed for maintenance and will thus reduce the scope for growth. This, in turn, will affect the ability of the organism to be active, grow, or reproduce and, where the energy available is insufficient even for maintenance, the survival of the organism is at risk.

Evolutionary adaptation to factors that change with latitude may involve adaptations of cellular and molecular mechanisms, as discussed above, as well as higher-level control mechanisms, size, and morphology. Such adaptations may not be complete in that adaptation or acclimation may exact a metabolic cost (thus, increasing the standard metabolic rate). As well, organisms in higher temperatures may be closer to the physical limits of survival than those at a lower temperature. For instance, declining oxygen saturation levels with temperature in some aquatic environments may limit oxygen availability to an extent that no amount of adaptation can overcome. In this general sense, then, tropical organisms may in some circumstances be closer to the limits of adaptation mechanisms, genetic or phenotypic, than those in higher latitudes and may thus respond differently to externally imposed stresses.

Apart from the direct stresses imposed by high temperatures, other stressors may be magnified by altered temperature conditions. For ectothermic aquatic organisms particularly, temperature increases generally increase the chemical toxicity of a range of compounds because of increased diffusion or active uptake across gill membranes. While detoxification mechanisms and excretory processes are also likely to increase with temperature[117] and thus may cancel out some of the temperature effects, it seems likely that a temperature rise invariably increases toxicity.[118,119] As a result of such increases, the minimal lethal concentration of toxic chemicals may be lowered, or the survival time at a particular concentration may be shortened.[120]

For example, the effects of temperature on copper accumulation and depuration in the tissues of the tropical gastropod *Cerithidea cingulata* were studied at 18, 28, and 38°C.[117] The tissue accumulation of copper was found to be more at high temperature than at the ambient and low temperatures although the depuration rate was highest at the high temperature. At ambient and low temperatures, Cu accumulation was slow and prolonged, and recovery from sublethal concentrations at ambient and low temperatures was 48 and 72 h, respectively, compared with recovery after 36 h at high temperature.

The shrimp *Palaemon elegans* is capable of regulating the body concentration of zinc to a relatively constant level over a wide range of dissolved zinc concentrations.[39] However, this regulatory capacity is markedly diminished with increasing temperatures, resulting in a rapid net accumulation of zinc.[39] The rate of accumulation has a Q_{10} of 1.8 from 10 to 20°C for an exposure concentration of 562 μg/L. Similarly, for the mussel *Mytilus edulis* zinc uptake was not only enhanced with increasing temperatures, but appeared to increase mortality during later exposures to higher temperatures.[120]

In reviews of the effects on fish and other aquatic organisms,[119,121] considerable variation was found among species, but increases in toxicity with temperature were generally most pronounced in relation to ammonia, cyanide, trace metals, chlorinated hydrocarbons, organophosphorus compounds, phenols, chlorine, detergents, and Kraft mill effluents. For microorganisms, such temperature effects were less clear-cut.[119,121]

4. Growth Rate, Longevity, and Morphology

Many aquatic invertebrates attain a larger final body size in the colder parts of their distributional area than in parts with normal or supranormal temperatures[41] even though growth rates and development may be more rapid at high temperatures.[4,123,124] Large size, thus, may indicate long slow growth rather than rapid growth and therefore implies longer life spans for species in the colder parts of their range.

The variability of growth rates in tropical marine invertebrates[4] suggests that some tropical species have become specialized for this environment and that this has involved both a reduction in growth rate as well as an inability to live outside the tropics. Those tropical species with a high growth rate appear also to have the metabolic flexibility to allow them to range into cooler waters.

In those habitats where desiccation is significant, size may be further influenced as small organisms are less resistant to water loss because of the greater ratios of surface area to volume. In turn, this suggests that tropical intertidal organisms would be smaller than intertidal organisms in temperate or polar regions where evaporation is relatively reduced.

In general, growth rate and longevity are inversely related. In mollusks, longevity is greater and growth rate less in cold waters. In the tropics, life spans ranged around 50 months, whereas 120 to 200 months was typical at temperatures around 8°C.[4]

Studies of the effect of temperature on body size of homometabolous aquatic insects (predominantly Chironomidae) showed that adult body size was consistently smaller in heated water compared with the ambient treatment.[125] For five of the seven species investigated, females decreased in size more than males. This reduction in growth occurred despite abundant autochthonous food resources, particularly in the heated water. In turn, they suggested that in this group of insects, somatic growth may be acting independently of changes in the food resources and that growth and development was more a function of temperature than food supply. This supports other findings[126] that temperature had more importance than diet for larval growth and adult size of Plecoptera and that even small changes in temperature during larval growth could change the magnitude of larval growth.

5. Behavior

Some ectotherms show behavioral thermoregulation; the ability of reptiles to control their body temperatures by altering their orientation to the sun has long been known, but behavior that has the effect of regulating temperature has been reported in few marine invertebrates. Some tropical fiddler crabs not only blanch in high temperatures, thus presumably reducing their absorption of heat from sunlight,[127] but they also return to their burrows more frequently, allowing some reduction of body temperatures.[51] Several intertidal organisms have been shown to regulate temperature by evaporative losses,[24] and the extent of evaporation could be controlled by alteration of shell gape and orientation.

Many teleost fish are known to show some form of thermoregulatory behavior by selection of water of a preferred temperature (e.g., juvenile salmon).[128] Such selective behavior is known in tropical, as well as temperate, fish; tropical marine reef fish from the Indo-West Pacific region have been shown to prefer mean temperatures between 20 and 30°C.[129] Species varied in the thermoregulatory precision of their behavior, but some forms could regulate precisely. In striped bass at 15°C the preference for a higher temperature could override the avoidance response to total residual chlorine.[130]

The metabolic implications of higher temperatures may also be manifest in behavioral modification. Thus, at increasing temperatures the water stick insect *Ranatra dispar* shows a linear increase in the capture rate and an exponential decrease in prey-handling rates to compensate for the increased food requirement.[131]

The evolutionary modification of behavior in relation to diadromous fish appears to have been the result of the differential availability of food resources in ocean and freshwater habitats.[132] As shown in Figure 1, oceans are more productive than freshwaters in temperate latitudes, and anadromy predominates. Catadromy, on the other hand, is more frequent in the tropics where freshwater productivity exceeds that of the ocean.

C. REPRODUCTIVE EFFECTS

1. Reproductive Onset

The onset of spawning in aquatic organisms is generally[4,133] but not always[134] triggered by temperature, usually by rising temperature. Such a critical rise in temperature will be attained in spring or early summer where there is a large seasonal temperature range in midlatitudes, while in polar or tropical regions, with smaller temperature ranges, such critical temperature rises may not be attained until midsummer. Observed spawning periods in polar, temperate, and tropical waters reflect this.[4] The degree of synchronization of spawning appears to depend on the extent of the annual temperature range[133] so that where little seasonal variation in temperature occurs, spawning is asynchronous. However, where the spawning season is long, secondary regulators such as lunar or tidal rhythms[135] or photoperiod[136] may be present.

Synchronization of spawning is generally advantageous to a species, ensuring that spawning occurs under favorable environmental conditions, that the probability of fertilization is high, and that gamete or larval losses to predators are minimized. Given these advantages, the frequency (ubiquity) of synchronized spawning could be expected to increase as periods of favorable conditions become shortened at increasing latitudes. Generally, the extent of synchronous spawning decreases toward the equator,[4,137-139] but a notable exception is the

TABLE 3
Summary of the Reproductive Trends of Tropical, Temperate, and Polar Fishes

Feature	Polar	Temperate	Tropical
Onset of sexual maturity	Late	Earlier	Early
Egg size	Large	Medium	Small
Demersal eggs	Rare	Rare without brood protection	Common with brood protection
Vitellogenesis	Short	Medium	Long
Oocyte development	Mostly synchronous	Variable	Mostly asynchronous
Spawning season	Short	Longer	Variable
Spawning	Total	Total or partial	Intermittent
Gonadosomatic index	c. 50%	<25%	<10%

Based on References 138, 141—144.

synchronous release of gametes by many species of coral in mass spawning episodes on the Great Barrier Reef and parts of the Indo-West Pacific.[140] This phenomenon, possibly triggered by temperature and lunar cycles, apparently has evolved to maximize fertilization and outbreeding and may minimize predation.

A summary of some early life history features of teleost fishes from freshwater and marine, tropical, temperate, and boreal habitats is presented in Table 3. In general, marine species have smaller eggs and larvae than freshwater species at similar temperatures, and cold water species tend to have larger eggs and larvae than warm water species.[141] Fishes from cold latitudes can be characteristically separated from warm water species (Table 3) by such reproductive traits as the onset of sexual maturity, the length of vitellogenesis and spawning season, and their gonadosomatic indices.[142]

In comparison with fish from temperate latitudes, marine tropical fishes from reefs, mangroves, and seagrass beds possess smaller pelagic eggs and show an increase in absolute fecundity and the relative frequency of intermittent spawners. It has also been suggested that as predator pressure is more intense in the tropics, active brood protection in fish with demersal eggs can provide higher progeny survival rate.[138,143]

2. Sex Ratio

Where the sex ratio of a species is not determined entirely by genetic factors, it may be affected by temperature or day length. For example, the percentage of females of the burrowing estuarine prawn *Upogebia africana* was 50.0% at 17.5°C, 50.4% at 19°C, 65.3% at 21°C, and 67.5% at 24°C.[134] The sex ratio of the Atlantic silverside, *Menidia menidia,* is also highly temperature dependent with the percentage of females 85% at 15°C, 20% at 19°C, and <20% at >23°C.[145] The sex ratio of white bream *Blicca bjoerkna* is altered significantly with temperature changes.[146] On the other hand, the sex ratio of the amphipod *Gammarus duebeni* is determined by the length of the photoperiod.[147]

While insufficient data are available for generalizations to be made, the differing thermal and light regimes of polar, temperate, and tropical regions may have a significant influence on the reproductive potential of organisms through the alteration of the sex ratio.

3. Life History Regulation

For many aquatic plants, the rate of spore or seed germination, and thus, the generation time of that species, is generally related to the ambient temperature under which development occurs. van den Hoek[1] has classified the algae in relation to the temperature regulation of their life histories and recognized a series of biogeographic zones in which they occur. In aquatic macrophytes such as the emergent *Ludwigia peploides,* the rate of seed germination increases with increasing temperatures.[148] Germination was most rapid and successful at 30°C. Both at the lower temperatures and at 40°C, germination was delayed and less complete, and at 10°C no germination had occurred within 27 d. At 40°C, the seedlings did not develop beyond the emergence of the radicle while at 15, 20, and 30°C, normal seedling development occurred.

The rates of egg and larval development of aquatic invertebrates also increase as temperature rises to a a certain point, while the range of temperatures tolerable to the development process varies with the species and is often correlated with the thermal characteristics of the habitat.[30,87,149,150] Even small increases (2.5 to 4°C) resulted in a more rapid development time for aquatic insects such as Diptera,[125] Hemiptera,[151] and Plecoptera.[126] This enhanced population survivorship in that development was rapid, but less somatic growth occurred in summer. Sacrificing somatic growth to maximize reproductive potential during the summer may be an important thermal strategy which could explain the ubiquitous presence of these aquatic insect groups[125] in many stream communities. Similarly, increased temperatures were found to shorten the incubation time for eggs of burrowing shrimp *Upogebia africana* from 67 d at 15°C to 26 d at 25°C, and this could be regressed as

$$\log Y = 4.037 - 1.881 \log X \tag{4}$$

where Y is the estimated development time in days and X the temperature in degrees celsius.[134] Because of the shortened incubation time, the maximum number of batches of eggs also increased from three batches at 17.5°C to seven batches at 24°C. Similar sets of power function curves that accurately predict the estimated development time from environmental temperature have been derived for marine harpacticoid copepods[152] and cyprinid fish.[153]

Obversely, low temperatures extend the period of egg maturation as, for example, in decapods.[22,154,155] In tropical portunid species such as *Scylla serrata, Portunus pelagicus, P. sanguinolentus,* and *Charybdis feriatus,* incubation periods became protracted (i.e., >70 d) at temperatures below 17 or 18°C.[155,156] In contrast, temperate portunid species had protracted egg development periods below 12°C;[154] clearly, the effects of temperature on egg maturation varies between populations of crabs separated geographically in temperate and tropical areas.

The generally shortened development times, together with increased growth rates and feeding rates, with increasing temperatures should combine to allow organisms in warmer waters to increase the number of annual cohorts. Correlation analysis[157] of 164 invertebrate populations in 51 lakes ranging from the equator to 74°N has allowed the calculation of the number of nonoverlapping generations per year (G) for various temperatures (T) for invertebrate populations of known mean population biomass (B) and maximum body mass (W_m) as follows:

$$\log (G) = -0.21 \log (B) + 0.05 (T) - 0.16 \log (W_m) - 0.64 \tag{5}$$

As shown in Figure 5 for a benthic population with a mean annual biomass of 0.1 $g \cdot m^{-2}$, the number of generations per year should increase with temperature and decrease with body mass. In turn, it suggests that for tropical waters, small organisms will have short generation times while for high latitude waters, long life cycles can be expected in the same size organism.

The effects of temperature on various physiological processes suggest that there may be differing thermal limits for general and reproductive processes. If there were significant differences between somatic and reproductive thermal limits, an organism living in a habitat

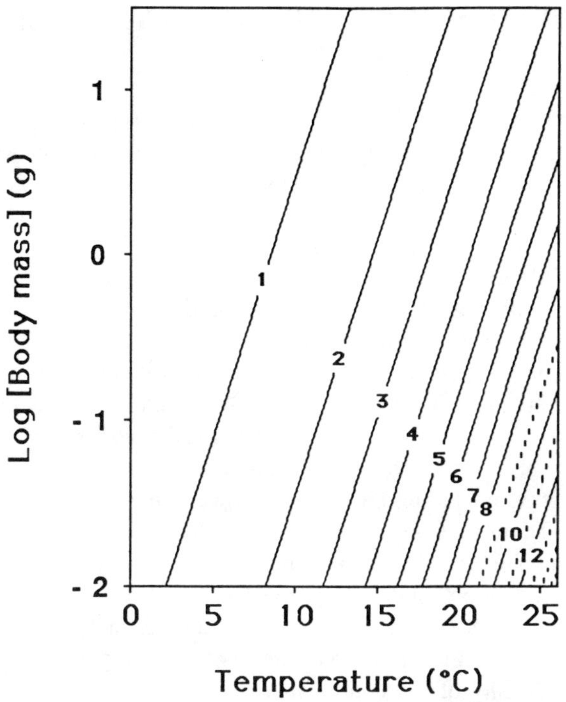

FIGURE 5. Number of nonoverlapping generations per year for a benthic population with a mean biomass of 0.1 g·m^{-2}. (Calculated from Equation 5 in Reference 157.)

with a narrow temperature range unsuited for reproduction might have to migrate to a thermally different habitat to reproduce.[85] Clearly, such migration requirements have considerable ecological significance and are more frequent in temperate and polar species than in tropical ones.

D. GENETIC CONSIDERATIONS

Genetic control is important in many of the processes described above including the ability to acclimate, the possession of certain metabolic pathways, the sex ratio, or the particular life history strategy of an organism. For example, protein synthesis is a fundamental component of physiological acclimation to change in temperature[158] and is genotype dependent.[159] In studies of the rates of protein synthesis and breakdown in *Mytilus edulis* subjected to increases in temperature above 10°C, individuals expressing high rates of protein turnover were more sensitive to change in temperature and died more rapidly at upper lethal temperatures.[160] By elevating the metabolic demand, high rates of protein synthesis conferred a significant disadvantage compared with individuals with lower and more efficient rates of protein synthesis. In general, organisms with genotype-dependent rates of growth (or protein synthesis) may have a differential fitness under conditions of temperature stress.[27] As genetic heterozygosity is correlated with low maintenance costs,[160,161] heterozygous individuals and genetically variable populations are likely to be more stress tolerant.

As with temperature, the resistance to pollution appears to be inversely related to the genetic diversity of species which, in turn, is correlated with their niche-width variation.[26]

While insufficient data are available to generalize, the genetic variability of fish populations appears to be high in tropical and temperate-pelagic species[162] but low in temperate-demersal ones.

IV. IMPLICATIONS FOR AQUATIC POLLUTION MANAGEMENT

Given the differences between polar, temperate, and tropical organisms and their environments, what are the implications of these features in relation to aquatic pollution management? The potential responses of aquatic organisms to pollutants must be evaluated against an ecological background.[163] Are there any features of adaptation to a tropical environment, in a physiological, genetic, or ecological sense, that make tropical organisms more or less vulnerable to pollutants or other forms of stress or disturbance?

It is worthwhile to distinguish between stress and disturbance.[164] ''Stress'' is caused by an environmental change or factor that reduces productivity, whereas ''disturbance'' results in destruction of biomass. These usages are not identical with those elsewhere but have been found useful in illuminating ecological strategies in different circumstances.[164,165] In general terms, temperate and polar environments may be seen as the more stressful, while disturbance may be more important in the tropics, as discussed below.

In temperate zones temperatures are generally lower, but seasonality is more pronounced, than in the tropics. Temperate organisms must be physiologically adaptable to cope with major seasonal changes in at least temperature, and possibly also in other factors such as light or salinity, depending on habitat. They must also be ''adapted'' in an ecological sense to seasonal cycles of productivity, nutrient cycles, and other factors. Such adaptability must be operative over the entire life of an organism, as there are year-to-year variations in the precise timing and intensity of seasonal changes. Interyear variation will be mediated not by changes in solar energy input, which varies only slowly with the years, but by climatic factors, such as changes in jet stream patterns and other aspects of the primary circulation of the atmosphere-ocean system. Occasional and unpredictable disturbances, such as destriction by storms, are superimposed on this pattern of more or less wide seasonal changes.

In temperate environments, then, seasonal changes in light, temperature, etc. may be regarded as stress factors, controlling the rate of primary and secondary productivity. In polar regions, the constant low temperatures, as well as wide variation in light intensity, would act as stressors. Such stress factors are predictable, in contrast to the destructive disturbances caused by storms. Note, however, that occasional cold winters or warm summers in temperate habitats may result in considerable loss of biomass in localized areas, so that greater than usual fluctuations in stress factors may themselves constitute a disturbance.

In the tropics, where seasonality is reduced (although rarely absent), within-year changes in ecological activity, such as production, seem to be less pronounced and ''favorableness'' may be more constant; i.e., stress is generally lower than in temperate areas. Intense localized disturbances due, for instance, to tropical cyclones, are unpredictable and more extreme than the more normal patterns of seasonal change. They are also less easily countered by evolutionary adaptation than the more predictable stresses of seasonal change in temperate habitats. These arguments do not imply that habitats in the tropics are absolutely more ''stable'' than those in temperate latitudes, merely that the scales and intensities of changes may differ.

The ecological implications of these contrasts are discussed in detail elsewhere but it is clear that, wherever they have been examined in the tropics, process rates such as respiration, decomposition, and nutrient cycling have been found to be higher than in the temperate zone.[166-169] This accelaration of ecological processes in the tropics is accompanied by a potential for shorter generation times (Figure 5) and thus, by extension, for more rapid evolutionary adaptation to habitat conditions. As discussed above, the greater ''favorableness'' of tropical environments (as contrasted with, for instance, polar habitats) suggests that tropical populations may be constrained more by biological interactions, such as competition, parasitism, and predation, than by the difficulty of individual adaptation to environmental conditions. This view accords well with the idea that favorable environments may support a greater variety of niches, many of which depend on biological construction of habitat space (e.g., coral reefs) as well as provision of other resources (such as food or nutrients). In this regard tropical habitats often support a greater variety of species than equivalent habitats in higher latitudes.

These arguments are importnt in modeling the effects of pollution and other disturbances on ecosystem resources in tropical areas. However, impacts on organisms, populations, and species, as opposed to ecosystem-level processes, are also important in tropical resource management. The important question here seems to be, "are individuals of tropical species necessarily more vulnerable to stress or disturbance than those of temperate forms?" It lacks a useful general answer. This chapter has emphasized that individuals in fluctuating environments must be able either to withstand changing conditions or to adapt to them. Short-term changes, such as occur seasonally, are too rapid to allow significant genetic adaptation to each episode of changed conditions. Instead, it seems more likely that it has been the ability to compensate physiologically that has evolved.

The process of physiological adaptation, whether at a molecular or a whole-organism level, has the effect of loosening the shackles of the environment on the life of an organism, of reducing the deterministic effect of external factors on biological performance. In general, temperate organisms seem to express a greater capacity for such adaptation than do tropical forms. In that sense, then, it may be said that tropical forms may in some circumstances be more vulnerable to physical and chemical stress but more resilient in response to biotic and abiotic disturbance. That, however, is about as far as the argument can usefully go. One reason is that the capacity to adapt to seasonal changes in temperature, light, or the supply of requisites, such as dissolved oxygen or nutrients, does not imply a similar capacity to adapt to novel chemicals, or to changes in ionic concentrations or oxygen tension, that are very different from the scale of changes seen in unimpacted environments. Furthermore, organisms do not exist as isolated entities reacting only to the physical and chemical features of their habitat. The ecological milieu within which they live may be as important as direct chemical or physical impacts in determining their functioning and survival.

REFERENCES

1. **van den Hoek, C.,** The distribution of benthic marine algae in relation to the temperature regulation of their life histories, *Biol. J. Linn. Soc.,* 18, 81, 1982.
2. **Blumenstock, D. L.,** Distribution and characteristics of tropical climates, *Proc. Pac. Sci. Congr.,* 20, 3, 1958.
3. **Fitzpatrick, E. A., Hart, D. and Brookfield, H. C.,** Rainfall seasonality in the tropical southwest Pacific, *Erdkunde,* 20, 181, 1966.
4. **Moore, H. B.,** Aspects of stress in the tropical marine environment, *Adv. Mar. Biol.,* 10, 217, 1972.
5. **Vernberg, F. J.,** Comparative studies of tropical and temperate zone coastal systems, *Bull. Mar. Sci.,* 31, 801, 1981.
6. **Turner, J. T.,** Latitudinal patterns of calanoid and cyclopoid copepod diversity in estuarine waters of eastern North America, *J. Biogeogr.,* 8, 369, 1981.
7. **Taylor, E. W. and Innes, A. J.,** A functional analysis of the shift from gill- to lung-breathing during the evolution of land crabs (Crustacea, Decapoda), *Biol. J. Linn. Soc.,* 34, 229, 1988.
8. **Hawker, D. W. and Connell, D. W.,** An evaluation of the tolerance of corals to nutrients and related water quality characteristics, *Int. J. Environ. Stud.,* 34, 179, 1989.
9. **Bakus, C. J.,** Energetics and feeding in shallow marine waters, in *International Review of General and Experimental Zoology,* Vol. 4., Felts, W. J. L. and Harrison, R. J., Eds., Academic Press, New York, 1969, 275.

10. **Paltridge, G. W. and Proctor, D.,** Monthly mean solar radiation statistics for Australia, *Sol. Energy,* 18, 235, 1976.
11. **Turton, S. M.,** The relationship between total irradiation and sunshine duration in the humid tropics. *Sol. Energy.,* 38, 353, 1987.
12. **McAlpine, J. R.,** Climate and water balance, in *Lands of the Alligator Rivers Area,* C.S.I.R.O. Aust. Land Res. Ser. No. 38, 1976, 35.
13. **Jones, K. and Betaieb, M.,** Nitrogen-fixing heterotrophic bacteria in coastal waters: a comparison of the contribution made by sewage effluent in a temperate and a tropical environment, in *Water and Wastewater Microbiology (Water Science and Technology,)* Jenkins, D. and Olson, B. H., Eds., 1988, 323.
14. **Duarte, C. M. and Kalff, J.,** Latitudinal influences on the depths of maximum colonization and maximum biomass of submerged angiosperms in lakes, *Can. J. Fish. Aquat. Sci.,* 44, 1759, 1987
15. **Barko, J. W., Hardin, D. G., and Matthews, M. S.,** Growth and morphology of submersed freshwater macrophytes in relation to light and temperature, *Can. J. Bot.,* 60, 877, 1982.
16. **Bulthuis, D. A.,** Effects of temperature on photosynthesis and growth of seagrasses, *Aquat. Bot.,* 27, 27, 1987.
17. **Taylor, J. A. and Tulloch, D.,** Rainfall in the wet-dry tropics: extreme events at Darwin and similarities between years during the period 1870-1983 inclusive, *Aust. J. Ecol.,* 10, 281, 1985.
18. **Doty, M. S.,** Antecedent event influence on benthic marine algal standing crops in Hawaii, *J. Exp. Mar. Biol. Ecol.,* 6, 161, 1971.
19. **Clarke, A. and Holmes, L. J.,** Lipid content and composition of some midwater crustaceans from the Southern Ocean, *J. Exp. Mar. Biol. Ecol.,* 104, 31, 1986.
20. **Saenger, P., Moverley, J., and Stephenson, W.,** Seasonal and long-term patterns in the macrobenthos versus benthic stability in a subtropical estuary, *Proc. Ecol. Soc. Aust.,* 15, 229, 1988.
21. **Southwood, T. R. E.,** Habitat, the templet for ecological strategies?, *J. Anim. Ecol.,* 46, 337, 1977.
22. **Clarke, A.,** On living in cold water: k-strategies in Antarctic benthos, *Mar. Biol. (Berlin),* 55, 111, 1979.
23. **Hynes, H. B. N.,** The ecology of stream insects, *Annu. Rev. Entomol.,* 15, 25, 1970.
24. **Garrity, S. D.,** Some adaptations of gastropods to physical stress on a tropical rocky shore, *Ecology,* 65, 559, 1984.
25. **Diehl, W. J. and Koehn, R. K.,** Multiple-locus heterozygosity, mortality, and growth in a cohort of *Mytilus edulis, Mar. Biol. (Berlin),* 88, 265, 1985.
26. **Nevo, E., Noy, R., Lavie, B., Beiles, A., and Muchtar, S.,** Genetic diversity and resistance to marine pollution, *Biol. J. Linn. Soc.,* 29, 139, 1986.
27. **Koehn, R. K. and Bayne, B. L.,** Towards a physiological and genetic understanding of the energetics of the stress response, *Biol. J. Linn. Soc.,* 37, 157, 1989.
28. **Ahlgren, G.,** Temperature functions in biology and their application to algal growth constants, *Oikos,* 49, 177, 1987.
29. **Vernberg, F. J.,** Latitudinal effects on physiological properties of animal populations, *Annu. Rev. Physiol.,* 24, 517, 1962.
30. **Vernberg, W. B. and Vernberg, F. J.,** *Environmental Physiology of Marine Animals,* Springer-Verlag, New York, 1972, 346.
31. **Kristensen, E.,** Ventilation and oxygen uptake by three species of *Nereis* (Annelida: Polychaeta). II. Effects of temperature and salinity changes, *Mar. Ecol. Progr. Ser.,* 12, 299, 1983.
32. **Emmerson, W. D.,** Oxygen consumption in *Palaemon pacificus* (Stimpson) (Decapoda: Palaemonidae) in relation to temperature, size and season, *Comp. Biochem. Physiol. A,* 81, 71, 1985.
33. **Mauro, N. A. and Maugum, C. P.,** The role of blood in the temperature dependence of oxidative metabolism in decapod crustaceans. II. Interspecific adaptations to latitudinal changes, *J. Exp. Zool.,* 219, 189, 1982.
34. **Burggren, W. W. and McMahon, B. R.,** Oxygen uptake during environmental temperature change in hermit crabs: adaptation to subtidal, intertidal, and supratidal habitats, *Physiol. Zool.,* 54, 325, 1981.
35. **Sukhtin, A. A.,** Respiration in cultured White Sea mussels (in Russian), *Ehkologiya,* 2, 55, 1988.
36. **Sinha, D. P. and Poandey, B. N.,** Effect of ambient water temperature on oxygen uptake in some freshwater fishes, *Indian J. Acad. Ichthyol. (Modinagar),* 6, 31, 1985.
37. **Nagata, W. D.,** Long-term acclimation of a parthenogenetic strain of *Branchionus plicatilis* Mueller to subnormal temperatures. II. Effect on clearance and ingestion rates, *Bull. Fac. Fish. Hokkaido Univ.,* 36, 1, 1985.
38. **Grant, J.,** Sensitivity of benthic community respiration and primary production to changes in temperature and light, *Mar. Biol. (Berlin),* 90, 299, 1986.
39. **Nugegoda, D. and Rainbow, P. S.,** The effect of temperature on zinc regulation by the decapod crustacean *Palaemon elegans* Rathke, *Ophelia,* 27, 17, 1987.
40. **Prosser, C. L.,** Temperature, in *Comparative Animal Physiology,* 2nd ed., Prosser, C. L. and Brown, F. A., Eds., W. B. Saunders, Philadelphia, 1961, 688.

41. **Kinne, O.,** The effects of temperature and salinity on marine and brackish water animals. I. Temperature, *Oceanogr. Mar. Biol. Annu. Rev.,* 1, 301, 1963.
42. **Vigh, L. and Joo, F.,** Modulation of membrane fluidity by catalytic hydrogenation affects the chilling susceptibility of the blue-green alga, *Anacystis nidulans, FEBS Lett.,* 162, 423, 1983.
43. **Chapelle, S., Brichon, G., and Zwingelstein, G.,** Effect of environmental temperature on the phospholipid metabolism of gill mitochandria of *Carcinus maenas, Biochem. Syst. Ecol.,* 9, 333, 1981.
44. **Haschemeyer, A. E. V. and Persell, R.,** Comparative analysis of leucine transport in temperate fish liver in vivo, *Comp. Biochem. Physiol. A,* 78, 381, 1984.
45. **Southward, A. J.,** The relationship between temperature and rhythmic cirral activity in some Cirripedia considered in connection with their geographical distribution, *Helg. Wiss. Meeres.,* 10, 391, 1964.
46. **Burton, D. T.,** Ventilation frequency compensation responses of three eurythermal estuarine fish exposed to moderate temperature increases, *J. Fish Biol.,* 15, 589, 1979.
47. **McLeese, D. W.,** Effects of temperature, salinity and oxygen on the survival of the American lobster, *J. Fish Res. Bd. Can.,* 13, 247, 1956.
48. **Fry, F. E. J., Hart, J. S., and Walker, K. F.,** Lethal temperature relations for a sample of young speckled trout, *Salvelinus fontinalis, Univ. Toronto Stud. Biol. Ser.,* 54, 35, 1946.
49. **Cocking, A. W.,** The effects of high temperature on roach *(Rutilus rutilus).* II. The effects of temperatures increasing at a known constant rate, *J. Exp. Biol.,* 36, 217, 1959.
50. **Somero, G. N. and Hochachka, P. W.,** The effect of temperature on catalytic and regulatory functions of pyruvate kinases of the rainbow trout and the Antarctic fish *Trematomus bernacchii, Biochem. J.,* 110, 395, 1968.
51. **Gilles, R.,** Mechanisms of thermoregulation, in *Marine Ecology,* Vol. 2 (1), Kinne, O., Ed., John Wiley & Sons, Chichester, 1975, 251.
52. **Thebault, M. T. and Bernicard, A.,** Bases moleculaires de l'adaptation thermique chez les invertebres marins, in *Biochemical and Biophysical Processes of Physiological Adaptation in Marine Organisms According to Environmental Factors,* Oceanis, Marseilles, 1981, 613.
53. **Crawford, D. L. and Powers, D. A.,** Molecular basis for equal enzymatic rates in populations from different thermal environments, *1st Int. Symp. Marine Molecular Biology,* Baltimore, MD, Abstr. No. 44, 1988.
54. **Christopherson, J.,** Basic actions of temperature action in microorganisms, in *Temperature and Life,* Precht, H., Christopherson, J., Hensel, H., and Larcher, H., Eds., Springer-Verlag, Berlin, 1973, 3.
55. **Morris, I. and Glover, H. E.,** Questions on the mechanism of temperature adaptation in marine phytoplankton, *Mar. Biol. (Berlin),* 24, 147, 1974.
56. **Li, W. K. W. and Morris, I.,** Temperature adaptation in *Phaeodactylum tricornutum* Bohlin: photosynthetic rate compensation and capacity, *J. Exp. Mar. Biol. Ecol.,* 58, 135, 1982.
57. **Peckol, P.,** Seasonal physiological responses of two brown seaweed species from a North Carolina continental shelf habitat, *J. Exp. Mar. Biol. Ecol.,* 72, 47, 1983.
58. **Daugherty, B. K. and Bird, K. T.,** Salinity and temperature effects on agar production from *Gracilaria verrucosa,* strain G-16, *Aquaculture,* 75, 105, 1988.
59. **Fry, F. E. J. and Hochachka, P. W.,** Fish, in *Comparative Physiology of Thermoregulation,* Vol. 1, Whittow, G. C., Ed., Academic Press, New York, 1970, 79.
60. **Sidell, B. D. and Stone, B. B.,** Metabolic responses of striped bass *(Morone saxatilis)* to temperature acclimation. I. Alterations in carbon sources for hepatic energy metabolism, *J. Exp. Zool.,* 218, 371, 1981.
61. **Winkler, P.,** Persistent differences in thermal tolerance among acclimation groups of a warm spring population of *Gambusia affinis* determined under field and laboratory conditions, *Copeia,* 2, 456, 1985.
62. **Dheer, J. M. S.,** Haematological, hematopoietic and biochemical responses to thermal stress in an air-breathing freshwater fish, *Channa punctatus* Bloch, *J. Fish Biol.,* 32, 197, 1988.
63. **Johnston, A. A. and Altringham, J. D.,** Muscle contraction in polar fishes: experiments with demembranated muscle fibres, *Comp. Biochem. Physiol. B,* 1988, 547, 1988.
64. **Tetens, V. and Wells, R. M. G.,** Antarctic fish blood: respiratory properties and the effects of thermal acclimation, *J. Exp. Biol.,* 109, 265, 1984.
65. **D'Avino, R. and Di Prisco, G.,** Hemoglobin from the Antarctic fish *Notothenia coriiceps neglecta.* I. Purification and characterization, *Eur. J. Biochem.,* 179, 699, 1989.
66. **De Vries, A. L.,** Effect of temperature on freezing avoidance and levels of antifreeze in Antarctic fishes, *Antarct. J. U.S.,* 15, 149, 1981.
67. **Ortega, M. M., Iglesias, J. I. P., and Navarro, E.,** Acclimation to temperature in *Actinia equina* L.: effects of season and shore level on aquatic oxygen consumption, *J. Exp. Mar. Biol. Ecol.,* 76, 79, 1984.
68. **Watts, S. A. and Lawrence, J. M.,** Seasonal effects of temperature and salinity on the organismal activity of the seastar *Luidia clathrata* (Say) (Echinodermata: Asteroidea), *Mar. Behav. Physiol.,* 12, 161, 1986.
69. **Brown, A. C. and da Silva, F. M.,** Effects of temperature on oxygen consumption in two closely-related whelks from different temperature regimes, *J. Exp. Mar. Biol. Ecol.,* 84, 145, 1984.

70. **Hall, L. W., Burton, D. T., and Abell, P. R.,** Thermal responses of Atlantic silversides *(Menidia menidia)* acclimated to constant and asymmetric fluctuating temperatures, *Arch. Hydrobiol.*, 94, 318, 1982.
71. **Refinetti, R. and Refinetti, K. Z.,** Absence of metabolic acclimation to cycling temperature conditions in the goldfish, *Physiol. Behav.*, 43, 121, 1988.
72. **Harborne, J. B.,** Constraints on the evolution of biochemical pathways, *Biol. J. Linn. Soc.*, 39, 135, 1990.
73. **Specht, R. L.,** Major vegetation formations in Australia, in *Ecological Biogeography of Australia,* Keast, A., Ed., Junk, The Hague, 1981, 163.
74. **Jones, M. B.,** Photosynthetic responses of C_3 and C_4 wetland species in a tropical swamp, *J. Ecol.*, 76, 253, 1988.
75. **Cowles, R. B. and Bogert, C. M.,** A preliminary study of the thermal requirements of desert reptiles, *Bull. Am. Mus. Nat. Hist.*, 83, 265, 1944.
76. **Lowe, C. H. and Vance, V. J.,** Acclimation of the critical thermal maximum of the reptile, *Urassaurus ornatus, Science,* 122, 73, 1955.
77. **Hutchinson, V. H.,** Critical thermal maxima in salamanders, *Physiol. Zool.*, 34, 92, 1961.
78. **Gonzalez, J. G.,** Critical thermal maxima and upper lethal temperatures for the calanoid copepods *Acartia tonsa* and *A. clausi, Mar. Biol. (Berlin),* 27, 219, 1974.
79. **Potter, M. A. and Hill, B. J.,** Heat mortality in the Sydney rock oyster, *Saccostrea (Crassostrea) commercialis* and the effectiveness of some control methods, *Aquaculture,* 29, 101, 1982.
80. **Menasveta, P.,** Lethal temperature of marine fishes of the Gulf of Thailand, *J. Fish Biol.*, 18, 603, 1981.
81. **Burton, D. T., Abell, P. R., and Capizzi, T. P.,** Cold shock: effect of rate of thermal decrease on Atlantic menhaden, *Mar. Pollut. Bull.*, 10, 347, 1979.
82. **Brett, J. R.,** Temperature: fishes, in *Marine Ecology,* Part 1, Vol. 1, Kinne, O., Ed., John Wiley & Sons, Chichester, 1970, 515.
83. **Kiyono, M. and Shimshima, K.,** Thermal tolerance of stone flounder eggs, *Bull. Jpn. Soc. Sci. Fish.*, 49, 701, 1983.
84. **Hodson, R. G., Fechhelm, R. G., and Monroe, R. J.,** Upper temperature tolerance of spot, *Leistomus xanthurus,* from the Cape Fear River estuary, North Carolina, *Estuaries,* 4, 345, 1981.
85. **Vernberg, W. B.,** Responses to estuarine stress, in *Ecosystems of the World, Estuaries and Enclosed Seas,* Vol. 26, Ketchum, B. H., Ed., Elsevier Scientific, Amsterdam, 1983, 43.
86. **Vernberg, W. B.,** Adaptations of host and symbionts in the intertidal zone, *Am. Zool.*, 9, 357, 1969.
87. **Vernberg, F. J. and Vernberg, W. B.,** Adaptations to extreme environments, in *Physiological Ecology of Estuarine Organisms,* Vernberg, F. J., Ed., The Belle W. Baruch Library in Marine Science, North Carolina, 1975, 165.
88. **Mundahl, N. D. and Benton, M. J.,** Aspects of the thermal ecology of the rusty crayfish *Orconectes rusticus* (Girard), *Oecologia,* 82, 210, 1990.
89. **Vaeille-Morel, C.,** Resistance of the prosobranch *Potamopyrgus jenkinsi* (E. A. Smith, 1889) to decreasing temperatures: an experimental study, *Ann. Limnol.*, 21, 221, 1985.
90. **Thompson, R. J. and Newell, R. I. E.,** Physiological responses to temperature in two latitudinally separated populations of the mussel, *Mytilus edulis,* in *Proc. 19th Eur. Marine Biology Symp.*, Gibbs, P. E., Ed., 1985, 481.
91. **Stirling, H. P.,** The upper temperature tolerance of prosobranch gastropods of rocky shores at Hong Kong and Dar es Salaam, Tanzania, *J. Exp. Mar. Biol. Ecol.*, 63, 133, 1982.
92. **Broekhuysen, G. J.,** A preliminary investigation of the importance of desiccation, temperature, and salinity as factors controlling the vertical distribution of certain intertidal marine gastropods in False Bay, South Africa, *Trans, R. Soc. S. Afr.*, 28, 255, 1940.
93. **Fraenkel, G.,** The heat resistance of intertidal snails at Shirahama, Wakayama-ken, Japan, *Publ. Seto Mar. Biol. Lab.*, 14, 185, 1966.
94. **Fraenkel, G.,** Resistance to high temperatures in a Mediterranean snail, *Littorina neritoides, Ecology,* 42, 604, 1961.
95. **Ansell, A. D., Barnett, P. R. O., Bodoy, A., and Masse, H.,** Upper temperature tolerances of some European molluscs. I. *Tellina fabula* and *T. tenuis, Mar. Biol. (Berlin),* 58, 33, 1980.
96. **Ansell, A. D., Barnett, P. R. O., Bodoy, A., and Masse, H.,** Upper temperature tolerances of some European molluscs. II. *Donax vittatus, D. Semistriatus* and *D. trunculus, Mar. Biol. (Berlin),* 58, 41, 1980.
97. **Chung, K. S. and Acuna, A.,** Upper temperature tolerance limit of mussel, *Perna perna, Bull. Jpn. Soc. Sci. Fish.*, 47, 411, 1981.
98. **Thibault, Y. and Couture, R.,** Resistence thermique superieure de *Gammarus fasciatus* Say (Crustacea, Amphipoda) et son utilisation en situation de rejets thermiques, *Can. J. Zool.*, 60, 1326, 1982.
99. **Coles, S. L.,** A comparison of effects of elevated temperature versus temperature fluctuations on reef corals at Kahe Point, Oahu, *Pac. Sci.*, 29, 15, 1975.

100. **Bourget, E.,** Seasonal variation of cold tolerance in intertidal mollusks and relation to environmental conditions in the St. Lawrence estuary, *Can. J. Zool.,* 61, 1193, 1983.
101. **Crisp, D. J. and Lewis, A. H.,** Factors controlling cold tolerance and breeding in *Balanus balanoides, J. Mar. Biol. Assoc. U. K.,* 64, 125, 1984.
102. **Murphy, D. J.,** Freezing resistance in intertidal invertebrates, *Annu. Rev. Physiol.,* 45, 289, 1983.
103. **Aarset, A. V.,** Freezing tolerance in intertidal invertebrates (a review), *Comp. Biochem. Physiol.,* 73, 571, 1982.
104. **Levitt, J.,** Winter hardiness in plants, in *Cryobiology,* Meryman, H. T., Ed., Academic Press, London, 1966, 495.
105. **Migata, S.,** Freeze preservation of *Porphyra* thalli in a viable state. II. Effect of cooling velocity and water content of thalli on frost resistance, *Bull. Fac. Fish. Nagasaki Univ.,* 21, 131, 1966.
106. **Rochet, M., Legendre, L., and Demers, S.,** Acclimation of sea-ice microalgae to freezing temperature, *Mar. Ecol. Progr. Ser.,* 24, 187, 1985.
107. **Scholm, H.,** Untersuchungen zur Hitze- und Frostresistenz von Tiefalgen, *Bot. Mar.,* 9, 55, 1966.
108. **Biebl, R.,** Uber die Temperaturresistenz von Meeresalgen verscheidener Klimazonen und verschieden tiefer Standorte, *Jahr. Wiss. Bot.,* 88, 390, 1939.
109. **Gessner, F.,** Temperature: plants, in *Marine Ecology,* Environmental Factors, Part 1, Vol. 1, Kinne, O., Ed., John Wiley & Sons, Chichester, 1970, 363.
110. **Bolton, J. J. and Luning, K.,** Optimal growth and maximum survival temperatures of Atlantic *Laminaria* species (Phaeophyta) in culture, *Mar. Biol. (Berlin),* 66, 89, 1982.
111. **Saenger, P. and Moverley, J.,** Vegetative phenology of mangroves along the Queensland coastline, *Proc. Ecol. Soc. Aust.,* 13, 257, 1985.
112. **Smillie, R. M. and Hetherington, S. E.,** Stress tolerance and stress-induced injury in crop plants measured by chlorophyll fluorescence in vivo. Chilling, freezing, ice cover, heat and high light, *Plant Physiol.,* 72, 1043, 1983.
113. **Hutchings, P. and Saenger, P.,** *Ecology of Mangroves,* University of Queensland Press, Brisbane, 1987, 388.
114. **Fry, F. E. J.,** Effects of the environment on animal activity, *Univ. Toronto Stud. Biol. Ser.,* 55, 62, 1947.
115. **Fry, F. E. J.,** The aquatic respiration of fish, in *The Physiology of Fishes, Metabolism,* Vol. 1, Brown, M. E., Ed., Academic Press, New York, 1957, 63.
116. **Bayne, B. L. and Newell, R. C.,** Physiological energetics in marine molluscs, in *The Mollusca,* Vol. 4, Saleudden, A. S. M. and Wilbur, K. M., Eds., Academic Press, New York, 1983, 407.
117. **Prabhakara-Rao, Y., Uma-Devi, V., and Prasada-Rao, D. G. V.,** Influence of temperature on copper accumulation and depuration in a tropical cerithiid *Cerithidea (Cerithideopsis) cingulata* (Gmelin 1790), *Water Air Soil Pollut.* 37, 365, 1988.
118. **Vernberg, W. B. and Vernberg, F. J.,** The synergistic effects of temperature, salinity, and mercury on survival and metabolisms of the adult fiddler crab, *Uca pugilator, Fish. Bull.,* 70, 415, 1972.
119. **Cairns, J., Heath, A. G., and Parker, B. C.,** The effects of temperature upon the toxicity of chemicals to aquatic organisms, *Hydrobiologia,* 47, 135, 1975.
120. **Cotter, A. J. R., Phillips, D. J. H., and Ahsanullah, M.,** The significance of temperature, salinity and zinc as lethal factors for the mussel *Mytilus edulis* in a polluted estuary, *Mar. Biol. (Berlin),* 68, 135, 1982.
121. **Hodson, P. and Blunt, B.,** Temperature-induced changes in pentachlorophenol chronic activity to early life stages of rainbow trout, *Aquat. Toxicol.,* 1, 113, 1981.
122. **Rehwoldt, R., Menapace, L. W., Nerrie, B., and Alessandrello,** The effect of increased temperature upon the acute toxicity of some heavy metal ions, *Bull. Environ. Contam. Tox.,* 8, 91, 1972.
123. **Dehnel, P. A.,** Rates of growth of gastropods as a function of latitude, *Physiol. Zool.,* 28, 115, 1955.
124. **Marchant, R.,** Some quantitative aspects of the life history of aquatic insects in temperate Australian rivers, in *Limnology in Australia,* De Deckker, P. and Williams, W. D., Eds., CSIRO Australia, Melbourne, 1986, 151.
125. **Rempel, R. S. and Carter, J. C. H.,** Temperature influence on adult size, development and reproductive potential of aquatic diptera, *Can. J. Fish. Aquat. Sci.,* 44, 1743, 1987.
126. **Sweeney, B. W. and Vannote, R. L.,** Growth and production of a stream stonefly: influences of diet and temperature, *Ecology,* 67, 1396, 1986.
127. **Wilkins, J. L. and Fingerman, M.,** Heat tolerance and temperature relationships of the fiddler crab *Uca pugilator* with reference to body coloration, *Biol. Bull. (Woods Hole),* 128, 133, 1965.
128. **Brett, J. R.,** Energetic responses of salmon to temperature. A study of some thermal relations in the physiology and freshwater ecology of sockeye salmon *(Oncorhynchus nerka), Am. Zool.,* 11, 99, 1971.
129. **Reynolds, W. W. and Casterlin, M. E.,** Behavioral thermoregulatory abilities of tropical coral reef fishes: a comparison with temperate freshwater and marine fishes, *Environ. Biol. Fish.,* 6, 347, 1981.
130. **Hall, L. W., Margrey, S. L., Burton, D. T., and Graves, W. C.,** Avoidance behavior of juvenile striped bass, *Morone saxatilis,* subjected to simultaneous chlorine and elevated temperature conditions, *Arch. Environ. Contam. Toxicol.,* 12, 715, 1983.

131. **Bailey, P. C. E.,** The effect of water temperature on the functional response of the water stick insect *Ranatra dispar* (Heteroptera: Nepidae), *Aust. J. Ecol.,* 14, 381, 1989.
132. **Gross, M. R., Coleman, R. M., and McDowall, R. M.,** Aquatic productivity and the evolution of diadromous fish migration, *Science,* 239, 1291, 1988.
133. **Pearse, J. S.,** Patterns of reproductive periodicities in four species of Indo-Pacific echinoderms, *Proc. Indian Acad. Sci.,* 68, 247, 1968.
134. **Hill, B. J.,** The effect of heated effluent on egg production in the estuarine prawn *Upogebia africana* (Ortmann), *J. Exp. Mar. Biol. Ecol.,* 19, 291, 1977.
135. **Neumann, D.,** Lunar and tidal rhythms in the development and reproduction of an intertidal organism, in *Physiological Adaptation to the Environment,* Vernberg, F. J., Ed., Intext Educ. Publishers, New York, 1975, 451.
136. **Segal, E.,** Light, in *Marine Ecology,* Kinne, O., Ed., Wiley-Interscience, New York, 1970, 159.
137. **Lake, P. S., Barmuta, L. A., Boulton, A. J., Campbell, I. C., and St Clair, R. M.,** Australian streams and northern hemisphere stream ecology: comparisons and problems, *Proc. Ecol. Soc. Aust.,* 14, 61, 1985.
138. **Clarke, A.,** Temperature, latitude and reproductive effort, *Mar. Ecol. Progr. Ser.,* 38, 89, 1987.
139. **Dean, J. C. and Cartwright, D. I.,** Trichoptera of a Victorian forest stream: species composition and life histories, *Aust. J. Mar. Freshwater Res.,* 38, 845, 1987.
140. **Harrison, P. L. and Wallace, C. C.,** Reproduction, dispersal and recruitment of scleractinian corals, in *Coral Reef Ecosystems: Ecosystems of the World,* Vol. 25, Dubinsky, Z., Ed., Elsevier Science, Amsterdam, chap. 7, in press.
141. **Bagarinao, T., Chua, T. E., and Dizon, L. B.,** Egg size and larval size among teleosts: implications to survival potential, in *Proc. 1st Asian Fisheries Forum,* Maclean, J. L. and Hosillos, L. V., Eds., Manila, 1986, 651.
142. **GarciaCagide, A., Claro, R., and Kosheliev, B. V.,** Peculiaridades de los ciclos reproductivos de los peces de differentes latitudes, *Rep. Invest. Inst. Oceanol. Acad. Cienc. Cuba,* 14, 1983.
143. **Billard, R.,** Le cycle reproducteur chez les poissons teleosteens, *Cah. Lab. Hydrobiol. Montereau,* 12, 43, 1981.
144. **Reshetnikov, Y. S.,** Reproduction peculiarities in tropical fishes, in *Peculiarities of Reproductive Cycles of Fishes in Water Bodies from Different Latitudes* (in Russian), Koshelev, B. V., Ed., 1985, 12.
145. **Conover, D. O. and Heins, S. W.,** Adaptive variation in environmental and genetic sex determination in a fish, *Nature (London),* 326, 496, 1987.
146. **Volodin, V. M. and Grechanov, I. G.,** Effect of temperature on gonadal development and sex ratio in the progeny of a pair of white bream *Blicca bjoerkna, J. Ichthyol.,* 25, 52, 1985.
147. **Bulnheim, H. P.,** Ueber den Einfluss der Photoperiode auf die Geschlechtsrealisation bei *Gammarus duebeni, Helg. Wiss. Meeresunters.,* 15, 69, 1967.
148. **Yen, S. and Myerscough, P. J.,** Co-existence of three species of amphibious plants in relation to spatial and temporal variation: investigation of plant responses, *Aust. J. Ecol.,* 14, 305, 1989.
149. **Vannote, R. L. and Sweeney, B. W.,** Geographic analysis of thermal equilibria: a conceptual model for evaluating the effect of natural and modified thermal regimes on aquatic insect communities, *Am. Nat.,* 115, 667, 1980.
150. **Ward, J. V. and Stanford, J. A.,** Thermal responses in the evolutionary ecology of aquatic insects, *Annu. Rev. Entomol.,* 27, 97, 1982.
151. **Park, S. O.,** Effects of temperature on the development of the water strider, *Gerris paludum insularis* (Hemiptera: Gerridae), *Environ. Entomol.,* 17, 150, 1988.
152. **Webb, D. G. and Parsons, T. R.,** Empirical analysis of the effect of temperature on marine harpacticoid copepod development time, *Can. J. Zool.,* 66, 1376, 1988.
153. **Herzig, A. and Winkler, H.,** The influence of temperature on the embryonic development of three cyprinid fishes, *Abramis brama, Chalcalburnus chalcoides mento* and *Vimba vimba, J. Fish Biol.,* 28, 171, 1986.
154. **Wear, R. G.,** Incubation in British decapod crustacea, and the effects of temperature on the rate of success of embryonic development, *J. Mar. Biol. Assoc. U.K.,* 54, 745, 1974.
155. **Campbell, G. R. and Fielder, D. R.,** Egg extrusion and egg development in three species of commercially important portunid crabs from S. E. Queensland, *Proc. R. Soc. Qd.,* 99, 93, 1988.
156. **Heasman, M. P. and Fielder, D. R.,** Laboratory spawning and mass rearing of the mangrove crabs *Scylla serrata* (Forskal), from first zoea to first crab stage, *Aquaculture,* 34, 303, 1983.
157. **Plante, C. and Downing, J. A.,** Production of freshwater invertebrate populations in lakes, *Can. J. Fish. Aquat. Sci.,* 46, 1489, 1989.
158. **Hazel, J. R. and Prosser, C. L.,** Molecular mechanisms of temperature compensation in poikilotherms, *Physiol Rev.,* 54, 620, 1974.
159. **Hawkins, A. J. S., Bayne, B. L., and Day, A. J.,** Protein turnover, physiological energetics, and heterozygosity in the blue mussel, *Mytilus edulis:* the basis of variable age-specific growth, *Proc. R. Soc. London Ser. B,* 229, 161, 1986.

160. **Hawkins, A. J. S., Wilson, I. A., and Bayne, B. L.,** Thermal responses reflect protein turnover in *Mytilus edulis, Functional Ecol.,* 1, 339, 1987.
161. **Koehn, R. K. and Gaffney, P. M.,** Genetic heterozygosity and growth rate in *Mytilus edulis, Mar. Biol. (Berlin),* 82, 1, 1984.
162. **Smith, P. J. and Fujio, Y.,** Genetic variation in marine teleosts: high variability in habitat specialists and low variability in habitat generalists, *Mar. Biol.,* 69, 7, 1982.
163. **Johannes, R. E. and Betzer, S. B.,** Introduction: marine communities respond differently to pollution in the tropics than at higher latitudes, in *Tropical Marine Pollution,* Ferguson Wood, E. J., and Johannes, R. E., Eds., Elsevier Scientific, Amsterdam, 1975, 1.
164. **Grime, J. P.,** *Plant Strategies and Vegetation Processes,* John Wiley & Sons, Chichester, 1979.
165. **Grime, J. P.,** Evidence for the existence of three primary strategies in plants and its relevance to ecological and evolutionary theory, *Am. Nat.,* 111, 1169, 1977.
166. **Jenny, H., Gessel, S. P., and Bingham, F. T.,** Comparative study of decomposition rates of organic matter in temperate and tropical regions, *Soil Sci.,* 61, 419, 1949.
167. **Nye, P. H.,** Organic matter and nutrient cycles under moist tropical forests, *Plant Soil,* 13, 333, 1961.
168. **Olsen, J. S.,** Energy storage and the balance of producers and decomposers in ecological systems, *Ecology,* 44, 322, 1963.
169. **Jordan, C. F. and Herrera, R.,** Tropical rainforests: are nutrients really critical? *Am. Nat.,* 117, 167, 1981.

Section III: Biological Effects of Pollution in Major Tropical Ecosystems

Chapter 4

OIL SPILLS IN THE TROPICS AND SUBTROPICS

Anitra Thorhaug

TABLE OF CONTENTS

I. INTRODUCTION

This chapter will review major impacts of oil spills in the tropics and the techniques for managing those oil spills, including recent dispersed-oil data. It will also include cost and policy issues, as well as recommendations. The information is meant to bring the reader to a more comprehensive view of a subject frequently treated in most books as individual case studies, which are of very limited value. Predictive use of data for decision-makers and implementors is essential. Thus, a synoptic understanding is important.

Much of the tropics lies in the third world. The petroleum industry has production and shipping ports throughout the subtropical and tropical world: the Red Sea, Arabian (Persian) Gulf, Indonesia and Brunei, Venezuela, Ecuador, Mexico, Nigeria, Angola, and the Gulf of Mexico. A map of major shipping lanes (from the Middle East, Nigeria, Indonesia, Mexico, and Venezuela) by the International Maritime Organization (IMO) shows that many tropical nations, which themselves do not produce oil, are on major shipping routes between producers and consumers. (Bilging brings 7.2 million bbl of oil into the environment per year.) All of the East and West African and Caribbean nations fall on major oil tanker routes, as do Pakistan, India, Sri Lanka, and Southeast Asian countries. Other nations, such as South Pacific nations, must realize that all ships carry petroleum. Local oil depots, refineries, and ports and harbors have higher accident rates than do petroleum transport ships.

Yearly input of petroleum hydrocarbons, from transportation sources, includes the following totals: 126 million gal from accidents, 300 million gal from bilging; 320 million gal from municipal wastes; 12 million gal from offshore production.[1] Bilging and accidents are chiefly on tanker routes.

This chapter will deal mainly with the 126 million gal spilled from accidents and 12 million gal from offshore production. A large amount of the 300 million gal from bilging — if reported and treated rapidly enough — can also be handled in the same manner. Municipal wastes are a question of storm water runoff and hazardous materials disposal, and must be treated separately. We do not address that issue here.

A. DECISION-MAKING REALITIES FOR ENVIRONMENTAL SCIENTISTS AND ENVIRONMENTAL MANAGERS

Oil spills present a different problem to the environmental manager and to the environmental scientist than do most pollution sources. The spill is already an ongoing fact. An unfortunate accident has occurred which no one wanted, no one chose. The sole problem for the environmentalist is, "Which strategy would be the most realistic to minimize environmental impact?" There are few choices, none ideal. To do nothing may result in a great catastrophe. The management process is something like a forest fire, without any good biological results at the conclusion. Also, there will be a great many players intensely involved in oil spills, including those from the military, the petroleum industry, and politicians.

B. CLEANUP METHODS

Despite more than 50 years of research, there are only four major methods of cleaning spills: (1) mechanical cleanup, (2) shoreline cleanup, (3) no spill control, and (4) dispersants.

Mechanical cleanup had 20 years during which it was the only acceptable or preferred method. A great deal is known about its limits. Much environmental damage has been done because it was the accepted method. It works in small discrete spills quickly reported and responded to, near the site of the equipment where an experienced team can immediately surround the spill and where the seas and winds are fairly calm. Frequently mechanical cleanup fails. For a developing nation it is seen as advanced technology which is expensive to prepare for and difficult to perform well. At the end of the cleanup, there is the question of land disposal of the used oil. The situations under which mechanical cleanup is not likely to work are the following:

1. When seas are rough; tide currents or winds are high.
2. When the site is distant from the equipment; the time to surround the spill is critical to calculate ahead of time.
3. The shallow areas or sensitive bottom types can be damaged during deployment, since the boats must maneuver.
4. The boom must be in good working order; breakages have been known (Panama), and then the cleanup results can be very difficult; therefore a backup position is always necessary. Secondary spills are common.
5. It is rare in a large spill (50,000 gal or more) that all the oil is mechanically contained. A second choice frequently must be made for the noncontained remaining oil.

The environmentalist must face the fact that the "mechanical option" is likely to result in much unmanaged oil. It is far easier and safer to make the decision to cleanup by mechanical means than it is to implement this decision, particularly in second- and third-world nations. The final problem is disposal of waste oil which must enter as a part of the "mechanical decision". Many countries do not have places to dispose of waste oil.

The choice of shoreline cleanup has several environmental hazards, if the shoreline surface — in contact with the oil — has coral or vegetation, mangroves, marshes, algae, and animals. Intertidal flats, especially coral reefs within 2 ft of low tide, are frequently the tropical shorelines in which that oil strands. Mangroves are definitely killed by oil when the oil hits their roots. Corals are definitely killed by oil. Animals are killed by oil. Thus, living shorelines, which concentrate large amounts of oil, are not good places for oil to strand. Beaches can be scraped to remove oil. This is costly, and it also can lead to beach erosion. Cleanup on a coral reef or in mangroves can cause harm.

The "do nothing" decision is usually best when oil is headed out to the deep sea and/or when the spill is in the deep sea and not coming to shore. It is the worst decision nearshore, when coral reefs and/or mangrove swamps are going to be affected. Frequently, "do nothing" is what occurs due to bad preparations and lack of adequate decision-making ahead of time. Hundreds of miles of mangroves around the world have been damaged, due to this decision (Indonesia, Nigeria, Puerto Rico, as examples). The dispersant decision is discussed in this chapter in detail, as well as recent work in the 1980s on dispersants which had not previously been analyzed in great detail. Dispersant solution was shelved for 20 years. The problem was that the older generations of dispersants, as exemplified by Torrey Canyon, were highly toxic. Thus, two decades have passed during which time the dispersant option has not been used in large-scale tropical field situations, although in the Temperate Zone alone, the author can name 50 examples of dispersant use in the last 5 years with no negative effects. This lack has provided large numbers of mechanical cleanup, beach cleanup, and "do nothing" examples. Many of these have been highly resource damaging. The dispersant option for temperate and arctic organisms has been discussed by the U.S. National Academy of Science,[41] and they seemed to agree that dispersants are a viable option. They emphasize the need for tropical information and quote some of the older tropical work included here.

Dispersants have the ability to flow with the tides and currents, whereas normal oil flows with wind vectors. Usually the oil alone ends up, in good part, stranded onshore, whereas dispersed oil tends to be carried out to sea.

The other options are, at this point, mostly theoretical, i.e., not tested and not known to be either effective or nontoxic, except under specific conditions.

1. Flaring is rarely done because it needs a certain thickness of oil and needs to be away from the vessel for safety factors.

2. Sinking compounds coat or smother organisms, can re-release oil, and blobs of oiled compounds in the environment, which must be dealt with. Obviously, they are very bad for infauna.
3. The gels are not yet commercially on the same order of magnitude with sorbant pads or dispersants and are practically out of the question, due to cost. Some have low toxicity.
4. The bacterial-activated compounds have not been tested, to our knowledge, in large scale on mangrove forests, coral reefs, or in seagrass beds. Their work on beaches, rocks, and the deep sea looks promising, but expensive. If developing nations must clean up the spills themselves, cost is of the essence frequently.
5. Other new products continually come on the market, and one must keep up to date on these products.

No doubt 5 years from now, the subject will have dramatically advanced. It is necessary to keep up with research in this field.

C. TYPES OF BIOLOGICALLY IMPORTANT OIL SPILLS IN THE TROPICS

Another important set of factors to remember is the types of spills and the priority to biological protection. There are offshore spills which can be from moving vessels or from fixed points. If the spill moves offshore or out to sea, the biological protection priority is fairly low. If it is moving toward land, it is high. The second case is coastal shoreline spills. This case usually needs biological protection, especially with onshore winds and sensitive shorelines (mangroves, corals, seagrass, etc.).

The third case is in an estuary. This type of spill almost always needs much protection. The unfortunate part of the estuarine type is that there is seldom much time between the spilling and the resource being hit. The time to mobilize mechanical cleanup crews and boats can be 4 to 6 h. Oil from a port or harbor collision can be in the mangroves in 1/2 h. The realities of the cleanup operation and its timing should be clearly understood by the biologist. If the mechanical cleanup tools are available, but there is no chance of their protecting the estuarine mangroves before 6 h to mobilize crew, obtain boom from storage, and put the boom in place (the fairly simple projection of the winds pushing the rapidly leaking oil from a barge that has collided with an object midestuary predicts 40,000 gal in the mangroves in 1.5 h), the biologist would be highly irresponsible to stick with the high risk of mechanical equipment not adequate in timing for protection. The management Plan B (nontoxic dispersant sprayed right away on the leading edge) should be called into immediate action, because unless that operation is mobilized, the mangroves will be killed as the tide goes low.

The key here is clearly and realistically seeing the options and risks, then generating predecisions so that the command person does not need to struggle through a postdecision, as in the Exxon Valdez incident. The command person will have enough difficulty implementing the plan, whatever it is, even if a predecision has been made.

A quantitative picture, rather than a qualitative picture, of the exact dimension and trajectory of the spill is of greatest help to all concerned.

D. COMPUTER MODELS FOR DECISION-MAKERS REGARDING OIL SPILLS AND DISPERSED-OIL IMPACT ON SPECIES

One of the best uses of concentration mortality data for species has been seen in the simulations of oil spills and dispersed oil spills. Much of the above-mentioned data have been used in a Gulf of Mexico simulation by Ross et al.[2] for the U.S. Gulf Coast. The field experiments done on spill trajectories and dispersed oil trajectories, along with currents, winds, tides, and chemical data, are programmed into a geographic information system with extensive spatial analytic capabilities. The environmental regulators in the Gulf States chose 90 species, including important commerical fish and invertebrates, critical habitat organisms (corals, seagrass, mangroves, marshes), endangered species, and a few others. The algorithms assessed for resource-specific impact linking oil or dispersed-oil concentrations to organism tolerance can compute mortality from the spill point as the spill is diluted in time and space. The readout is a map with spill trajectory over time as concentration dilutes. The impact on all resources and location of impact is calculated and read out specifically such as "percent mangrove mortality". The program is run a second time with the same oil source and spill volume, but dispersed. The mortality is again computed, readout, and this time compared with "oil alone". Frequently, there are large resource impact differences between the two spills. The quantitative visualization of the potential impact over time, space, and important species can wipe away "fuzzy thinking" and allow detailed predecisions about whole sets of spill scenarios. Many biological managers would be better off to have a visual image of spill trajectory over time and concentration.

I recommend this technology for predetermination of operational techniques with all parties concerned with studying the options and making decisions.

Up to now, no tropical nation has this capability. South Florida is the only tropical region that has this.

E. COSTS

Cost accounting must consider short-term cleanup costs and long-term resource value costs. When all options are available and it is the taxpayers or citizens of the area who will bear the expense of cleanup (Bahrain's catastrophic spill, for example), then cost becomes a major factor to be taken into consideration. Long-term as well as direct costs must be considered. If the cost will be borne by the insurance company of the spiller or by the spiller, then the manager must deal with these parties.

There is the hidden cost of taking away the oil removed from beach cleanup and disposing of it properly. If the developing nation has no appropriate oil dump so the oil will not enter into the ecosystem through storm runoff from dumps, then the shoreline cleanup option must be reappraised due to risk from the ultimate dumping point of oil. The Jamaican government now faces a fisherman's legal action for such a strategy.

The cost to the ecosystem of tens of miles of dead mangroves which could take 25 years to reestablish the forest properties for primary productivity, fish nurseries, etc. is a large amount of money. A coral reef is worth more than \$200,000/ha, according to United Nations Environment Program (UNEP). Tourism loss where a spill covers a tourist beach would be a direct cost to take into account.

The socioeconomic costs of disruption of the lives and diet of communities dependent on artesanal fisheries, especially children and pregnant women, is another hidden cost. The direct loss of tourist industry revenue for exchange dollars should be carefully thought out during plans for oil-spill contingencies in areas such as the Caribbean, East Africa, and Pacific nations. (See Reference 26 for a full explanation of environmental cost accounting.)

II. RESULTS

A. COMPARATIVE TOXICITY OF OIL ALONE ON VARIOUS TROPICAL ECOSYSTEMS

1. Mangroves

There has been a wide array of mangrove oil spills reported to kill mangroves in Nigeria, Indonesia (on several occasions), Panama,[6] Kenya, Puerto Rico,[8] and other areas. Several oil spills which did not sink into the root systems, but only oiled the bark of mangroves at high tide, were reported not to kill the mangroves.[24,42,43] A mesoscale field experiment with mature red mangroves in the Caribbean[13] indicates that dispersed oil Corexit 9527 does not kill the red mangrove, even at high concentrations, but oil alone at that same concentration kills the red mangrove *Rhizophora mangle*. If the dispersant was placed on the mangrove after the oil reached the roots, the mangrove died as if no dispersant had been added, but oil alone had spilled. The field experiments of Getter et al.,[9] Ballou et al.,[2] and Cubit et al.[21] on red mangroves in Panama reestablished that dispersed oil (Corexit 9527) does not kill mangroves.

Our experiments[3] in the laboratory with very young (6 months to 1 1/2 years) red, black, and white mangroves showed that the oil concentration between 1 and 125 ppm had no effect on red and very little on black and white mangroves. The higher concentrations (1250 and 12,500 ppm) did have effects on black and white mangroves (Tables 1 and 2).

2. Corals

The coral, *Diploria strigosa*, in laboratory and field conditions, has been shown by Knap et al.[4,5] to have been affected by oil alone. Oil spills in Egypt,[36] in Indonesia, in Curacao,[7] and in Panama[6] showed a series of species of corals in various oceans to be sensitive to oil.

Emerged coral reefs are clearly more vulnerable to oil than those submerged. Bak[7] has shown oil effect down to 125 ft on high energy coastlines in Curacao (Table 3).

3. Fish

Fish have been extensively tested in temperate and subarctic waters, but tropical oil tests and reports are far less available.[1] Several large tropical oil spills enumerate fish and invertebrate kills such as Puerto Rico[6,8] and Indonesia and Singapore government reports.

The data in Thorhaug, et al.[3] show four important tropical commercial fish to be very sensitive to oil spills. Frequently, fish are far less sensitive to pollutants than tropical seagrass and corals, but this is not apparent in these data (Table 4).

4. Seagrasses

Tropical and subtropical seagrasses have been tested between 12.5 and 125 ppm oil. The results show that lethal concentrations are greater for *Thalassia testudinum* and less for *Halodule* and *Syringodium*. The type of oil was important to the toxic effect on seagrasses (Figure 2). In field situations, real spill results have been far less accurate for seagrass than any other critical habitats (Tables 5, 6, and 7).

5. Comparison

When the three sets of dominant matrix organisms were tested using the same relative methods and the same oil at the same location, the mangroves were approximately 100 times more tolerant than the seagrasses. The seagrasses were somewhat more tolerant than the corals at high concentrations of oil (Tables 5 and 6). (These experiments used oil at 10 times the concentration expected in a normal open water spill and 6 to 10 times longer exposure time). The tropical commercial fish had tolerances similar to corals.

Now, why do we have so many reports of mangrove deaths if this is so? First, the oil grows in concentration along a shoreline, which acts as a sorbant. Mangroves have been found by Teas et al.[13] in the field and in the literature to have oil levels 5 to 10 l/m^2. Second, oil spreads from the source where it could be 0.3 mm thick to a thinner and thinner film. Thus, the corals and seagrass, unless in very shallow water, would never experience more than 50 ppm oil whereas mangroves might experience orders of magnitude higher, especially in the holes in the sediment, where oil might accumulate near root structures of mangroves.

TABLE 1
Dispersed Oil and Oil Effects on Mangroves

Location	Type	Dispersant used & dilution	Type of oil	Amount of spill	Date	Resource affected	Impact	Dispersant effect	Ref.
Panama	Field	Corexit 9527, 24 h, 1:20	50 ppm Prudhoe Bay crude	Exp.[a]	1984	Mangroves	Defoliation, death	Dispersed oil before it reached mangroves	9
Coast on Caribbean side of Panama	Accidental	Corexit 9527, ca. 21,000 l, 1:20	Medium weight crude	55,000—60,000	Apr 27, 1986	*R. mangle*	Defoliation, death		21
Coast on Caribbean side of Panama	Experimental	Corexit 9527, 1:20	Prudhoe Bay crude 50 ppm	Exp.	1985	*R. mangle*	28% trees defoliated	No defoliation at sites with dispersant	9
South Florida Turkey Pt., Biscayne Bay, FL	Field	Corexit 9527, 1:20	LA Crude concentrated	Exp.	1982—1986	*R. mangle*			13
Panama	Field spill	Corexit 9527, 1:20	Medium weight crude		Fall, 1986	Mangroves *R. mangle*	Observed mangrove death	If dispersed before oil on mangroves, less mortality	34
Jamaica	Lab	11 dispersants	Venezuelan	Exp.	1988—1989	*Rhizophora Avicennia Laguncularia*	Defoliations, death of root	Various at 1250 ppm not low	3

[a] Exp. = experimental.

TABLE 2
The Mortality Percent of Jamaican Mangroves Exposed for 10 h at 1250 ppm Dispersed Oil

Dispersants	Rhizophora (red) (%)	Avicennia (black) (%)	Laguncularia (white) (%)
Conco K	28.6	14.3	80
OFC D609	42.9	14.3	25
Corexit 9527	71.4	0	40
Corexit 9550	0	0	20
Wonder-O	42.9	14.3	50
ADP-7	28.6	28.6	50
Jansolv	14.3	0	60
Cold Clean	28.6	14.3	20
Finasol	0	14.3	40
V-25	71.4	14.3	40
LTX	nd	nd	nd
Oil only	14.3	28.6	40
Control	0	14.3	14.3

From Thorhaug, A., Carby, B., Reese, R., Rodriguez, M., McFarlane, J., Teas, H., Sidrak, G., Anderson, M., Aiken, R., McDonald, F., Miller, B., Gordon, V., and Gayle, P.., in *Proc. 1991 Oil Spill Conf.*, American Petroleum Institute, Washington, D.C., 1991.)

B. COMPARATIVE DISPERSED-OIL EFFECTS ON VARIOUS GROUPS OF TROPICAL ORGANISMS

Tables 1 to 4 show dispersant effects available from the literature. Seagrasses, mangrove, and corals are shown separately to compare dispersant effects within each critical habitat group (Table 8, A and B). Unfortunately, scientists and government workers in the tropics do not publish material at the same rate as those in the first world, so there may be dispersed oil spills with unpublished results. For instance, the Philippine Coast Guard was funded for a large dispersant toxicity testing program, but results were not published nor are they available (fish were the major test organism).

In general, there are more detailed data in laboratory experiments about seagrass than others: Fish data are abundant from temperate zones, but sparsely available in the tropics. There is more field testing on corals than mangroves, seagrasses, or tropical fish. A wider range of dispersants has been tested on seagrasses than on mangroves or corals. More real-life spills have been documented on mangroves than seagrasses or corals.

C. DIFFERENCES OF DISPERSED-OIL TOXICITY EFFECTS AMONG VARIOUS DISPERSANTS

The U.S. Environmental Protection Agency (EPA) toxicity tables indicate large differences between the 20 dispersant products listed with the agency. A biological indicator organism test was done by a scientific group using a temperate Pacific estuarine species. Indeed, large differences highly correlated with the EPA information were found.[10] Countries in temperate zones have found differences between their organisms.

1. Seagrasses

A set of laboratory experiments using three subtropically grown (but distributed throughout subtropics and tropics) seagrasses and seven dispersants was run.[11,12] The seagrasses responded generally in the same relative tolerance order to each of the dispersants. Dispersants could be clumped into three groups: low, medium, and high toxicity. Finasol OSR-7 was

TABLE 3
Dispersed Oil and Oil Effects on Corals

Location	Type	Dispersant used & dilution	Conc. of Dispersant	Amount of spill	Date	Resource affected	Impact	Dispersant effect	Ref
Bermuda	Field & lab	Corexit 9527 BP 1100 WD	1:20 1:10	Arabian light crude	1981—1986	Corals	6—24 h after, 1—50 ppm on *Diploria strigosa*	No effect to brief exposures; when oil dispersed 20 ppm polychaetes, bivalves crustacea intolerant unclear after 9 mo. whether dispersant had effect or not	4,5
Arabian Gulf	Field	Corexit 9527 20:1		Arabian light crude experiment	1980	Corals	No impact immediately, some death after 6 mo. during winter cold		28
Panama	Field	Corexit 9527	50 ppm 20:1	Prudhoe Bay crude experiment	1985	Corals, seagrasses mangroves	No coral death at 24 h exposure	No death of corals with dispersant	9
Panama	Spill	Corexit 9527	20:1	50,000 med. wt. crude	1986	Corals, seagrasses mangroves	Coral death	Reports intertidal reefs extensive mortality, subtidal to 2 m mortality	12
Jamaica	Lab	10 dispersants	1:10	Venezuela light	1988—1989	Corals	Various	3 nontoxic, 5 hightly toxic	3

TABLE 4
Jamaican Coral Mortality and Seagrass Mortality, Dispersed Oil and Oil at Various Concentrations and Times

	125 ml, 6 h			75 ml, 10 h	
	Por. por.	**Mont. an.**	**Ac. pl.**	**Por. por.**	**Mont. an.**
Conco	100	100	100	100	100
OFC D609	100	100	100	91	91
Corexit 9527	90	86	100	88	76
Kemarine	90	57	100	85	90
ADP 7	90	72	100	85	90
Corexit 9550	64	14	100	0	12
Jansolv	0	0	50	0	0
Elastosol	15	0	4	0	0
Cold Clean	15	0	15	8	0
Finasol	8	0	15	0	0
Oil only	52	52	100	12	0
Control	0	0	29	0	0

Note: Por. por. = *Porietes porietes;* Mont. an. = *Montastrea annularis*; Ac. pl. = *Acropora palmata.*

From Thorhaug, A., McDonald, F., Miller, B., McFarlane, J., Carby, B., Anderson, M., Gordon, V., and Gayle, P., in *Proc. Int. Oil Spill Conf.*, American Petroleum Institute, Washington, D.C., 1989, 455.

least toxic; Conco K and OFC-D-609 were most toxic; Corexit 9550, Jansolv 60 had medium toxicity; Cold Clean 500 and Corexit 9550 had low toxicity. The differences between dispersants was much greater than differences between species (Figures 1, 3 and Tables 5 to 7). This correlated well with animal toxicity data.[10]

The differences between dispersants was also determined on tropical seagrasses. The ranking was precisely the same as the subtropical seagrasses from Florida. The locally made products V-25, Kemarine (Jamaica), Wonder-O (Jamaica) were all very toxic, especially at higher concentrations.[19]

2. Mangroves

A series of comparative mangrove seedling experiments were performed by Thorhaug et al.[3] These showed that at concentrations of 12.5 and 125 ppm, typical of open ocean spills, that no difference between dispersants or between dispersants and oil or dispersants and control appear. All are healthy. The next group of experiments showed that at 1250 ppm and 12,500 ppm, large differences in effect between dispersants occurred. Corexit 9527, Conco K, V-25, ADP7 were highly toxic. Jansolv, Cold Clean, Corexit 9550, and Finasol OSR-7 were far less toxic. The others fell in between (Tables 1 and 2).

3. Corals

Most field experiments, as well as laboratory tests, have dealt with one dispersant product, Corexit 9527. One study tested Exxon 9527 (1:20 dilution) and BP 1100 WD (1:10 dilution) and found no statistical difference in the ranges 1 to 50 ppm, 6 to 24-h exposure with light Arabian crude oil.[5] A second studied LTX and found 162 ppm LC50 for *Madracis mirabilus* on Indo-Pacific species.[14]

Knap et al.[5,6] discussed differences between coral toxicity and BP 1100 WD vs. Corexit 9527. There seemed little difference in response.

Thorhaug et al.[3,19] have investigated three major reef species from the Central Caribbean reef crest *(Acropora palmata),* reef face *(Montastrea annularis),* and reef back *(Porites*

TABLE 5
Tropical and Subtropical Seagrass Dispersant Oil and Oil Effects on Seagrasses

Location	Type	Dispersant used & dilution	Type & Conc. of dispersed oil	Amount of spill	Date	Resource affected	Impact	Dispersant effect	Ref
	Lab out doors	Corexit 9527 1:20	50 ppm oil, 1:20, 24 h	50 ppm oil, lab	1984	*Thalassia testudinum*	LD_{50} 12 & 96 hr bioassays, oil & dispersed oil	Oil with dispersant has lower toxicity than without dispersant	
Miami, FL	Lab out doors	Corexit 9527, 1:20	LA crude, Murhan	Lab	1983—1984	*Thalassia* *Halodule* *Syringudium*	LD_{50} vs. time & conc. at 5 to 100 h	At medium conc. High High	11
Miami, FL	Lab. out doors	Arco D-609, 1:10	LA crude, Murban	Lab		*Thalassia* *Halodule* *Syringudium*	LD_{50} 5 h 100 h	Low to medium Low to medium Low to medium at 75 & 125 ml	12
Miami, FL	Lab. out doors	Conco K (K), 1:10	LA crude, Murban	Lab		*Thalassia* *Halodule* *Syringudium*	LD_{50} at 5 & 100 h	Medium to high High High	12
Panama	Field	Corexit 9527	Prudhoe Bay crude, 50 ppm @ 24 h	Lab	1985	*T. testudinum*	None to *Thalassia*	No effect on Thalassia	20
Miami, FL	Lab out doors	Corexit 9550 1:20	LA crude 125 & 75 ml oil 1:20 disp. in 100,000 cc SW	Lab	1986	*Thalassia* *Halodule* *Syringudium*	LD_{50} at 100 h	Low Medium Low to medium	12
Miami, FL	Lab out doors	OFC-D-607 1:10	LA crude 75 &	Lab	1986	*Thalassia* *Halodule*	LD_{50} at 100 h	Low Low	12

TABLE 5 (continued)
Tropical and Subtropical Seagrass Dispersant Oil and Oil Effects on Seagrasses

Location	Type	Dispersant used & dilution	Type & conc. of dispersed oil	Amount of spill	Date	Resource affected	Impact	Dispersant effect	Ref
			125 ml oil LA crude			*Syringudium*		Medium	
Miami, FL	Lab out doors	Cold Clean 500 1:10	LA crude 75 & 125 ml in 100,000 cc SW	Lab	1986	*Thalassia* *Halodule* *Syringudium*	LD_{50} 100 h	Low Low to medium Low	12
Miami, FL	Lab	Finsol OSP-7 1:10	LA crude 75 & 125 ml in 100,000 cc SW	Lab	1986	*Thalassia* *Halodule* *Syringudium*	LD_{50} 100 hr	Medium Low Low to medium low	12

From Thorhaug *et al.* lab 10 dispersants Venezuelan lab 1988-89 Thai, Syr., Hal LD_{50} 3 low, 3 med., 4 high.

TABLE 6
A Comparison of the Effect of Murban Oil vs. Louisiana Crude Plus Conco K(K) Dispersant on *Thalassia*, *Halodule*, and *Syringodium* at a Variety of Exposure Times and Volumes in 100 l Seawater

Dosage (ppm)	Volume of dispersant (ml)	Exposure (hr)	Dispersant/ oil	*Thalassia testudinum* n	G 95% C.L.	CV%	% mortality	*Halodule wrightii* n	G 95% C.L.	CV%	% mortality	*Springodium filiforme* n	G 95% C.L.	CV%	% mortality
Murban oil															
75.0	7.5	5	1/10	15	3.80 ± 0.21	42	13	15	3.34 ± 0.21	34	13	15	3.25 ± 0.61	31	13
75.0	7.5	100	1/10	15	2.78 ± 0.33	61	40	15	1.11 ± 0.25	29	87	15	1.35 ± 0.38	25	100
125.0	12.5	5	1/10	15	3.21 ± 0.46	38	20	15	3.04 ± 0.61	58	20	15	3.15 ± 0.43	61	27
125.0	12.5	100	1/10	15	1.21 ± 0.65	29	73	15	0.68 ± 0.21	30	100	15	0.93 ± 0.25	44	100
0	0	100	—	15	4.15 ± 0.21	18	0	15	4.51 ± 0.65	24	0	15	4.15 ± 0.31	25	0
Louisiana crude															
75.0	7.5	5	1/10	15	4.21 ± 0.11	28	0	15	3.74 ± 0.23	32	0	15	3.55 ± 0.28	24	7
75.0	7.5	100	1/10	15	3.10 ± 0.61	41	33	15	1.98 ± 0.23	41	73	15	1.65 ± 0.88	45	87
125.0	12.5	5	1/10	15	3.98 ± 0.25	13	7	15	3.25 ± 0.45	35	7	15	3.10 ± 0.36	31	13
125.0	12.5	100	1/10	15	2.51 ± 0.41	37	40	15	1.66 ± 0.41	48	100	15	1.05 ± 0.47	52	100
0	0	100	—	15	4.25 ± 0.21	21	0	15	4.10 ± 0.21	21	0	15	3.99 ± 0.24	21	7

Note: Mean specific growth rates (G % p d, 95% confidence limits, and coefficient of variation in growth rates (CV%) are given on number of blades measured (n) for each species of seagrass.

From Thorhaug, A. and Marcus, J., in *Proc. Oil Spill Conf.*, American Petroleum Institute, Publ. 4452, Washington, D.C., 1987, 223.

TABLE 7
Seagrasses in Jamaica vs. Toxicity

Dispersant product	Dispersability ratio[a]	Cost[b]	Mortality 125 ml; 6 h *Thal*	*Hal*	Mortality 75 ml; 6 h *Thal*	*Hal*	Mortality 12.5 ml; 6 h *Thal*	*Hal*
Conco	0.580	0.59	100	82	48	70	17	35
OFC D609	0.007	0.08	70	63	25	70	20	25
Corexit 9527	0.009	0.11	89	93	42	72	10	22
Kemrarine	—	—	63	68	42	68	17	28
ADP 7	—	—	50	46	18	68	21	30
Corexit 9550	0.009	0.11	40	46	15	20	7	7
Jansolv	—	—	0	0	0	0	0	0
Elastosol	—	—	15	46	10	11	8	10
Cold Clean	—	—	0	10	0	8	0	5
Finasol	0.038	0.28	0	8	0	0	0	0
Oil only	0	0	30	28	10	16	10	12
Control	0	0	11	9	7	5	0	5

[a] Dispersability is the ratio of dispersant to oil required to disperse 90% of the oil.[44]
[b] Cost is the relative effective cost of sufficient dispersant to disperse 90% of 1 gal of oil under the conditions of the Mackay apparatus.
[c] Concentrations in 100,000 cc of seawater were 10:1 of oil to dispersant product.

From Thorhaug, A., Carby, B., Reese R., Rodriguez, M., McFarlane, J., Teas, H., Sidrak, G., Anderson, M., Aiken, R., McDonald, F., Miller, B., Gordon, V., and Gayle, P., in *Proc. 1991 Oil Spill Conf.*, American Petroleum Institute, Washington, D.C., in press.

porites). They were all very sensitive to Conco K, Corexit 9527, V-25, ADP7, and Wonder-O (local product). The low toxicity comparisons were Cold Clean, Finasol OSR-7, Corexit 9550, and Jansolv, in that order. Other dispersants fell in between (Table 4, Figure 2).

4. Fish

Laboratory tests[3] of dispersants on four species of tropical commercial fish showed Conco K, Kemarine, V-25, Wonder-O, ADP-7, and Jansolv to be very toxic. Cold Clean, OFC D609 were least toxic. Fish were *Holocentrus rufus, Acanthuris spp., Haemulon spp.,* and *Archosargus rhonboddales*. Some subtropical Gulf of Mexico species have been investigated by Shuba and Heikamp[16] (two shrimp, blue crab, oyster, and redfish) and temperate fish and invertebrates by Anderson *et al.*[10] and by McAuliffe[15] showing clear differences between dispersant toxicities. Cold Clean and Corexit 7664 (a fourth-generation dispersant) were of low toxicity to temperate species. Of high toxicity were Corexit 9527, OSP-20 (Table 8).

5. Comparative Dispersant Effects On Tropical Matrix Species

Recently, there has been a comparison of important tropical species to the same or similar subtropical species using the same methods and involving the same investigators.[19,29,45] The tropical and subtropical species and phyla had the same ranking in sensitivity to dispersed oils. That is, the species behaved the same, whether subtropical or tropical, to the toxicity level of the dispersed oil. The quantitative levels of sensitivity were approximately the same to a given oil or dispersed oil. Tolerance levels were the following: mangroves > seagrasses > fish > corals.

TABLE 8A
The Mean Mortality of 3 Jamaican Fish Species, 3 Jamaican Seagrasses, and 3 Jamaican Coral Species

	Fish[a]				Seagrasses[b]				Coral[c]			
	1	2	3	X̄	1	2	3	X̄	1	2	3	X̄
Conco	100	100	100	100	100	82	100	94	100	100	100	100
OFC D609	40	100	100	80	70	63	nd[d]	67	100	100	100	100
Corexit 9527	40	100	100	80	89	93	nd	91	72	100	100	91
V-25	100	100	100	100	nd	nd	nd	nd	100	100	100	100
Wonder-O	100	100	100	100	100	100	100	100	100	100	100	100
Kemarine	nd	nd	nd	nd	63	68	nd	66	100	100	100	100
ADP-7	100	100	100	100	50	46	nd	48	100	100	100	100
Jansolv	100	100	100	100	0	0	0	0	0	0	73	24
LTX	100	100	100	100	nd	nd	nd	nd	nd	nd	nd	nd
Corexit 9550	0	60	0	20	40	46	nd	43	43	0	100	47
Cold Clean	0	60	80	47	0	0	0	0	0	0	21	7
Finasol	0	0	20	7	0	0	0	0	0	0	11	4
Oil	0	0	0	0	30	28	30	29	52	52	58	54
Control	0	0	0	0	11	9	10	10	0	0	0	0

Note: At 6-h exposure of 125 ppm dispersed oil for the seagrasses and coral; at 3-h exposure of 125 ppm dispersed oil for the fish. X̄ is mean of toxicity of three species.

[a] Fish 1, *Holocentrus rufus;* fish 2, *Acanthurus sp.;* fish 3, *Haemulon sp.*
[b] Seagrass 1, *Thalassia testudinum;* seagrass 2, *Halodule wrightii;* seagrass 3, *Syringodium filiforme.*
[c] Coral 1, *Porites porites;* coral 2, *Montastrea annularis;* coral 3, *Acropora palmata.*
[d] nd = no data.

From Thorhaug, A., Carby, B., Reese, R., Rodriguez, M., McFarlane, J., Teas, H., Sidrak, G., Anderson, M., Aiken, R., McDonald, F., Miller, B., Gordon, V., and Gayle, P., in *Proc. 1991 Oil Spill Conf.*, American Petroleum Institute, Washington, D.C., 1991, 142.

D. DIFFERENCES OF DISPERSED-OIL TOXICITY EFFECTS AMONG SPECIES WITHIN PHYLA

1. Seagrasses

In the subtropical Atlantic, the seagrass *Thalassia testudium* is the most tolerant to high concentrations of oil and dispersed oil.[12,13,17] The least tolerant is *Syringodium filiforme. Halodule wrightii* usually is similar, but slightly more tolerant to oil than *Syringodium filiforme.* These differences are amazingly constant between concentrations, oil, and dispersant type. We have no data yet on the Indo-Pacific species, but would generally expect cogenetors to behave similarly based on comparative Atlantic-Pacific work.[18] (See Figures 1 to 4.)

The tropical species investigated by Thorhaug et al.[19] had the same order of sensitivity ranking as subtropical species determined by Thorhaug and Cruz.[18] It may be interesting to note from a large range of field experiments in the Philippines (central in distribution of Indo-Pacific species) that the sensitivity from most sensitive to least sensitive ranked *Syringodium* fairly sensitive, whereas *Halodule* was tolerant of most substances.[18] In the Indo-Pacific, one would expect *Enhalus* and *Thalassia* to be the most tolerant, and *Syringodium* and *Thallasodendron* least tolerant.

2. Mangroves

Mangroves *(Rhizophora mangle)* were dispersant tested in experimental mesoscale setups by Teas et al.[13] Field situations consistently have given similar results for red mangroves.[2,9,21]

Experiments by Thorhaug et al.[3] showed that the red mangrove *(Rhizophora mangle)*

TABLE 8B

Survival (%) and Growth (Shoots m^{-2}) of Five Species of Seagrass Tested Vs. Pollutant Type in Philippine Test Plots

	Enhalus		*Thalassia*		*Halodule*		*Syringodium*		*Cymodocea*	
Pollutant	**Sprigs**	**Seeds**	**Sprigs**	**Plugs**	**Sprigs**	**Plugs**	**Sprigs**	**Plugs**	**Sprigs**	**Plugs**
Dredge & fill										
Marinduque	78 (1.9)	80 (2.7)	78 (2.4)	38 (4.3)	36 (16.5)	80 (40)	30 (6.1)	28 (3.8)	20 (6.7)	38 (8.9)
Manila Bay	60 (2.64)	62 (2.22)	80 (1.8)	78 (9.2)	18 (5.4)	82 (12.3)	n.p.	n.p.	20 (2.3)	50 (8.2)
Urban Outfall										
Manila Bay	38 (1.8)	n.p.	0	0	0	0	n.p.	n.p.	0	0
Mine tailings										
Marinduque										
East (high energy)	68 (2.3)	8 (1.0)	n.p.	40 (2.6)	16 (11.8)	32 (42)	n.p.	n.p.	40 (1.5)	52 (3.1)
West (low energy)	80 (1.3)	60 (1.2)	84 (2.4)	100 (14.4)	84 (43.5)	96 (60.8)	24 (14.3)	44 (5.2)	4 (1.0)	60 (4.4)
Chemical factory										
(Phosphate fertilizer)	0	0	0	0	4 (3.5)	0	0	0	0	0
Power plant										
Bataan	60 (16.2)	n.p.	52 (12.2)	28 (3.7)	24 (2.2)	24 (35)	0	0	60 (3.7)	60 (4.3)
High current										
Tip of Marinduque	4 (2)	n.p.	n.p.	0 (0)	n.p.	0 (0)	n.p.	n.p.	n.p.	n.p.

Note: Number without parenthesis is survival %. The number in parenthesis is number of shoots per square millimeter. n.p. = not planted.

From Thorhaug, A. and Cruz, R., in *Restoring the Earth*, Berger, A., Ed., Elsevier, Amsterdam, 1989, 21. With permission.

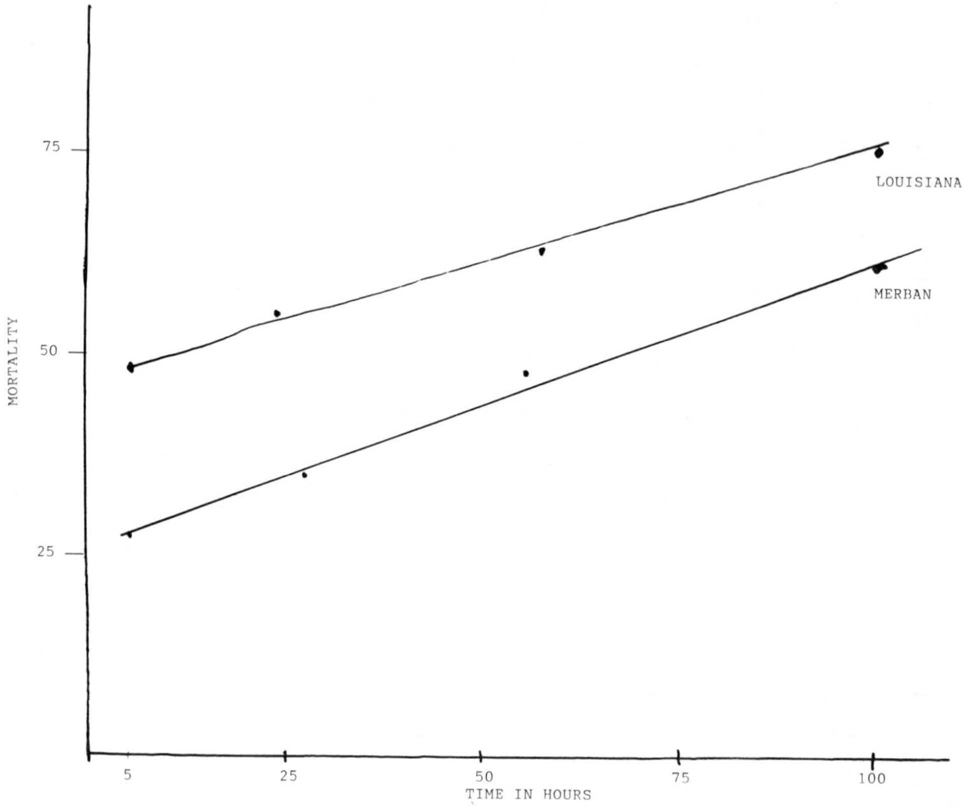

FIGURE 1. Laboratory experiment: time vs. mortality of subtropical *Thalassia testudinum* and Thalassia seagrass.

was far more tolerant than the black *(Avicennia niger)* and black more tolerant than white *(Lagungularia sp.)* to oil and dispersed oil. This was done over four orders of magnitude of concentration, the red and black with 25 dispersants, the red vs. white with 13 dispersants (Table 2).

Field results from the Atlantic would indicate *Rhizophora mangle* had higher tolerance than the other two species. Other parts of the world are yet to be tested.

3. Corals

Several coral species have been employed in dispersant experiments. LeGore et al.[22] mention several *Acropora* species. Although several sets of field experiments have been carried out which utilized various corals, there was no particular attempt to increase concentration until toxicity differences appeared to occur. Thus, at present there is not an established hierarchy of tolerance.

The experiments of oil alone in Bermuda (at the far upper edge of corals anywhere in the Atlantic) done by Knap et al.[4] *in situ* and in laboratory experiments showed a fairly high level of control conditions and excellent chemistry. The species used were *Diplora strigosa,* brain coral, dominant to Bermuda reefs.

The work of Ballou et al.[2] later reported by Cubit et al.[21] from the Caribbean (Panama) did not seem to indicate differences between the ubiquitous *Porites porites* (transplants to site) and *Agaricia sp.* found in back reef (also transplanted to site). Both of these species are estuarine, as well as reef.

In the report of LeGore et al.[22] in the Arabian Gulf on several *Acropora* species, the corals were exposed for 1 to 5 d. Differences between species were not noted.

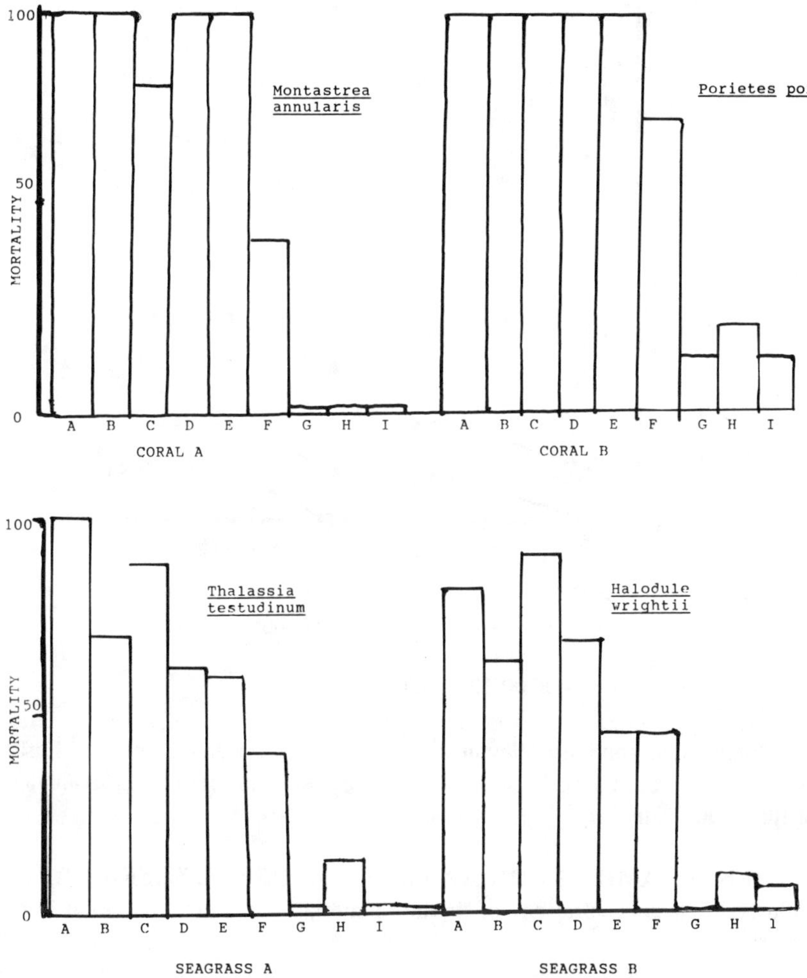

FIGURE 2. Laboratory results of nine types of dispersed oil with two seagrasses, and two coral species. (From Thorhaug, A., Carby, B., Reese, R., Rodriguez, M., McFarlane, J., Teas, H., Sidrak, G., Anderson, M., Aiken, R., McDonald, F., Miller, B., Gordon, V., and Gayle, P., in *Proc. 1991 Oil Spill Conf.*, American Petroleum Institute, Washington, D.C., 1991, 142.)

The work of Thorhaug et al.[19] shows strong differences between species. *Porites porites* was, by far, the most tolerant of all substances, and *Acropora palmata* (crest of reef) was by far the most sensitive to all substances. The reef-building *Montastrea annularis* was in the middle, but closer to *Porites* in response to changes than to *Acropora*. The tolerance to other substances or changes of *Porites* is great while *Acropora* is sensitive to changes and substances (Table 6).

4. Fish

There were differences in tolerance to dispersant and concentration in the work of Aiken et al. (See Reference 9.) The fish were ranked: squirrel fish *(Holocentrus rufus)* > grunts *(Haemulon sp.)* > sea bream *(Archosargus rhonboddales)* > doctor fish *(Acanthurus sp.)* for tolerance. Not a great deal has been done with tropical commercial fish and tolerances to pollution (Table 8).

5. Invertebrates

For the more than 100 tests of various invertebrates vs. dispersed oil, a few in the Gulf

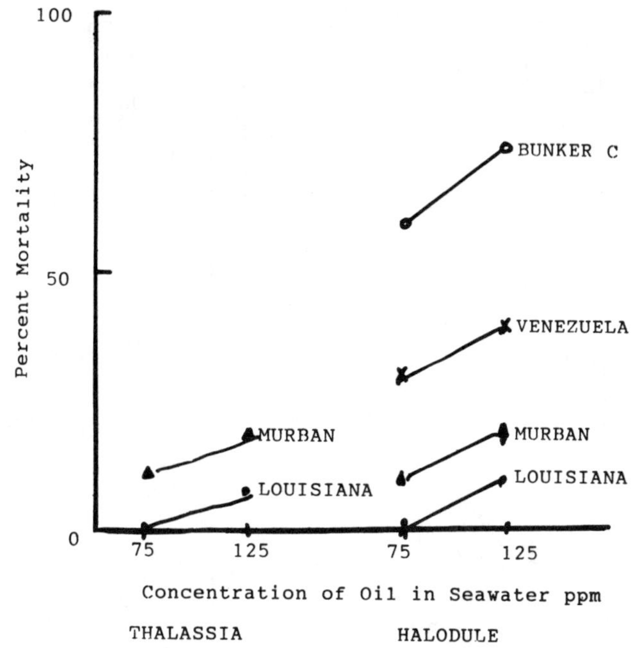

FIGURE 3. Oil type vs. mortality.

of Mexico[16] temperate zone are relevant to subtropical and tropical species. Early shrimp and crab stages appear relatively sensitive, but less so than corals, as evaluated by the National Research Council.[1]

E. DIFFERENCES AMONG CONCENTRATIONS OF DISPERSED OIL

Obviously, high concentrations of organic materials are more toxic than low concentrations.

1. Seagrasses

Large toxicity tolerance differences appeared as increased concentrations of oil dispersants raised percent mortality for a species. In each case seen in Tables 6 and 7, a general concentration range exists where some species show less than 50% mortality and others *(Syringodium* and *Halodule)* are greater. Time of exposure also influences these sets of curves (Figure 1). Statistically significant differences occurred between species at higher concentrations. These ranges are at least an order of magnitude above advised usage levels for dispersants. The differences hold for subtropical and tropical seagrasses of the same species, as evaluated by Thorhaug.[3]

2. Mangroves

Our studies[3] showed that mangroves do not respond to the levels of concentration that fish corals and seagrasses do, that is, 12.5 to 125 ppm. No differences between controls, oil, or concentrations existed. Thus, we attempted higher concentrations (one and two orders of magnitude greater) to obtain differences showing the red mangrove *(Rhizophora mangle)* most resistant to oil and dispersed oil, the white *(Laguncularia)* the least tolerant, black *(Avicennia niger)* medium tolerant. One should keep in mind the high concentrations found by the scientists who measured buildup of oil in mangroves. (Open ocean concentrations would be 1 to 50 ppm.) Reports of liters of oil per square meter are not uncommon in mangrove spills. The time of exposure for mangroves at an interface of oil is often longer than organisms found further seaward or in the ocean itself.

3. Corals

There is experimental work by Knap and colleagues[4, 5] using concentrations in the range 1 to 50 ppm. The higher concentrations showed sublethal effect of behavioral treatment (tissue contraction, tentacle retraction, and localized tissue rupture); lower concentrations did not show behavior effects.

The work by Thorhaug et al.[3, 19] shows oil only has some effect on *Acropora palmata* at 12.5 ppm, far greater effect (100% mortality) at ten times the level (i.e., 125 ppm). *Porites porites* and *Montastrea annularis* go from 1 and 12% death, respectively, at 75 ppm oil to 100% at 125 ppm oil. The Pacific species *Madricis mirabilus* has LC_{50} of 162 ppm to Shell LTX.[14] BP and Corexit 9527 showed no mortality in the concentration range 1 to 20 ppm on *Diploria strigosa*.[5] Corexit 9527 was not lethal in mesoscale field experiments at 50 ppm on *Porites porites* and *Agaricia sp.*[21] The work of Thorhaug et al.[19] shows different responses due to concentrations. At 125 ppm, many coral have a very toxic response to all but four dispersants. At 12.5 ppm, there was a wide variety of responses ranging from 0 to 50% mortality for 6-h exposure. Corals tested were *Porites porites, Montastrea annularis,* and *Acropora palmata*. The mortality differences between concentrations were seen for all species.

4. Fish

Four tropical fish were tested by Thorhaug, Aiken, Walker et al. (see Reference 9.) They all were very concentration sensitive, the doctor fish *Acanthurus sp.* being the most sensitive and the squirrel fish *Holocentrus rufus* the least. At 12.5 ppm, they were not as affected for short exposure times (1 to 3 h, which the National Academy of Sciences[1] has as the time period a fish would encounter an oil spill).

F. DIFFERENCES BETWEEN OIL TYPES

In general, fresh oil vs. "aged oil" and various weights might make a difference in toxicity. To date, no comprehensive program on any of the three critical habitat types has been carried out using various ages and types of petroleum products.

1. Seagrasses

Figure 3 shows the effect of different oils on seagrass toxicity.

2. Mangroves

Prudhoe Bay crude, Louisiana crude, and Venezuelan middle weight has been used for mangrove results. No differences have been pointed out by authors or are apparent from results.

3. Corals

Arabian light crude, Prudhoe Bay, and Venezuelan middle weight have been tested on corals. The authors do not point out any differences nor are they apparent from results.

G. DIFFERENCES IN TIMES OF EXPOSURE

There are several times of exposure which are relevant to tropical coastal oil spills: (1) very short times (1 to 2 h) where a parcel of dispersed or oiled water might flow over a habitat; (2) 6 to 12 h, where a tidal cycle could wash an oiled parcel onto a habitat, especially in an estuary; and (3) longer times where oil might sit in an estuary or be trapped behind a reef or strand in a mangrove swamp. The EPA standard of 96 h (nearly 100) is relevant to this; however, this standard has been applied mostly to fish, which are the least likely to have to tolerate 96 h of exposure in the same way that plants or corals would.

1. Seagrasses

Three subtropical and tropical Atlantic seagrasses have been tested for time of exposure response to various dispersed oils. Figure 1 and Tables 6 and 7 show these results. Clearly, the longer exposure times have a greater toxic effect, especially at higher concentrations of dispersant and oil.

2. Mangroves

Mangroves appear not to have had systematic experiments on time of exposure to dispersed oil. Thorhaug et al.[3] used 12 h (Table 2).

3. Corals

Corals were exposed from 6 to 24 h to dispersed oil in a continuous flow experiment.[4] No toxicity appeared within this time frame. Field experiments were 24 h[20] and 6 h.[22]

H. SEASONAL EFFECTS

1. Seagrasses

Tests were run throughout the year in the subtropics. No differences were found.

2. Mangroves

There are no data for mangroves.

3. Corals

Seasonal differences in response to BP 1100 WD were seen in winter.[4] Differences of corals exposed to Corexit 9527 treatments vs. controls were seen in winter in the Arabian Gulf.[22] Both the above studies are at the furthest limit of corals in the respective ocean basin.

I. TEMPERATURE OF SPILL

In general, the warmer the temperature, the more toxic the compound's effect for physiological processes. The temperate and arctic work has corroborated this. However, there is a small window of temperatures available for tropical organisms: 25 to 30°C are the continental shelf temperatures for tropical organisms. The estuarine temperatures can reach 34°C and fall to 20°C in shallow waters, but these are diurnal variations with low tides and flush back to the central temperatures within hours.

The work of Thorhaug et al.[3,17] was carried out at the 25 to 30°C temperature range, Knap et al.[4] at 22 to 28°C, Cubit et al.[21] at 27°C, and Teas et al.[13] in the 25 to 30°C range.

J. MITIGATION AFTER OIL SPILLS

There have been three notably successful sets of attempts to mitigate after oil spills. Two were in temperate areas on marshes and the third on mangroves in the tropics. The tropical one is handled in detail.

What we see from all three sets of authors is the following: (1) the physical-chemical conditions of the sediment are altered by the spill, even after extensive cleaning (in the French case); (2) recurrences of the spilled oil may occur if oil is not all cleaned up; (3) some species and some elevations produce better results than others. The higher or lower above the tidal zone, the better the result.

In the case of Teas et al.[34] the sediment of the mangroves was noted as altered. The first attempt to restore mangroves after the spill produced a low survival rate of planted red mangrove propagules *(Rhizophora mangle)*. The second attempt several months after the spill was more successful with intermediate survival. The third attempt, many months after the spill, was the one which the authors recommend. Plants higher in elevations survived better than lower, an example of the system restoring itself.

The attempt in France to replant marsh by Seneca[35] was very well executed and was differentially successful. The upper elevation plants survived better than lower elevation plants. One portion of the problem was that the oil came back with tides which recirculated it after plantings had been done. A great deal of steam and other rock cleaning was done which evidently resulted in oil being carried out to sea and then back in some time later. Some of the lower tidal levels were not restored. The higher portion could be restored.

Some careful studies have been done on temperate marshes after mesocosm spills in Great Britain.[33] The studies showed restoration of marshes after heavy oiling was possible, but not a simple endeavor.

Seagrass, corals, and fish stocks have not been attempted to be restored after oil spills. Techniques for restoring these under a variety of other impacts have been successful in Atlantic and Pacific Basins.[18,28-30]

K. METHODS OF MANAGING SPILLS

1. Mechanical Cleanup

These methods include booms to contain oil, skimmers, and sorbants. The performance of any of these methods so as to have low biological damage is dependent on the speed of transporting them to the site and having them operational in time to function on the spilled oil. Physical conditions at the site frequently make this option nonviable: (1) current must be less than 1 kn; (2) wind should be less than 6 kn; (3) wave action must be very low, less than the freeboard of the barrier; (4) the nature of the oil slick itself. The stress on the boom is appreciable, and the booms frequently break, demanding a second strategy be employed. Skimmers decrease their recovery rate at decreasing thickness of oil. Below 1 mm, the rate is almost negligible.

The distance the spill site is from the stockpile of sufficient mechanical equipment and the time for deployment are major factors as to whether "mechanical" will be an adequate combat tool. In many cases, even "optimal" time of setup of mechanical equipment precludes use of mechanical as an option. Estuarine spills are a chief example of this. There is almost no time (minutes) between a port spill and mangroves in many areas. In shallow, ecologically sensitive habitats, it is a particular problem to setup a boom. Mechanical cleanup is a far more highly skilled technology, dependent on speed, accuracy, maintenance of equipment, and logistic coordination. The author has many reports of failed attempts to contain oil in less-developed countries and developed nations. Mechanical cleanup frequently fails due to operator errors or inability. In the U.S., only 15% of mechanical cleanup occurs, leaving 85% uncleaned.

If the mechanically retained oil cannot be recovered, then the waste oil must be taken ashore and disposed of correctly. This author has made a survey. Most nations in Central America, South America, the Caribbean, and East or West Africa do not have any central disposal sites for waste oil.

Mechanical cleanup costs \$700 to \$7000 per barrel[38,39] (Tables 9 and 10).

2. Shoreline and Reef Cleanup

If the winds are onshore at points during the life of the spill and if proximity to shore is small, some of the oil will land on shore. If mangroves are hit by the spill, they will die (unless it is a high tide and the oil is carried out before the tide goes low).

Coral reefs may also be a site of shoreline circulation, which at high oil concentration will kill corals.

Beach cleanup includes removing sand which can lead to accelerated erosion. In a tourist-oriented location, beach cleaning is not appropriate.

Hard surfaces, such as rocks, can be sandblasted or steam cleaned. This reinjects oil into the water for further stranding and must be done along with mechanical barriers.

These shoreline methods are all expensive (see Tables 9 and 10).

TABLE 9
Costs of Oil Spill Cleanup Techniques

Type	Cost per barrel	Ref.
Mechanical	\$65—5000	38
Dispersant	\$15—65	38
Shoreline	\$650—7000	39
Gels	Above \$7000	39

TABLE 10
Short-Term Oil Spill Cleanup and Indirect Long-Term Costs

	Capitalization costs	Direct costs	Indirect waste removal	Indirect environmental costs	Socioeconomic costs
Mechanical	High	\$1.50—0.60/gal	\$2/gal		None if successful, high if not successful
Beach cleanup	Low	\$8—17/gal	\$4/gal	Very high in productive ecosystems (mangroves, reefs)	High if tourism is involved or if mangroves/coral dies; low if rock or sand not used
No action	0	0	0	About \$100,000/acre if mangroves, seagrasses corals are killed	Very high if fisheries or tourism is affected or if mangroves or corals die
Dispersant	Medium to low	\$0.06—0.30/gal	0	None if nontoxic dispersants used, some if toxic dispesants used	Low

3. Other Techniques

If the slick is thick enough and human life and property are not at risk, flaring may be possible. Substantial air pollution results. Ixtoc in Mexico used flaring.

Gels are still in experimental stages or very expensive. No known tropical field testing has been done.

Bottom sinking agents: seagrasses, corals and meiofauna are notably unable to deal with these substances and can be smothered. Sinking agents can releak oil.

Bacterial agents are still very expensive although the 1990 MegaBorg spill in open waters far off Galveston, TX used them. Bacterial agents may be better for remediation.

The field is fast moving. Keeping abreast of technology is critical.

4. Dispersant Cleanup

The real spills for which dispersants have been used are far less in the tropics than in the temperate zone. However, in the many reports of use of dispersant, a report of the dispersed oil killing a population of any organisms could not be found. There were reports where one portion of a spill was not dispersed while another was, the nondispersed portion causing harm.

The Ixtoc and Yum spills, both drilling rig blowout spills, offer an interesting comparison. The Ixtoc spill was the largest spill in the tropics with the largest amount of dispersants used up to that time. Jernalov and Linden[32] reported that no statistical difference in total fish landing occurred after the spill started flowing from the year before. The Yum spill which flowed from approximately the same area on the northern shelf of the Yucatan for only 4 months compared with Ixtoc (more than 1 year). The volume of flow was approximately the same. The UNEP regional headquarters in Mexico and the fisheries report about 50% reduction in fish catch in Yum. No dispersant was used in Yum.

The reports on the Panama Texaco spill where the dispersant Corexit 9527 was used are difficult to separate out due to legal procedures which inhibit information. However, there are reports that a portion of the spill which occurred in the mangroves did get dispersed by the dispersant airplane of the Clean Caribbean Cooperative, which sprayed dispersant on a sizable portion of the spill. There are also reports of nondispersed portions of the spill being carried by winds into mangroves, killing them. Some controversy remains as to exactly where the dispersed oil on the leading edge was.

The reports of other spills in Belize, Jamaica, and Egypt, which were dispersed without fish or mangrove kills, have been reported to international governments.

The preliminary data analysis[1] has not revealed any tropical "Torrey Canyon", where large-scale dispersants were used to the detriment of the environment. This is an important setting in which to place the laboratory and mesoscale data.

L. DISPERSABILITY OF OIL

One important point for a commander of an oil spill is the effectiveness of whatever technique he is going to use on the oil cleanup. The problem at present is that the effectiveness of the dispersants have not been tested at higher temperatures relevant to tropical seas and estuaries, that is, from 22 to 35°C. Most data are from 1 to 15°C; some data go to 20°C. Certain dispersant manufacturers claim they are presently testing at higher temperatures and salinities.

Effectiveness of emulsions are strongly affected by temperature, and, in general, the warmer the temperature, the greater the immiscibility.

Effectiveness is also a function of salinity. There are no effectiveness tests for the very salty waters in the Red Sea, Gulf of Arabia, or Gulf of Aquaba, for example. Some dispersants are effective over a larger range of salinities than others (Corexit 9550). There are very few tests for low salinity or brackish water situations.

Various oils are differently dispersed. Generally, the heavier the oil, the more difficult to disperse; the lighter the oil, the easier.

We must now have results of studies of major oils from major shipping routes vs. the more-toxic dispersants at high temperatures to understand their effectiveness and efficiency. We must have the effectiveness from 0 to 40‰ salinity at higher temperatures.

III. RECOMMENDATIONS

A. POLICY

The policies most useful are those which allow the commander of the oil spill to predetermine implementation during pre-oil spill planning sessions, finding the least resource-damaging option, and comparing the spill with the predetermined model to implement it immediately at the time of spill. Unless the spill is far offshore, time is of the essence for the spill commander to do his best for a good result. Any delay whatsoever may be to the detriment of the very resource the environmentalist wishes to protect.

Nations and states should have oil-spill policies which allow them to use cost-accounting recovery for their damaged coastal resources. This means the nation should carry out "pricing" of coral reefs, mangroves, and sustainable fisheries.

The 1980 IMO policy about mangroves was cautious when formalized, but is out of date because it would cause mangrove death directly by not having dispersed oil as it entered the mangroves.

The Bonn Agreement may be a good policy entry point for developing nations without a present policy. The central focus of these European policies are the ''least impact''.

B. PREPAREDNESS

Successful management of oil spills has preparedness of all parties as the key. Any and all decisions which can be made ahead of time should be. All decisions in which resources are the highest priority should be made ahead of time. As much experience and coordination as possible should be made as in preplanning. The luxury of offshore spills, like the 1989 Morocco spill (of over 1 million gal) which took days to come ashore, are few. Most spills (55%) take place within several miles of shore. In Florida, a spill can move 60 mi in 24 h. Thus, most spills near shore take only hours to hit living resources which might be irreparably damaged.

C. STOCKPILING CORRECT EQUIPMENT

When the spill occurs, either the equipment of choice is there or difficult arrangements to get it flown in must be made. For some African countries, it is easier to get it flown in from Europe than from a neighbor.

If dispersants will be useful, be sure the nontoxic ones are stockpiled in your area so the temptation to use the toxic ones will not occur. This must be done ahead of time.

D. DEPENDENCE ON OTHERS' EQUIPMENT

The problem with not having adequate equipment is that the neighbors' equipment may not work. In regions with many small nations (East and West Africa, Caribbean, Pacific Islands), the expenditure for protecting the nation from a major spill is always seen as a trade-off from some other project. The regional center now being founded by UNEP is one alternative.

IV. DISCUSSION

A. MANAGEMENT PRINCIPLES FOR USE OF DISPERSANTS IN TROPICAL HABITATS

It is now apparent that nontoxic dispersants at concentrations recommended by the manufacturer can be used under various sets of emergency conditions for oil spills which frequently occur in the tropics as they are being used by the Bonn Agreement nations: (1) estuarine spills where or when mechanical means are inadequate to control oil from impacting one or more types of habitat, especially mangroves; (2) nearshore coastal spills where environmental conditions are rapidly transporting spill toward one or more critical habitats; (3) in weather conditions when mechanical cleanup is ineffective and there is a danger of impacting corals, mangroves, or seagrasses.

All parties must stop using the generic term ''dispersants'' within the oil spill cleanup plans. Specific tested and *non-toxic* dispersants must be named for spills on each habitat type with the upper concentration limits for use described.

Further laboratory tests must be done by nations in tropical areas to test their commonly used and stockpiled dispersants for toxic effects on their critical habitat organisms such as various species of corals, mangroves, seagrasses, and marshes. It is unrealistic to imagine small third-world nations will find this a priority. Regional multinational and industrial funds should be found to do this. It must be done according to international or accepted standards.

A network of rapid information dissemination to industry, environmental management, government spill cleanup managers should be organized so that whatever information is derived can be disseminated.

Resource maps which must include the *exact* species of mangrove, seagrass, or corals should be included in oil-spill contingency plans. Since the toxic effects differ by more than an order of magnitude between species, present ''lumping'' resource maps are inadequate. Dominant species must be available on sensitivity maps.

ACKNOWLEDGMENTS

The Office of Science Advisor, U.S. Agency for International Development (AID) has generously supported research on toxicity limits to oil and various dispersants. The Jamaica Mission of U.S. AID also supported the project as did the Office of Disaster Preparedness, the Government of Jamaica, and Florida International University. The author wishes to acknowledge Mobil Oil Foundation for a grant to test dispersants on seagrasses, corals, and fish. Mr. S. Horne, Dr. J. Hildrew, and Dr. Marum of Mobil were most helpful in establishing guidelines for testing relevant to industry. M. Flaherty of the U.S. EPA (ret.) is thanked for helpful discussions about differences among dispersants. Captain Charles Corbett and the U.S. Coast Guard oil spill (cleanup) staff in Washington, D.C., Miami, and in Puerto Rico Matt Dillon are thanked for discussions on cleanup problems.

A variety of members of IPEICA and APS members are thanked for the industry viewpoint. Members of the United Nations Environmental Programme office such as Dr. Tolba, Dr. Keskes, Arthur Dahl, Macrum Gerges, and other offices (Dr. Noel Brown, North American office), especially Franklin McDonald, United Nations Disaster Relief Organization, Mrs. Beverly Miller, and Salvino Bresenio, Dr. S. Lizarraga of UNEP are also thanked. Natural Resources Conservation Department, Jamaica, Venezuelan environmental staff, Mrs. Veronica Villavicencio, Philippines, Drs. Badawi and Esaway, Prof. M. Kassas, Egypt.

Dr. Howard Teas is especially thanked as mentor on oil spill and physiology, particularly during the preliminary stages of our experiment. Dr. Jeffrey Marcus and Dr. Fitz Booker were involved in early years of laboratory experiments on seagrasses, and Dr. Jeffrey Marcus is co-author of much of the subtropical seagrass results quoted here. Barbara Carby, John McFarlane, Richard Reese from the Jamaica Office of Disaster Preparedness, Mike Rodriquez from Jamaica Defense Department Coast Guard, Marcel Anderson and staff, NRCD, Beverly Miller and Valerie Gordon, IMERU, Prof. George Sidrak, Dr. Karl Aiken, and Peter Gayle and Wendy Walker, UWI, were all involved in phases of the Jamaica data and model project. The Oil Spill Committee of Jamaica is thanked for its input as is the Jamaican Coast Guard. Investigators and cleanup regulators around the world have contributed experiences and ideas to this document. Dr. G. Kollenberg and Dr. Ibe of IOC (UNESCO) for sending me to the Persian Gulf to report on the oil spill.

REFERENCES

1. **National Research Council,** *Using Oil Spill Dispersants on the Sea,* National Academy Press, Washington, D.C., 1989.
2. **Ballou, T. G., Dodge, R., Hess, S., and Knap, A.,** Tropical oil pollution investigations in coastal systems (tropics): the effects of untreated and chemically dispersed Prudoe Bay crude oil on mangroves, seagrasses, and corals in Panama, in *Oil Dispersants; New Ecological Approaches,* Flaherty, L. M., Ed., ASTM, Philadelphia, 1989, 229.
3. **Thorhaug, A., Carby, B., Reese, R., Rodriguez, M., McFarlane, J., Teas, H., Sidrak, G., Anderson, M., Aiken, R., McDonald, F., Miller, B., Gordon, V., and Gayle, P.,** Dispersant use for tropical nearshore waters: Jamaica, in *Proc. 1991 Oil Spill Conf., American Petroleum Institute,* Washington, D.C., 1991, 142.

4. **Knap, A. H., Wyers, S. C., Dodge, R. E., Sleeter, T. D., Frith, H. R., Smith, S. R., and Cook, C. B.,** The effects of chemically and physically dispersed oil on the brain coral *Diporia strigosa* (Dana) — A summary review, in Proc. Oil Spill Conf., American Petroleum Institute, Washington, D.C., 1985, 547.
5. **Knap, A. H. et al.,** Oil spills research in tropical environments and contingency planning in the wider Caribbean, International Petroleum Industry Environmental Conservation Association, 1986, Venezuela, 1985, 24.
6. **Jackson, J. B. C., Cubit, J. D., Keller, B. D., Balista, V., Burns, K., Caffey, H. M., Caldwell, R. L., Garrity, S. D., Getter, C. D., Gonzalez, C., Guzman, H. M., Kaufmann, K. W., Knap, A. H., Marshall, J., Steger, R., Tompson, R. C., and Weil, E.** Ecological effects of a major oil spill on Panamanian coastal marine communities, *Science,* 1989.
7. **Bak R.,** Effects of oil on coral reefs, in West Indies Island Mar. Lab. Proc., 1981, 31.
8. **Lugo, A., Teas, A. H., and Lewis, R.,** Effect of oil on mangroves and restoration, in *Restoration and Management of Marine Ecosystems Impacted by Oil,* Cairns, J. and Buikema, A., eds., Butterworths, New York, chapt 5, 1984.
9. **Getter, C. D., Ballou, T. G., Hess, S. C., Knap, A., Dodge, R., and Sleeter, R.,** Final results of the API tropics oil spill and dispersant use experiments in Panama, in Proc. Oil Spill Conf., American Petroleum Institute, 1987, 634.
10. **Anderson, J. W., McQuerry, D. L., and Kiessere, S. L.,** Laboratory evaluation of chemical dispersants for use on oil spills at sea, *Environ. Sci. Technol.,* 19(5), 454, 1985.
11. **Thorhaug, A. and Marcus, J.,** Effects of dispersant and oil on tropical and subtropical seagrasses, in Proc. Oil Spill Conf., American Petroleum Institute, 1985, 497.
12. **Thorhaug, A. and Marcus, J.,** Effects of dispersant and oil on subtropical and tropical seagrasses, in *Proc. Oil Spill Conf.,* American Petroleum Institute Publ. 4452, Washington, D.C., 1987, 223.
13. **Teas, H., Duerr, E. O., and Wilcox, J. R.,** Effects of dispersed and non-dispersed oil on the red mangrove, *Mar. Pollut. Bull.,* 18(3), 122, 1985.
14. **Elgershuizen, J. H. W. B. and deKruijk, H. A. M.,** Toxicity of crude oils and a dispersant to the stoney coral *Madracis mirabilis, Mar. Pollut. Bull.,* 7(2), 22, 1976.
15. **McAuliffe, C. D.,** Organism exposure to volatile soluble hydrocarbons from crude oil spills — a field and laboratory comparison, in *Proc. Oil Spill Conf., Washington, D.C.,* American Petroleum Institute, 1987, 275.
16. **Shuba, R. J. and Heikamp, A. J.,** Toxicity test on biological species indigenous to the Gulf of Mexico, in *Proc. Oil Spill Conf.,* American Petroleum Institute, Washington, D.C., 1989, 249.
17. **Thorhaug, A. and Marcus, J.,** Effect of seven dispersants on three subtropical seagrasses, *Mar. Pollut. Bull.,* 18(3), 124, 1987.
18. **Thorhaug, A. and Cruz, R.,** Restoration of Pacific seagrasses, in *Restoring the Earth,* Berger, A., Ed., Elsevier, Amsterdam, 1989, 21.
19. **Thorhaug, A., McDonald, F., Miller, B., McFarlane, J., Carby, B., Anderson, M., Gordon, V., and Gayle, P.,** Dispersed oil effects on tropical habitats, in *Proc. Int. Oil Spill Conf.,* American Petroleum Institute, Washington, D.C., 1989, 455.
20. **Getter, C. D.,** Oil spills research in tropical environments and contingency planning in the wider Caribbean, in Proc. Int. Petroleum Industry Environmental Conservation Association, Venezuela, 1986, 29.
21. **Cubit, J. D., Getter, C. D., Jackson, J. B. C., Garrety, S. D., Caffey, A. M., Thompson, R. C., Weil, E., and Marshall, M. J.,** An oil spill affecting coral reefs and mangroves on the Caribbean coast of Panama, in *Proc. Oil Spill Conf.,* American Petroleum Institute, Washington, D.C., 1987, 401.
22. **LeGore, R. S., Marszalek, D. S., Hofman, J. E., and Cuddeback, J. E.,** A field experiment to assess the impact of chemically dispersed oil on Arabian Gulf corals, SPE 11444, Society of Petroleum Engineers, 1982, 51.
23. IMO/UNEP Guidelines on Oil Spill Dispersant Application and Environmental Consideration, International Maritime Organization, London, 1982.
24. **Chan, E. T.,** Oil pollution and tropical littoral communities, Biological effects of the 1975 Fla. Keys Oil Spill, in *Proc. Oil Spill Conf.,* American Petroleum Institute, Washington, D.C., 1977, 539.
26. **Daly, H. E. and Cobb, J. B.,** *For the Common Good,* Beacon Press, Boston, 1989.
27. **Thorhaug, A.,** Dispersant effects on tropical ecosystems, in *Oil Dispersants: New Ecological Approaches,* Flaherty, M., Ed., ASTM, Philadelphia, 1988, pp 249-265.
28. **Thorhaug, A. and Cruz, R.,** Restoration of Philippine seagrass, in Proc. LSD Philippines Conf. on Seagrass, 1987, 20.
29. **Thorhaug, A., Miller, B., Cruz, and Gayle, P.,** Comparison of Seagrass Restoration from Atlantic and Indo-Pacific Tropics, in Proc. Int. Ecological Conf., 1986, 312.
30. **Grigg, R.,** Restoration of coral reefs after lava flows, Report to University of Hawaii, *Mar. Sci.,* 22 pp., 1981.
31. **Hudson, H. and Diaz, R.,** Damage survey and restoration of M/U Wellwood, in Proc. 6th Int. Coral Reef Symp., Vol. 2, 1988, pp 231-236.

32. **Jernalov, N. and Linden, O. E.,** IXTOC, *Ambio,* 10(6), 299, 1981.
33. **Baker, J. and Goodman, K. S.,** A preliminary ecological survey of coastline, British Petroleum International, London, 1982.
34. **Teas, H., Lasday, A. H., Elias, L. L., Morales, R. A., DeDiego, M. E., and Baber, J. M.,** Mangrove restoration after the 1986 refineries Panama oil spill, in Proc. 1989 Oil Spill Conf., American Petroleum Institute, Washington, D.C., 1989, 427.
35. **Seneca, A. E.,** Restoration of marshes after the Amoco Cadiz oil spill, in Proc. Int. Bot. Soc. Congr., Vancouver, 1980.
36. **Hanna, R. G. and Ormond, R. F.,** Oil pollution, urchin, erosion, and coral reef deterrants in the Egyption Red Sea, *Iraqi J. Mar. Sci.,* 1(1), 35, 1982.
37. **Bak, R. P. M.,** The effect of chronic oil pollution on a Caribbean coral reef, *Mar. Pollut. Bull.,* 18(10), 534, 1987.
38. **Lasday, A. H.,** Economic evaluation of dispersants to combat oil spills, in *Oil Dispersants, New Ecological Approaches,* 1018, Flaharty, M., Ed., ASTM STP Philadelphia, 1989, 41.
39. **White, I. C. and Nichols, I. A.,** The cost of spills, in Proc. Oil Spill Conf., Washington, D.C., American Petroleum Institute, 1983, 541.
40. United Nations Environment Program Latin American Office, 1987, personal communication.
41. **Butler, et al.,** 1989.
42. **Curry,** personal communication.
43. **Teas,** personal communication.
44. U.S. Environmental Protection Agency, 1984.
45. **Thorhaug, A.,** Tropical and subtropical thermal pollution, *Thalassia Yugoslavia,* 20(2), 103, 1986.

Chapter 5

NUTRIENTS AND MANGROVES

Kevin G. Boto

TABLE OF CONTENTS

I. INTRODUCTION

Mangrove forests occupy about 11,500 km^2 of the Australian coastline, of which some 10,900 km^2 occurs in the tropical regions.[1] As such, mangroves occupy nearly 75% of the available intertidal coastal area in northern Australia. Similarly, and despite widespread destruction of these forest in many areas, mangroves also occupy some 52,700 km^2 of coastline throughout the Association of S.E. Asian Nations (ASEAN) region.[2] Although not confined entirely to tropical regions, there are only a very limited number of species which exist in the temperate zone, compared with the much greater species diversity found near the equator. As a brief example of the pronounced latitudinal gradient in Australia, over 40 species are found in northeastern Australia,[3] decreasing rapidly to four species near Sydney, with only one species extant near Melbourne. In addition, forest luxuriance and primary production (for undisturbed forests) is highest in the wet tropical region of Papua New Guinea and northeastern Australia.[4] The more arid conditions of the Gulf of Carpentaria and northwestern Australia appear to be far less favorable to mangrove forest development.[4,5]

Mangroves can, therefore, be classified as an almost exclusively tropical marine ecosystem with virtually no counterpart in temperate regions having anywhere near the same macrophyte species diversity. Apart from being well-represented, in areal extent, along tropical coastlines, mangroves are considered to play a very important role in controlling coastal hydrodynamics and sediment movements[6,7] and have long been known as a habitat for a wide range of dependent biota.[8] More recently, Robertson and Duke[9,10] have, for the first time, provided sound scientific evidence for the importance of mangroves as a "nursery ground" for many fish and crustacean species. These studies directly and quantitatively compared fish and crustacean population structures in both mangrove and nearby nonmangrove habitats (seagrasses, subtidal sandflats, and mudflats), clearly demonstrating that mangroves were the preferred habitat for the juvenile stages of many fish and prawn species. Similarly, these and other more specific studies[11,12] have shown that mangroves are a crucial habitat for a number of commercially important prawn species in northern Australia.

In Southeast Asia, mangrove forests are also utilized, on a large scale, as an important source of wood products, notably charcoal production. On a more local or village-level scale, mangroves provide an important source of fish, shellfish, and other foods as well as providing material for building, thatching, and furniture.[13]

Despite their importance and unique character, tropical mangrove systems face severe threats of destruction from a number of sources. This situation is increasingly evident in Australia, ironically coinciding with an increased public awareness of the importance of these forests which were until fairly recently regarded as swampy, mosquito-ridden wastelands.

One of the indirect threats to mangroves is through nutrient pollution. The major sources of this type of pollution are sewage effluent and aquaculture wastewater disposal. Tourism development in Australia and Southeast Asia is very rapidly increasing, along with other urban development. These pose a direct threat to mangroves through outright destruction along with a longer-term, less direct effect of increased demand for sewage effluent disposal. While the discharge of untreated sewage is rare nowadays, secondarily and even tertiarily treated effluent still constains highly elevated levels of organic and inorganic nutrients compared with oligotrophic tropical nearshore waters. Similarly, the rapid growth of aquaculture (particularly prawn farming) in the Australasian region threatens mangroves. Even in areas where mangroves are not directly cleared to establish such ventures, it is a common practice to discharge the very large volumes of pond effluents directly into mangrove waterways. In many ways, this type of pollution can be worse than sewage effluent as fish pond wastewaters often do not receive any type of treatment; hence, they contain considerable concentrations of nutrients.

This chapter attempts to predict the possible impacts of such inputs into mangrove soils and waterways on both the "terrestrial" and "aquatic" nutrient stocks and biota. The discussion is necessarily both brief and speculative as there is a paucity of information on nutrients in pristine mangrove systems and very few published case studies of nutrient enrichment effects. Nevertheless, the small amount of available data do allow some estimation of the possible degrees of such impacts as well as suggesting methods of disposal which would be likely to be far less destructive to these relatively fragile and important coastal ecosystems.

II. NUTRIENT STOCKS AND FLUXES

This section examines the available data on standing stocks and fluxes of organic and inorganic nutrients in mangroves, while Section III describes the influence of inorganic nutrients on primary production and growth of the various elements of the mangrove ecosystem. Discussion is limited to the major nutrient elements, nitrogen (N) and phosphorus (P), and, in most cases, is limited to the situation in undisturbed forests. Section IV then attempts to synthesize the available information to predict the likely impact of nutrient pollution in tropical mangroves.

A. SOIL AND FOREST NUTRIENT STOCKS AND FLUXES

As is typical for many unpolluted tropical systems,[14] the concentrations of free inorganic nutrient species, such as ammonium, nitrate, and phosphate ions, in the soils and waters are very low. Most of the nitrogen and phosphorus in mangrove forests is "bound", in various particulate and dissolved organic forms, in the soils and plant tissue.

Boto and Wellington[15] have studied the variation of weak acid-extractable ammonium, nitrate, nitrite, and soluble reactive phosphorus (hereafter referred to as "phosphate") with time (monthly, over 1 year) and elevation within the intertidal zone. This study was carried out for a mangrove forest in northern Australia dominated by *Rhizophora, Bruguiera,* and *Ceriops* species. Within the vegetated zone, concentrations of all inorganic nutrients were low. Ammonium, the major form of the dissolved inorganic nitrogen (DIN), did not vary significantly with elevation (Table 1), but was significantly lower during periods of rapid plant growth, ranging from 0.2 to 11.0 μg N per gram soil dry weight over a year. Similar values for soil DIN have been reported for temperate *Avicennia* forests.[16,17] Phosphate, on the other hand, decreased significantly with increasing elevation (Table 1) within the intertidal zone, but did not vary significantly with time at any given elevation, unlike ammonium, which showed a marked response to plant uptake. This latter observation probably is an artifact of the extraction method used in this study which very likely overestimates the amount of plant-available P. Nevertheless, the extractable P values were low throughout the study transect, ranging from about 15 to 5 μg P per gram dry soil, from the low to high intertidal sites. Total soil N and P were orders of magnitude greater, in the order of 1000 to 4000 μg N per gram dry weight and 100 to 500 μg P per gram dry weight, respectively, indicating that the vast majority of soil N and P is bound in organic forms.

This latter observation is in agreement with studies by Hesse[18] and Udo et al.[19] who found that 75 to 85% of total P in Sierra Leone mangrove soils was in the organic form. Indeed, most of the inorganic P in mangrove soils is adsorbed or strongly bound within hydrated iron and aluminium colloidal sesquioxides,[18-20] thus, severely limiting its availability to the plants.

A study of sediment pore waters under *Rhizophora mangle* and *Avicennia germinans* mangrove forests in Florida[21] also showed generally low concentrations of free dissolved inorganic N and P species, although concentrations of phosphate in the *Avicennia* sediment pore waters (12 to 176 μ*M*) appeared to be higher than in the *Rhizophora* sediments (12 to

TABLE 1
The Variation of Weak Acid-Extractable (pH 4.5 Acetate Buffer) Ammonium-N (EXN) and Phosphate-P (EXP) with Position along a Transect Which Traverses a Mangrove Forest between Two Tidal Channels in Northern Australia[15]

Station	Elevation (m vs. AHD)	EXN ($\mu g\ g^{-1}$ DW)	EXP ($\mu g\ g^{-1}$ DW)
1	0.15	4.6 (2.2)	14.2 (3.8)
2	0.75	5.0 (3.0)	10.2 (2.4)
3	1.10	5.6 (2.4)	5.9 (2.1)
4	1.25	4.8 (2.4)	5.0 (2.2)
5	1.35	4.3 (2.0)	5.0 (2.2)
6	1.20	5.7 (2.4)	8.0 (2.5)
7	1.10	6.0 (2.5)	12.1 (3.5)
8	0.60	7.3 (2.6)	20.1 (4.0)

Note: The values shown are means (and standard deviations) derived from monthly samples at each station and averaged over the 0- to 100-cm depth which was sampled with each core. The elevation of each site is expressed in relation to Australian High Datum (AHD) = +15 cm vs. Mean Sea Level at this location.

23 μM). Ammonium ranged from 1 to 23 μM over both sediment types, whereas, similar to Boto and Wellington's findings,[15] nitrate was invariably low (<2 μM) and did not contribute significantly to the soil DIN pool.

There are some data suggesting that, despite the generally low ambient levels of available inorganic N and P, turnover of the much larger soil organic nutrient pool is quite rapid. For example, Boon and Cain[22] have measured quite high rates of metabolism for a variety of organic nitrogen compounds in the sediments of a temperate *Avicennia* forest in Victoria, Australia, with rates of up to 23 $\mu M\ cm^{-3}\ d^{-1}$ being recorded for metabolism of dipeptides, for example. It is very likely that decomposition rates would be greater in the tropics. For example, very high rates of bacterial production and bacterial densities in mangrove sediments in northern Australia have been reported by Alongi,[23] with production rates and densities sometimes exceeding 4 g C $m^{-2}\ d^{-1}$ and 10^{11} cells per gram soil, respectively. It seems reasonable to assume that this generally high rate of overall microbial activity is also translated into high rates of bacterial mineralization processes. Partial evidence to support this assumption can be derived from the studies of Stanley et al.[24] and Boto et al.[25] They found that the bacterial populations at the sediment-water interface, in a northern Australian mangrove forest, were sufficient to intercept all amino acid and dissolved organic matter efflux from the deeper sediments, thus preventing any significant flux to the overlying waters. Alongi[70] has also recently measured very low to negligible efflux rates of dissolved inorganic nutrients from the same mangrove sediments.

These studies indicate the N turnover is finally balanced, with rapid rates of organic N mineralization being matched by rapid plant uptake and/or microbial utilization of the released DIN within the sediments.

Overall net losses of N from the forest can result from denitrification[26,27] or ammonia volatilization.[28] Very little information on the rates of either of these processes in mangrove sediments is available, although a laboratory study of ^{15}N-enriched ammonium loss, over a 64-d period, from a tropical mangrove sediment,[29] has indicated that N loss from either of these processes in an unpolluted system is probably small (i.e., <10% of forest primary production N requirements per year).

Soil phosphorus mineralization and utilization processes are probably also rapid, although the phosphate produced from mineralization is subject to "immobilization" through incorporation into iron and aluminium sesquioxides,[30] as discussed above. The flux of P within and from sediments, as well as the ability of the sediments to absorb chronic P loading, is also complicated by the close links between soil anaerobiosis and phosphate mobility. Generally, low redox potentials in these soils leads to significant phosphate release (especially at pH <7) through:[31] (1) Reduction of insoluble ferric phosphate to the more soluble ferrous form, (2) the reduction of the hydrated ferric oxide coating on clay particles resulting in the release of occluded phosphate, (3) phosphate displacement from ferric-aluminum phosphates by organic anions and hydrolysis reactions, and (4) phosphate exchange by organic anions on clay particle anion-exchange sites.

As will be discussed later, all of these processes make it difficult to predict the long-term fate of high phosphorus inputs to an anaerobic soil system.

B. WATER-BORNE NUTRIENT STOCKS AND FLUXES

The relatively little available data suggest that particulate and dissolved organic and inorganic N and P concentrations in tropical mangrove waterways are extremely low. This appears to be especially so for mangrove systems which are influenced almost entirely by tidal movements[32] and which have virtually no freshwater input from terrestrial riverine or groundwater sources.

Concentrations and net fluxes of particulate[33] and dissolved[32] organic and inorganic N and P compounds in a northern Australian mangrove have been found to vary significantly with time of year, but with concentrations of dissolved components always being low. Figure 1 shows some typical values for the concentrations of dissolved organic and inorganic nutrients at various times of the year (over a 2-year period). In particular, the dissolved inorganic nutrients concentrations are extremely low, in many cases decreasing to below analytical detection limits during the cooler parts of the year (from about May to September). The particulate and dissolved organic N and P were also generally much lower than has been recorded for many temperate and subtropical marsh and mangrove systems (see references quoted by Boto and Wellington).[32]

Other tropical mangrove estuarine waters which are influenced by significant freshwater input from their terrestrial hinterland have DIN and phosphate concentrations of up to an order of magnitude higher. Wong[34] has reported nitrate plus nitrite and phosphate concentrations of up to 6 μM in a Malaysian mangrove estuary. The significance of the fresh water input was clearly seen from the ammonium concentrations which ranged from 50 μM at the freshwater end of the estuary to below detection limits at the seawater end. Nixon et al.[35] have also reported similar levels of DIN and phosphate in a comparison of a mangrove and a nonmangrove estuarine system in Malaysia. Interestingly, they found that the DIN speciation differed considerably between the two systems with ammonium predominant in the mangrove estuary while nitrate was the major form in the other.

Similarly, the 2-year study of the tidally dominated Coral Creek system[32] revealed no net annual exchange of most dissolved components with the nearshore coastal waters (Table 2), with the exception of dissolved phosphorus (organic and inorganic forms) which showed a significant net *import* into the mangrove system. Dissolved nitrogen also showed a small net annual *import* but the degree of this exchange was not statistically significant when the errors of such estimates were taken into account. Table 2 also shows a small, but statistically significant net *export* of particulate N and P, mainly in the form of intact mangrove litter (leaves, fruits, etc.) has been estimated for this system.[36] The net exchange of N, in particulate and dissolved forms from the mangroves to the adjacent nearshore system, amounts to a

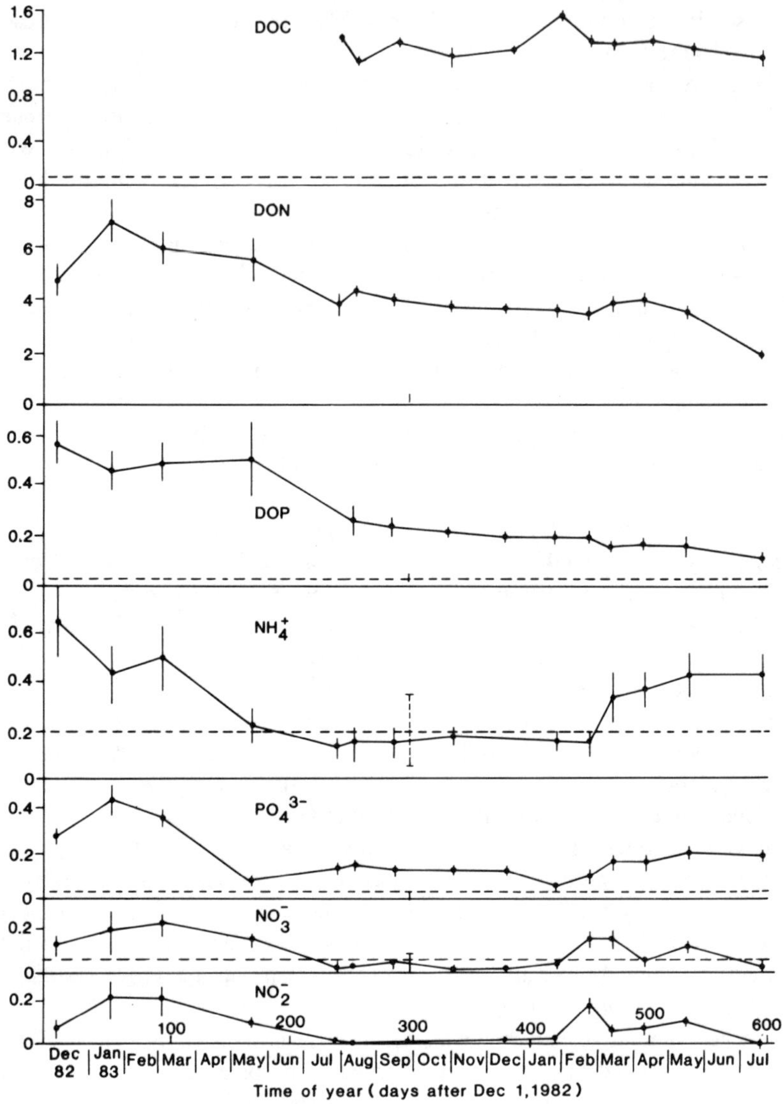

FIGURE 1. Variation of mean concentrations of dissolved materials in Coral Creek over a 20-month sampling period. Each mean was derived from 30 to 50 samples taken during each full tidal cycle on the sampling date shown. Vertical lines: 95% confidence intervals for each mean. Dashed horizontal lines: analytical detection limits and 95% confidence intervals for each component (where these limits have been established). Units: Dissolved organic C (DOC), mg C l^{-1}; all others, μM. (From Boto, K. G. and Wellington, J. T., *Mar. Ecol. Prog. Ser.*, 50, 151, 1988. With permission.)

very small net *export* amounting to about 5% of forest net primary production requirements per year. This net loss, along with losses via denitrification and ammonia volatilization (see before), is probably balanced by the small, but crucial amounts of nitrogen fixation which occur within mangrove sediments, algae, and other sources[37-39] within the forests. Total P exchange, taking into account particulate and dissolved organic-inorganic forms, amounts to a small net annual *import* equivalent to the supply of 12% of annual P requirements for forest primary production. This small degree of net import may not be sufficient to balance P loss through occlusion into sesquioxide minerals and subsequent burial and immobilization, as many areas of these forests have been shown to be phosphorus deficient (see later).

TABLE 2
Net Annual Fluxes of Particulate and Dissolved Carbon, Nitrogen, and Phosphorus from the Coral Creek System (Northeastern Australia) via Tidal Transport[32,33]

Component	Net annual exchange (g C, N or P m^{-2} $year^{-1}$)	Porportion of forest primary production requirements (%)
Particulate matter (mainly intact mangrove plant detritus)		
Particulate organic C	−327	−35.9
Particulate organic N	−3.7	−13.4
Particulate organic P	−0.25	−12.2
Dissolved materials		
Dissolved organic C	7.3	0.8
Dissolved organic N	1.3	4.7
Dissolved organic P	0.37	17.9
NH_4	0.15	0.6
NO_3+NO_2	−0.03	−0.1
PO_4	0.13	6.3
Total dissolved N	1.45	5.4
Total dissolved P	0.50	24.2

Note: The exchange rates are given in terms of forest area contained within the coral creek basin and in terms of net forest primary production requirements. A negative sign denotes net export.

The only other published study of water-borne nutrient fluxes in tropical mangroves was carried out by Nixon et al.[35] in Malaysia. This study was of a very preliminary nature only but did suggest a net export of most dissolved nutrients from both a mangrove-lined and nonmangrove estuary. As postulated for the nutrient concentrations, the apparent net export of dissolved nutrients from these systems is most likely from the surrounding terrestrial systems rather than from the mangroves. The profound influence of terrestrial runoff on nutrient export from temperate salt marsh systems is also apparent from the large body of literature on that subject, although only specifically recognized in a few of these studies.[40,41]

The degree to which net N and P loss occurs through tidal-borne export of particulate matter is dependent on local hydrodynamic and topographic factors. Woodroffe[42] has studied particulate transport from a temperate *Avicennia* forest in New Zealand, where the system effectively forms an enclosed basin at low stages of the tide. The data for this system show that a much higher proportion of the forest detrital production is retained within the forest compared with Boto and Bunt's[33,36] estimates for the northern Australian system. An even more extreme example has been described by Twilley[43] for a subtropical basin forest in Florida. This system has minimal direct tidal exchange with nearshore waters and very little net transport of particulate materials occur.

In the more usual situation, however, it appears that substantial export of mangrove detrital material occurs, although this does not lead to any substantial net loss of N and P owing to their low concentrations in the detrital material. When these minor losses are compared with the small to moderate net imports of dissolved N and P as found for the tidally dominated northern Australian system,[32] it appears that pristine tropical mangrove systems can be considered to be in a very finely balanced state for the major nutrient elements N and P. This has significant implications for situations in which substantial extraneous inputs of nutrients may occur, as will be discussed later.

III. EFFECTS OF NUTRIENTS ON PRIMARY PRODUCTION

A. FOREST GROWTH AND PRODUCTION

There is some evidence to suggest that mangrove forests are generally nutrient limited with respect to N, and in some cases, P. Some early studies in subtropical Florida mangroves[44-46] suggested this possibility, and further field and laboratory studies in tropical[4,15,47] and temperate[17,48] mangroves have lent some support to this premise.

The most direct study of nutrient limitation on mangrove forest growth was provided by a field experiment carried out over a 12-month period in northern Australia.[47] In this experiment, N (as ammonium sulfate) or P (as superphosphate) was added to the soil at 3-month intervals for a year and tree growth response monitored by measurement of rate of new leaf appearance. Each fertilized plot was quite large (225 m^2) in order that the results would be more likely to represent whole forest conditions and response. P and N were added to the soils in amounts totaling 400 kg P (or N) per hectare over the 12-month period. These studies strongly indicated that N limitation to forest growth was ubiquitous, while P limitation was also evident in the high-intertidal, but not in the low- to midintertidal areas of the forest. Where responses were significant, rates of new leaf appearance increased by about 35% over the 12-month study period. A concurrent study of undisturbed sites along the same 360-m transect showed that forest biomass was highly and significantly correlated ($r = 0.85$, $p < 0.01$) with soil-extractable P (acetate buffer pH 4.5 as extractant) averaged over a 12-month period for each of the eight sampling sites.

A regional scale survey of tropical mangrove forest primary production and soil nutrient status at 25 widely separated sites in northern Australia and Papua New Guinea[4] showed that many of the low productivity forests in northeastern and western Australia coincided with soils containing extremely low levels of soil-extractable P. This provided further indication that P limitation may be a general feature of the tropical mangroves of Australia. This contrasted with the highly productive forests in the Gulf of Papua[4] where soil P is much higher than in most Australian soils.[49]

In general, Australian soils tend to be P deficient[50] and, hence, the above conclusions are not surprising. It is probably unlikely that P deficiency is so widespread in other tropical mangrove systems, however, as exemplified by studies of mangrove soils in Sierra Leone[18] and Nigeria[19] where levels of soil total and extractable P were considerably higher than were recorded for the northern Australian mangrove soils.[4,15]

Table 3 shows a summary of the data obtained for forest primary production for various sites in the Coral Creek mangroves of northern Australia[51,52] and its variation in relation to elevation within the intertidal zone and soil extractable P[15] (averaged over a 12-month period). As for forest biomass, the close relationship between forest primary production (P_N) and soil extractable P (EXP) is evident.

Other studies of temperate or subtropical mangrove systems have been less direct, but nevertheless provide some evidence of general nutrient (N or P) limitation of forest growth. Onuf et al.[46] reported a significant enhancement of growth, and earlier onset of flowering, for *Rhizophora mangle* in response to nutrient enrichment. These authors compared the nutrient status and growth of mangroves on two islands near Fort Pierce, FL. One of these islands received a nutrient enrichment of about 3 kg m^{-2} per year from a bird colony. The authors found that the soil ammonium and plant N levels were also significantly higher at this site compared with the control site which received no nutrient enrichment.

Concentrations of soil extractable or interstitial ammonium in temperate mangrove forests near Sydney[17] and Melbourne[16] were similar to those recorded in tropical Australia,[4,15] that is, in the order of 1 to 10 μg N per cubic centimeter. These results,coupled with data from a laboratory experiment[48] which showed that *Avicennia marina* exhibits enhanced growth with added ammonium levels of up to 14 μg N per cubic centimeter, also point to these

TABLE 3
The Variation of Weak Acid (Acetate Buffer pH 4.5) Extractable Phosphorus (EXP) and Forest Primary Production Estimates (P_N) with Elevation within the Intertidal Zone for Various Sites within the Mangrove Forests at Coral Creek, Northeastern Australia[15,52]

Site No.	Elevation (m vs. AHD)	EXP (μg P g^{-1} DW)	P_N (kg C ha^{-1} d^{-1})
1	0.10	16.0	36.2
2	0.22	13.8	32.5
3	0.34	12.0	30.0
4	0.61	10.9	31.0
5	1.00	9.9	23.2
6	1.30	5.8	16.0

Note: The elevation of each site is expressed in relation to Australian High Datum (AHD), with AHD = 0 approximately equal to mean sea level.

forests being N limited. However, Cain and Boon[16] did not find any significant relationship between cellular osmotica of the mangroves and levels of inorganic N in the soils at their Westernport Bay study site. This result would tend to suggest that the mangrove growth in that area is limited more by soil salinity rather than available N.

Soil salinity variations with time, and elevation within the intertidal zone, may be a confounding factor in the control of mangrove forest growth and primary production. For example, while the results shown earlier for Coral Creek mangroves strongly suggest that soil P availability exerts a strong influence on net photosynthetic primary production, actual plant growth may well be controlled by salinity. The extent to which carbon assimilated during photosynthesis is used to build plant tissue or is lost in "maintenance respiration" and production of osmotica, will most likely be dependent on soil salinity.[53] Accordingly, the input of nutrients coupled with lowering of salinity would be expected to have a beneficial effect on mangrove photosynthesis and growth. This has important implications when considering forest response to extraneous nutrient inputs as will be discussed later.

B. ALGAL PRODUCTION

There is very little information available on algal production in mangrove forests. Alongi[23] and Kristensen et al.[54] have found that benthic microalgal primary production on the forest floor is negligible to low when compared with other intertidal areas. Kristensen et al.,[54] for example, found that benthic microalgal production in a *Rhizophora apiculata* forest in southern Thailand amounted to only 4 to 20% of forest production. They suggested that the algal production was limited by light and availability of DIN. Alongi,[23] however, reported very low levels of both benthic microalgal biomass (as chlorophyll *a*) and negligible rates of primary production in a northern Australian mangrove forest where light penetration through the dense canopy has previously been measured to be only 5% of external incident photosynthetically active radiation.[52] This suggests that light availability, rather than nutrient deficiency, is the major control of microalgal production in mangrove forests and, hence, added nutrients are unlikely to effect this production substantially.

An interesting study by Lapointe et al.[55] compared ambient levels of nutrient availability and the effects of nutrient (N and P) enrichment for three species of dominant macroalgae in waters adjacent to coral reef and mangrove forests in Belize. They found that the reef-associated algae responded significantly to nutrient enrichment (mainly by P), whereas the mangrove-associated algal response was not significant or even reduced. This suggests that macroalgae in mangrove waterways may also be limited by light, rather than nutrients, although it is impossible to generalize from the results of only one study.

C. WATER COLUMN PRIMARY PRODUCTION

Again, very little information is available on aquatic primary production by phytoplankton in mangrove waters. Gong[56] obtained low values for phytoplankton production in a Malaysian mangrove waterway and has suggested that mangrove aquatic primary production is generally light-limited owing to the usually high turbidity of these waters. Boto and Bunt[33] similarly reported low phytoplankton biomass in Coral Creek mangrove waters (measured as total chlorophylls which averaged 1.3 ± 0.3 $\mu g\ l^{-1}$ seawater). Conversely, mangrove waters in southern India[57] have been reported to contain up to 20 $\mu g\ l^{-1}$ of chlorophyll *a*. This suggests that water turbidity does not always limit aquatic primary production, and local factors such as ambient suspended sediment concentrations and hydrodynamic factors may be important.

A recent report of cyclone-induced sediment resuspension in nearshore waters in northern Australia[58] does suggest that nutrient input to mangrove waters may increase phytoplankton production. It was demonstrated that the sediment disturbance during the cyclone released large quantities of soil interstitial nutrients into the water column. This nutrient release was followed, within days, by massive phytoplankton blooms, with chlorophyll concentrations of up to three times the normal level being recorded. It is highly likely that nutrient input to mangrove waters would have a similar effect, especially during slack tide periods when water turbidity is significantly decreased,[33] and when light limitation may not be controlling.

There are often significant growths of algae associated with mangrove stilt roots or pneumatophores. Apart from their ability to fix nitrogen, however,[38,39] these have received little attention, and there are no available data on the biomass or productivity of such algal growths and whether they are nutrient limited.

IV. EFFECTS OF NUTRIENT POLLUTION

A. EFFECTS ON FOREST SOILS, PRIMARY PRODUCTION, AND GROWTH

The major sources of nutrient input to mangroves are likely to be (1) sewage effluent, (2) aquaculture pond wastewaters, and (3) runoff from agricultural activity. Unfortunately, there are very few documented case studies of such disturbances and their effects on mangroves, and, hence, we are basically forced to speculate based on our current knowledge of essentially undisturbed or slightly disturbed systems.

Substantial nutrient input can be expected to effect significantly the soil chemistry and microbiology as well as forest primary production and growth. These effects are most amenable to prediction based on currently available information, as limited as that may be. Effects on other aspects of the tropical mangrove ecosystem such as aquatic productivity, dependent biota, and ecosystem structure and diversity (see later discussion) can only be afforded very brief mention because of the paucity of relevant information.

1. Soil Chemistry and Microbiology

Soil chemistry and microbiology are likely to be profoundly affected by substantial and chronic nutrient inputs. When soluble reactive phosphate is added to soils, it is usually rapidly immobilized by adsorption reactions[30,59] depending to a large extent on the soil clay mineralogy, iron content, and redox status. A great deal of this absorbed P appears to remain in a plant-available pool. For example, in the large-scale fertilization experiment described earlier,[47] addition of 400 kg P per hectare (as superphosphate), over a 12-month period, was found to increase the amount of weak acid-extractable P from 5 to 65 μg P per gram dry weight.

However, the capacity of mangrove sediments to absorb and immobilize P is limited by availability of exchange sites and by redox status. As discussed above, highly reduced mangrove soils have far less capacity to absorb and occlude P because of the greater solubility of iron (II) sesquioxides in the reduced soils. Measurements of adsorption isotherms[18,60] for several undisturbed mangrove sediments from northern and southern Australia indicate adsorption maxima in the range 250 to 700 μg P per gram dry weight. In comparison, a lower adsorption maximum was reported[60] for a mangrove sediment near Darwin, northern Australia, which had received treated sewage effluent for about 20 years. In this sediment, total P (1720 μg P per gram dry weight) was much higher than recorded in most Australian mangrove sediments.[4,15] Holford and Patrick[30] have also cautioned that reduced sediments have a limited long-term capacity to adsorb and immobilize phosphate.

Inorganic nitrogen in mangrove sediments can be quite mobile, owing to the high concentrations of sodium ions which tend to swamp cation-exchange sites. Our own measurements, for example, have shown ammonium concentrations in the soil interstitial waters to be in the order of 1 μg NH_4-N per cubic centimeter. While this would suggest that ammonium would be readily lost through leaching and tidal exchange, measurements of ammonium fluxes from the sediments to the overlying tidal waters[24] and tidal export of ammonium from the forests as a whole[32] have been shown to be negligible. This apparent contradiction is most probably due to efficient interception of most dissolved nutrients at the sediment-water interface by the very high microbial densities and activities in the upper few centimeters of mangrove sediments.[23-25]

However, it is very unlikely that the sediments could similarly retain the very high concentrations which would result from nutrient pollution. Depending on the loading, substantial amounts of added ammonium, at least that which is not lost through denitrification reactions or increased plant uptake (see below), would be expected to be readily leached into the mangrove waterways.

As discussed in Section II, denitrification reactions in anaerobic soils can lead to substantial losses of nitrogen to the atmosphere, especially where soils have been amended with large amounts of nitrogenous fertilizers.[26] Sewage effluent in particular would be extremely susceptible to denitrification losses as a high proportion of the DIN is in the nitrate form.

2. Plant-Nutrient Interactions

Clough et al.[60] have estimated that nutrient uptake by mangroves can lead to the immobilization of significant amounts of N and P by incorporation into the plant tissue. Based on average leaf N and P contents of about 1 and 0.1%, respectively, and woody tissue content of about 0.5% N and 0.05% P, coupled with literature values for leaf[61] and wood production,[62] the authors estimated that mangrove forests could immobilize around 150 to 250 kg N per hectare and 15 to 20 kg P per hectare annually. Some of this immobilized N and P would be returned as plant litter; however, studies by Boto and Bunt[33] and Robertson[63] have shown that the plant litter is either consumed by sesarmid crabs (and thereby largely incorporated to crab biomass) or lost from the system through tidal flushing.

These estimates were conservative, being based on tissue nutrient concentrations and growth rates for undisturbed forests. The addition of N and P to the forests will have two likely effects both of which would lead to greatly increased nutrient immobilization through plant uptake. First, fertilization has been shown to lead to increased tissue levels of N and P. Clough et al.[60] report foliar N and P concentrations of 2.04 and 0.167%, respectively, for a mangrove forest which received long-term treated sewage effluent, compared with values of 1.15% N and 0.097% P at nearby undisturbed control sites.

Similarly, Boto and Wellington[47] observed significant increases in foliar N and P for new leaves at fertilized sites compared with adjacent untreated sites. Their samples, taken at the end of the 12-month fertilization period, showed that leaf N levels increased from 1.20 (SD = 0.05) to 1.43% (0.08) at the N-fertilized site. Leaf P increased from 0.088 (0.004) to 0.095% (0.003) at one P-fertilized site and from 0.113 (0.009) to 0.122% (0.005) at the other. As discussed above, Onuf et al.[46] also reported significant increases in *Rhizophoro mangle* foliar N and P at sites enriched by guano.

In addition to increased tissue nutrient content, the effects of fertilization would be expected to lead to increased mangrove growth in most situations, as discussed above. This increased growth, by up to 30%[47] coupled with the increased tissue nutrient levels, would lead to very substantial increases in plant uptake and immobilization of added nutrients compared with the relatively conservative initial estimates of Clough et al.[60]

These considerations imply that input of inorganic nutrients to mangrove forests are unlikely to be detrimental to the trees and in most circumstances are likely to be beneficial, leading to increased plant growth. In turn, the increased plant growth will lead to the immobilization of a substantial amount of the added nutrients, by up to at least 300 kg N per hectare and 30 kg P per hectare anually. This uptake and immobilization process will complement the capacity of the soil to sorb and immobilize P and to lose N via denitrification. In general, therefore, mangroves would be expected to have a very high capacity to process added nutrients.

A note of caution must be given here, however, as the long-term capacity of the system to process added nutrients and prevent their entering the waterways (where the effects are likely to be severe, discussed later) will greatly depend on the amounts and rate of nutrient addition. Based on the estimates given above, it is probably unlikely that mangrove forests could process and immobilize long-term nutrient inputs in excess of about 200 to 300 kg N per hectare or 15 to 30 kg P per hectare annually. This is, nevertheless, a fairly high input rate and could be accommodated by careful design of timing and placement of effluent discharge, as will be discussed below. A further note of caution, as pointed out by Clough et al.,[60] concerns the probable incapacity of mangroves to tolerate significant inputs of labile dissolved organic carbon. Such inputs would be likely to drive the soils and waters into a state of extreme anaerobiosis with which even the mangroves, with their adaptions to tolerate moderate degrees of anaerobiosis, would be hard-pressed to cope.

B. EFFECTS ON AQUATIC PRODUCTIVITY

As discussed above, there is conflicting evidence concerning the degree to which aquatic productivity in tropical mangroves is limited by light or nutrient availability. While there is some evidence that water turbidity restricts phytoplankton production,[56] it must also be considered that tropical mangrove waters are highly oligotrophic,[32] with dissolved inorganic nutrient levels commonly at or near analytical detection limits. In addition, Boto and Wellington[32] found that concentrations of dissolved nitrate and phosphate in Coral Creek waters showed significant day-night differences (lowest concentrations during the day), implying that phytoplankton activity could effectively "strip" these nutrients from the waters. This, in turn, suggests that added nutrients in high concentrations could lead to phytoplankton blooms and, hence, to eutrophic conditions which could eventually drive these waters into a highly anaerobic state. This "worst case" situation is very likely in mangrove waterways which are generally characterized by very slow tidal flushing rates and water residence times of days to weeks owning to salinity trapping effects.[64]

Other evidence, from the Furnas[58] study of cyclone effects, suggests that nutrient inputs to nearshore and shelf waters, following substantial bottom sediment resuspension, leads to phytoplankton bloom conditions. This study was carried out in northeastern Australia where the nearshore shelf waters are commonly as turbid as mangrove waters, even under normal climatic and hydrodynamic conditions. It, therefore, seems reasonable to assume that substantial and chronic nutrient input to mangrove and nearshore waters would be potentially catastrophic, giving rise, after the decay of algal blooms, to severely anaerobic conditions in waters which can be quite low in dissolved oxygen (ranging from about 10 to 90% saturation) under normal conditions.[65]

C. EFFECTS ON DEPENDENT BIOTA, ECOSYSTEM STRUCTURE, AND DIVERSITY

There is no published information on the effect of high loadings of inorganic nutrients on any of these aspects of mangrove ecosystems, and it is difficult to make predictions. As a first guess, it is probably reasonable to assume that moderate loadings of inorganic nutrients will not have any significant impact on dependent fauna, although very high concentrations of ammonium, in particular, would be expected to be toxic, especially to the burrowing infauna. However, if moderate levels of inorganic nutrient input were detrimental to such fauna, then this would almost certainly have severe effects on the food chains which operate within mangroves and on the forest structure. Studies by Robertson[63] and Robertson et al.[66] have clearly demonstrated the key trophic role of sesarmid crabs as major consumers of mangrove litter and their larvae as a possible major food source for many species of juvenile fish which utilize the mangroves as a habitat. Smith[67] has also shown that these crabs play a major role in determining the distribution and occurrence of a number of the commonly dominant tree species through their grazing on seeds and propagules. It is especially interesting to note that, in the tropical northern Australian forests examined by Smith,[67] the tree species *Avicennia marina* was shown to be virtually excluded from zones where sesarmid crab populations were high (usually the midintertidal zone) owing to the particularly voracious grazing of its propagules by these crabs. This species is probably the hardiest and among the fastest-growing of the mangroves, being able to tolerate high salinity ranges, highly reduced soils, and very wide temperature ranges.[68] It is reasonable, therefore, to speculate that drastic reductions in sesarmid crab populations would lead to this species becoming dominant in most areas. In this, and more subtle ways, it would be expected that any adverse effects of inorganic nutrient inputs (especially if such inputs were chronic) could be expected to lead to long-term changes to the structure and species diversity of tropical mangrove forests.

No information is available on the direct response of the various mangrove tree species to nutrient levels. Beadle[50] has postulated, on the basis of very large-scale surveys, that the high species diversity of many Australian terrestrial forests or vegetational associations is a consequence of the low P status of Australian soils in many areas. There is no good evidence yet available to suggest similarly that low nutrient status may contribute to high species diversity of tropical mangroves. Large-scale surveys of mangrove forests in northern Australia and Papua New Guinea[4] have shown that high-diversity forests do tend to be associated with areas of lowest soil P availability; however, these areas also coincide generally (and more consistently) with areas of low salinity. It is the latter factor which is more likely to be responsible for the higher species diversity of these forests.[69] It is, therefore, not possible to make any informed guesses on the (direct) effects of nutrient input on mangrove forest species diversity on the basis of currently available information.

As discussed above, the input of substantial amounts of organic nutrients would be expected to drive the mangrove soils and waters into extreme degrees of anaerobiosis which would certainly be highly detrimental to the burrowing fauna and to the trees themselves. This would presumably arise from the increased microbial activity which would be expected to result from the extra supply of labile carbon. Normal levels of bacterial densities and activity are quite considerable in tropical mangrove soils[23] and apparently controlled by diurnal (tidal inundation) and seasonal temperature variations rather than by carbon or other nutrient limitation. These observations are, however, probably only valid within the normal range of concentrations of dissolved organic nutrients in the soils and waters, and it is highly likely that large inputs of labile organic matter would greatly accelerate microbial activity accompanied by large changes in the bacterial population structures.

V. SUMMARY AND CONCLUSIONS

Nutrient limitation (N or P) is probably a general feature of tropical mangrove ecosystems and, hence, the input of inorganic nutrients would be expected to be beneficial to mangrove growth. In addition, because of the increased capacity for sediment denitrification reactions and the ability of the trees to take up and immobilize quite high amounts of added inorganic nutrients, it is reasonable to assume that mangrove forests would have a fairly good capacity to tolerate and to process such inputs. In accordance with the calculations described above, sustained inputs of around 300 kg N per hectare and 30 kg P per hectare annually should be capable of being processed by the forests. It is highly desirable that inorganic nutrients be immobilized within the plants and soil to prevent contamination of the waterways. In the waters, elevated nutrient levels would be expected to give rise to greatly increased phytoplankton-algal production which, during later decomposition, would greatly exacerbate the low oxygen conditions which are a common feature of many mangrove waters in their normal state.

In this regard, the placement and timing of effluents discharged into mangroves will play a very large part in determining the severity of effects.[60] Direct discharge into mangrove waterways is the least desirable approach, and it is recommended that effluents containing high levels of inorganic nutrients be discharged into the high intertidal zone of the mangrove forests during neap tide periods when tidal waters do not reach these zones. In this manner, the added nutrients would be largely absorbed within the sediments and the ability of the forests and soils to process the nutrients would be maximized, thus greatly limiting the input into the waterways. Effluents which are less saline than the ambient waters would help to ameliorate the salinity stresses which often limits mangrove growth in the upper intertidal areas, also helping to increase the uptake of nutrients by the trees.

The levels of organic matter input, in addition to the as yet unknown tolerance of mangrove infauna to extraneous nutrient input, may ultimately limit the ability of mangrove to tolerate nutrient pollution. A great deal of research is required to address these aspects properly.

REFERENCES

1. **Galloway, R. W.,** Distribution and physiographic patterns of Australian mangroves, in *Mangrove Ecosystems in Australia. Structure, Function and Management,* Clough, B. F., Ed., Australian Institute of Marine Science — Australian National University Press, Canberra, 1982, 31.
2. **Fortes, M. D.,** Mangroves and seagrass beds of east Asia: habitats under stress, *Ambio,* 17, 207, 1988.
3. **Bunt, J. S. and Williams, W.,** Vegetational relationships in the mangroves of tropical Australia, *Mar. Ecol. Prog. Ser.,* 4, 349, 1981.
4. **Boto, K. G., Bunt, J. S., and Wellington, J. T.,** Variations in mangrove forest productivity in northern Australia and Papua New Guinea, *Estuar. Coast. Shelf Sci.,* 19, 321, 1984.
5. **Wells, A. G.,** Mangroves of northern Australia, in *Mangrove Ecosystems in Australia. Structure, Function and Management,* Clough, B. F., Ed., Australian Institute of Marine Science — Australian University Press, Canberra, 1982, 55.
6. **Wolanski, E., Jones, M., and Bunt, J. S.,** Hydrodynamics of a tidal creek-mangrove swamp system, *Aust. J. Mar. Freshwater Res.,* 31, 415, 1980.
7. **Wolanski, E., Chappell, J., Ridd, P., and Vertessy, R.,** Fluidization of mud in estuaries, *J. Geophys. Res.,* 93 (C3), 2351, 1988.

8. **Milward, N. E.,** Mangrove-dependent biota, in *Mangrove Ecosystems in Australia. Structure, Function and Management,* Clough, B. F., Ed., Australian Institute of Marine Science — Australian University Press, Canberra, 1982, 121.
9. **Robertson, A. I. and Duke, N. C.,** Mangroves as nursery sites: comparisons of the abundance and species composition of fish and crustaceans in mangroves and other nearshore habitats in tropical Australia, *Mar. Biol.,* 96, 193, 1987.
10. **Robertson, A. I. and Duke, N. C.,** Mangrove fish communities in tropical Australia: spatial and temporal differences in standing stocks and community structure, *Mar. Ecol. Progr. Ser.,* in press.
11. **Robertson, A. I.,** Abundance, diet and predators of juvenile banana prawns *Penaeus merguiensis* in a tropical mangrove estuary, *Aust. J. Mar. Freshwater Res.,* 39, 467, 1988.
12. **Staples, D. H., Polzin, H. G., and Heales, D. S.,** Habitat requirements of juvenile penaeid prawns and their relationship to offshore fisheries, in *2nd Australian National Prawn Seminar,* Rothlisberg, P. C., Hill, B. J., and Staples, D. J., Eds., Cleveland, NPS2, 1985, 47.
13. **Field, C. D. and Dartnall, A. J., Eds.,** *Mangrove Ecosystems of Asia and the Pacific; Status, Exploitation and Management,* Australian Institute of Marine Science, Canberra, 1987.
14. **Alongi, D. M.,** The ecology of tropical soft-bottom benthic ecosystems, *Oceanogr. Mar. Biol. Annu. Rev.,* 28, 381, 1990.
15. **Boto, K. G. and Wellington, J. T.,** Soil characteristics and nutrient status in a northern Australian mangrove forest, *Estuaries,* 7, 61, 1984.
16. **Cain, S. and Boon, P. I.,** Cellular osmotica of plants in relation to sediment nitrogen and salt contents in mangroves and saltmarshes at Western Port, Victoria, *Aust. J. Mar. Freshwater Res.,* 38, 783, 1987.
17. **Clarke, P. J.,** Nitrogen pools and soil characteristics of a temperate estuarine wetland in eastern Australia, *Aquat. Bot.,* 23, 275, 1985.
18. **Hesse, P. R.,** Phosphorus fixation in mangrove swamp soils, *Nature (London),* 193, 295, 1962.
19. **Udo, E. J., Ibedu, M. A., and Fagbami, A. A.,** Phosphorus status of some saline and non-saline hydromorphic soils of the Niger Delta of Nigeria, *Plant Soil,* 77, 327, 1984.
20. **Attiwill, P. M. and Clough, B. F.,** Productivity and Nutrient Cycling in the Mangrove and Seagrass Communities of Westernport Bay, Ministry for Conservation, Victoria, 1978.
21. **Carlson, P. R., Yarbro, L. A., Zimmermann, C. F., and Montgomery, J. R.,** Pore water chemistry of an overwash mangrove island, *Estuaries,* 4, 282, 1981.
22. **Boon, P. I. and Cain, S.,** Nitrogen cycling in salt marsh and mangrove sediments at Western Port, Victoria, *Aust. J. Mar. Freshwater Res.,* 39, 607, 1988.
23. **Alongi, D. M.,** Bacterial productivity and microbial biomass in tropical mangrove sediments, *Micro. Ecol.,* 15, 59, 1988.
24. **Stanley, S. O., Boto, K. G., Alongi, D. M., and Gillan, F. T.,** Composition and bacterial utilization of free amino acids in tropical mangrove sediments, *Mar. Chem.,* 22, 13, 1987.
25. **Boto, K. G., Alongi, D. M., and Nott, A. L. J.,** Dissolved organic carbon-bacteria interactions at sediment-water interface in a tropical mangrove system, *Mar. Ecol. Prog. Ser.,* 51, 243, 1989.
26. **Tusneem, M. E. and Patrick, W. H., Jr.,** Nitrogen Transformations in Waterlogged Soils, Bull. No. 657, Department of Agronomy, Louisiana State University, Baton Rouge, 1971.
27. **Keeney, D. R.,** The nitrogen cycle in sediment-water systems, *J. Environ. Qual.,* 2, 15, 1973.
28. **Vlek, P. L. G. and Stumpe, J. M.,** Effects of solution chemistry and environmental conditions on ammonia volatilization from aqueous systems, *Soil Sci. Soc. Am. Proc.,* 42, 416, 1978.
29. **Boto, K. G., Saffigna, P., and Clough, B. F.,** Role of nitrate in the nitrogen nutrition of the mangrove *Avicennia marina, Mar. Ecol. Prog. Ser.,* 21, 259, 1985.
30. **Holford, I. C. R. and Patrick, W. H., Jr.,** Effects of redox potential and pH on phosphate removal from wastewater during land application, *Prog. Water Technol.,* 11, 215, 1979.
31. **Patrick, W. H., Jr., and Mahapatra, I. C.,** Transformation and availability to rice of nitrogen and phosphorus in waterlogged soils, *Adv. Agron.,* 20, 323, 1968.
32. **Boto, K. G. and Wellington, J. T.,** Seasonal variations in concentrations and fluxes of dissolved organic and inorganic materials in a tropical, tidally-dominated, mangrove waterway, *Mar. Ecol. Prog. Ser.,* 50, 151, 1988.
33. **Boto, K. G. and Bunt, J. S.,** Tidal export of particulate organic matter from a northern Australian mangrove system, *Estuar. Coast. Shelf Sci.,* 13, 247, 1981.
34. **Wong, C. H.,** Mangrove aquatic nutrients, in *Proc. Workshop on Productivity of the Mangrove Ecosystem: Management Implications,* Ong, J. E. and Gong, W. K., Eds., Universiti Sains Malaysia, Penang, 1984, 60.
35. **Nixon, S. W., Furnas, B. N., Lee, V., Marshall, N., Ong, J. E., Wong, C. H., Gong, W. K., and Sasekumar, A.,** The role of mangroves in the carbon and nutrient dynamics of Malaysia estuaries, in *Proc. Asian Symp. on Mangrove Environment: Research and Management,* Soepardmo, E., Rao, A. N., and Macintosh, D. J., Eds., University of Malaya, Kuala Lumpur, 1984, 535.

36. **Boto, K. G. and Bunt, J. S.,** Carbon export from mangroves, in *The Cycling of Carbon, Nitrogen, Sulfur and Phosphorus in Terrestrial and Aquatic Ecosystems,* Galbally, I. E. and Freney, J. R., Eds., Australian Academy of Science, Canberra, 1982, 105.
37. **Mann, F. D. and Steinke, T. D.,** Biological nitrogen fixation (acetylene reduction) associated with blue-green algal cyanobacterial communities in the beachwood mangrove nature reserve, South Africa. I. The effect of environmental factors on acetylene reduction activity, *S. Afr. J. Bot.,* 55, 438, 1989.
38. **Zuberer, D. A. and Silver, W. S.,** Biological dinitrogen fixation (acetylene reduction) associated with Florida mangroves, *Appl. Environ. Microbiol.,* 35, 567, 1978.
39. **Boto, K. G. and Robertson, A. I.,** The relationship between nitrogen fixation and tidal exports of nitrogen in a tropical mangrove system, *Estuar. Coast. Shelf Sci.,* 31, 531, 1990.
40. **Valiela, I., Teal, J., Volkmann, S., Shafer, D., and Carpenter, E. J.,** Nutrient and particulate fluxes in a salt marsh ecosystem: tidal exchanges and inputs by precipitation and groundwater, *Limnol. Oceanogr.,* 23, 798, 1978.
41. **Chrzanowski, T. H., Stevenson, L. H., and Spurrier, J. D.,** Transport of dissolved organic carbon through a major creek of the North Inlet ecosystem, *Mar. Ecol. Prog. Ser.,* 13, 167, 1983.
42. **Woodroffe, C. D.,** Studies of a mangrove basin, Tuff Crater, New Zealand. III. The flux of organic and inorganic particulate matter, *Estuar. Coast. Shelf Sci.,* 20, 447, 1985.
43. **Twilley, R. R.,** The exchange of organic carbon in basin mangrove forests in a southwest Florida estuary, *Estuar. Coast. Shelf Sci.,* 20, 543, 1985.
44. **Hicks, D. B. and Burns, L. A.,** Mangrove metabolic response to alterations of natural freshwater drainage to south-western Florida estuaries, in *Proc. Int. Symp. on Biology and Management of Mangroves,* Vol. 1, Walsh, G. E., Snedaker, S. C., and Teas, H. J., Eds., University of Florida, Gainesville, 1975, 238.
45. **Lugo, A. E., Sell, M., and Snedaker, S. C.,** Mangrove ecosystem analysis, in *Systems Analysis and Simulation in Ecology,* Vol. 4, Patten, B. C., Ed., Academic Press, New York, 1976, 113.
46. **Onuf, C. P., Teal, J. M., and Valiela, I.,** Interactions of nutrients, plant growth and herbivory in a mangrove ecosystem, *Ecology,* 58, 514, 1977.
47. **Boto, K. G. and Wellington, J. T.,** Phosphorus and nitrogen nutritional status of a northern Australian mangrove forest, *Mar. Ecol. Prog. Ser.,* 11, 63, 1983.
48. **Naidoo, G.,** Effects of salinity and nitrogen on growth and water relations in the mangrove, *Avicennia marina* (Forsk.) Vierh., *New Phytol.,* 107, 317, 1987.
49. **Viner, A. B.,** The status and transport of nutrients through the Purari River (Papua New Guinea). *Purari River (Wabo) Hydroelectric Scheme Environmental Studies,* No. 9, Office of Environment and Conservation and Department of Minerals and Energy, Papua New Guinea, 1979.
50. **Beadle, N. C. W.,** The edaphic factor in plant ecology with a special note on soil phosphates, *Ecology,* 34, 426, 1953.
51. **Boto, K. G.,** Nutrient status and other soil factors affecting mangrove productivity in north-east Australia, *Wetlands (Australia),* 3, 45, 1983.
52. **Bunt, J. S., Boto, K. G., and Boto, G.,** A survey method for estimating potential levels of mangrove forest primary production, *Mar. Biol.,* 52, 123, 1979.
53. **Clough, B. F. and Sim, R. G.,** Changes in gas exchange characteristics and water use efficiency of mangroves in response to salinity and vapour pressure deficit, *Oecologia,* 79, 38, 1989.
54. **Kristensen, E., Anderson, F. Oe., and Kofoed, L. H.,** Preliminary assessment of benthic community metabolism in a south-east asian mangrove swamp, *Mar. Ecol. Prog. Ser.,* 48, 137, 1988.
55. **Lapointe, B. E., Littler, M. M., and Littler, D. S.,** A comparison of nutrient-limited productivity in macroalgae from a Caribbean barrier reef and from a mangrove ecosystem, *Aqua. Bot.,* 28, 243, 1987.
56. **Gong, W. K.,** Mangrove primary productivity, in *Productivity of the Mangrove Ecosystem: Management Implications,* Ong, J. E. and Gong, W. K., Eds., Universiti Sains Malaysia, Penang, 1984, 10.
57. **Balasubramanian, T. and Venugopalan, V. K.,** Dissolved organic matter in Pitchawaram mangrove environment, Tamil Nadu, South India, in *Proc. Asian Symp. on Mangrove Environment: Research and Management,* Soepardmo, E., Rao, A. N., and Macintosh, D. J., Eds., University of Malaya, Kuala Lumpur, 1984, 496.
58. **Furnas, M. J.,** Cyclonic disturbance and a phytoplankton bloom in a tropical shelf ecosystem, in *Red Tides: Biology, Environmental Science and Toxicology,* Elsevier Science, Okaichi, T., Anderson, D. M., and Nemoto, T., Eds., New York, 1989, 273.
59. **Holford, I. C. R. and Mattingly, G. E. G.,** A model for the behaviour of labile phosphate in soil, *Plant Soil,* 44, 219, 1976.
60. **Clough, B. F., Boto, K. G., and Attiwill, P. M.,** Mangroves and sewage: a reevaluation, in *Tasks for Vegetation Science,* Vol. 8, Teas, H. J., Ed., Junk, The Hague, 1983, 151.
61. **Duke, N. C., Bunt, J. S., and Williams, W. T.,** Mangrove litter fall in north-east Australia. I. Annual totals by component in selected species, *Aust. J. Bot.,* 29, 547, 1981.
62. **Teas, J. H.,** Silviculture with saline water, in *The Biosaline Concept,* Hollaender, A., Aller, J. C., Epstein, E., San Pietro, A., and Zaborsky, O. R., Eds., Plenum Press, New York, 1979, 117.

63. **Robertson, A. I.,** Leaf-burying crabs: their influence on energy flow and export from mixed mangrove forests (*Rhizophora* spp.) in northeastern Australia, *J. Exp. Mar. Biol. Ecol.,* 102, 237, 1986.
64. **Wolanski, E.,** Circulation anomalies in tropical Australian estuaries, in *Hydrodynamics of Estuaries,* Kjerfve, B., Ed., CRC Press, Boca Raton, FL, 1988, 53.
65. **Boto, K. G. and Bunt, J. S.,** Dissolved oxygen and pH relationships in northern Australian mangrove waterways, *Limnol. Oceanogr.,* 26, 1176, 1981.
66. **Robertson, A. I., Dixon, P., and Daniel, P. A.,** Zooplankton dynamics in mangrove and other nearshore habitats in tropical Australia, *Mar. Ecol. Prog. Ser.,* 43, 139, 1988.
67. **Smith, T. J., III,** Seed predation in relation to tree dominance and distribution in mangrove forests, *Ecology,* 68, 266, 1987.
68. **Saenger, P.,** Morphological, anatomical and reproductive adaptations of Australian mangroves, in *Mangrove Ecosystems in Australia — Structure, Function and Management,* Clough, B. F., Ed., Australian Institute of Marine Science — Australian National University Press, Canberra, 1982, 153.
69. **Bunt, J. S., Williams, W. T., and Clay, J. S.,** The detection of species sequences across environmental gradients, *Mar. Ecol. Prog. Ser.,* 24, 197, 1985.
70. **Alongi, D. M.,** unpublished data.

Chapter 6

EFFECTS OF SALINITY ON CORAL REEFS

Stephen L. Coles and Paul L. Jokiel

TABLE OF CONTENTS

I. INTRODUCTION

The effects of salinity on reef corals and associated coral reef marine organisms have not been thoroughly studied, despite the importance of salinity to osmoregulation and other physiological processes necessary for reef organism survival. Part of the reason for this lack of attention to such an important physical factor may be because most coral reefs occur where salinity is relatively stable, and years can pass without significant variation from normal oceanic values. Perhaps because of this salinity stability scientists have often considered corals to be highly stenohaline organisms[1] with little capacity for survival or adaptation to salinity change. However, reef corals and coral reefs can encounter a large range of salinities.[2] Such exposures can have results ranging from indeterminate to devastating, depending on the extent of deviation from normal salinity and the duration of exposure.

Most of the limited experimental data on salinity tolerances of coral reef organisms come from experiments conducted 60 to 75 years ago and are mostly limited to reef corals. As scientific investigations have explored more regions of the world where coral reefs occur, salinity tolerance limits of corals have ranged beyond those shown by early studies, particularly at the upper end of the scale. The evidence suggests that corals and other reef organisms can adapt to salinity environments well beyond average oceanic salinity.

II. CORAL REEF SALINITY ENVIRONMENTS

A. NORMAL SALINITY RANGES ON CORAL REEFS

Unrestricted mixing of all the salts in the oceans of the world would yield an average salinity of 34.72‰.[3] However, land barriers, differences in water flux between the ocean and the atmosphere, and salt transport between oceans by currents prevent complete mixing, causing considerable salinity differences among the oceans. Open ocean salinities are maximal at about 20°N and S latitude because of high net evaporation and minimal in the region of the equator because of high net precipitation. Current systems and other factors modify these latitudinal belts of salinity, resulting in further differences in salinity between oceans and at different locations within the same ocean.

The maximum open ocean salinities occur in the Atlantic Ocean, where northern summer average values at 37 to 37.25‰ at 20 to 30°N and at 10 to 20°S.[4] Maximum salinities in the Indian and Pacific Oceans are lower, with mean open ocean values ranging up to 36.5‰ in three regions: west of the Indian subcontinent at latitudes 10 to 25°N, west of Australia at about 30°S, and east of the Society Islands and Tuamotos in the Pacific at 15 to 35°S.[4]

Minimum open ocean salinities occur in the Indian and Pacific Oceans. The lowest are in a triangular area enclosing the entire northeast region of the Indian Ocean and the South China Sea.[4] Northern summer mean open ocean values range down to 30‰ in this region of low salinity, which includes the offshore waters of Burma, Thailand, Malaysia, Indonesia, and the western Philippines.

Major coral growth and coral reef development occur in all these regions, which comprise the full range of surface salinities for the three major oceans.[3,4] Most of the significant coral reefs in the Atlantic Ocean occur in the Caribbean basin in the western North Atlantic, where mean salinities usually range above 36‰. Likewise, prolific coral and coral reef populations occur where open ocean salinities are low in the Indian and Pacific Oceans. Significantly, this region is also the world center of maximum coral diversity and abundance. Low salinity has therefore, not prevented the long-term survival of reef corals or the development of flourishing coral reefs.

Table 1 shows ranges of salinities which have been measured in or near areas of active coral growth and coral reef development. The worldwide ranges and patterns are similar to those of open ocean salinities. Most of the high values of around 37‰ occur in the western

TABLE 1
Salinity Ranges in Areas of Coral Growth Throughout the World

Location	Salinity range (‰)	Comment	Ref.
Atlantic			
Tortugas, Florida	32.1—36.3	Low value after rain	5
Biscayne Bay, Florida	35.2—35.7	Increase into S. Bay	5, 6
Bermuda	36.3—36.6	Open ocean	7
Florida Keys	35.6—36.6	Open ocean	7
Bahamas	36.0—36.4	Open ocean	7
N. Florida Keys	36.6—37.3	Drought conditions	8
Magueyez Is., P. Rico	31.9—37.5	Low during Hurr. Edith	9
Discovery Bay, Jamaica	33.8—34.9		10
Grand Bahamas Is.	31.0—36.6		11
Caribbean, Costa Rica	32.0—38.0	Low after heavy rain	12
Caribbean, Mexico	34.7—37.0		13
Bahia, Brazil	36.5—37.5		14
Indo-Pacific			
Waikiki, Hawaii	32.5—34.4	Offshore reef flat	15
	26.6—32.4	Nearshore reef flat	15
Kaneohe Bay	30.9—36.2	Semienclosed lagoon	16
Abu Dhabi, Arab. Gulf	42.0—48.0		17
Gulf of Aqaba, Red Sea	40.0—41.0		18
Christmas Is. Saline Lake	51.0—52.0	World max. for coral surviv.	19
Hollandes Cay, Panama	34.4—35.9	Sept.—Oct. 1970	20
Singapore-Malaya	28.5—31.3		21
Tahiti	35.0—36.0	Lagoon	22
Fiji Islands	33.0—34.0		22
Great Barrier Reef	25.8—35.9	Nearshore	23
	33.8—35.3	Offshore	23
Canton Island	35.0—39.0	Enclosed atoll lagoon	24, 25
Northern Philippines	32.2—34.7	1-year measurements	26
Phuket, Thailand	31.0—33.0		27
Kuwait, Arabian Gulf	38.6—42.4		28
Reunion Is., Ind. Ocean	34.5—35.4	Dec. 1984	29
Gulf of Thailand	25.0—30.0	Oct.-Nov. 1974	30
New Caledonia	35.0—35.6	1977-79	31
Gulf Of Suez, Red Sea	40.5—41.1		32
Saudi Arabia	39.5—41.0	Offshore reefs	33
Arabian Gulf	39.0—46.0	Nearshore reefs	33
Bahrain	41.0—50.0		34
Kenya, Indian Ocean	26.0—35.4	Low dur. S. E. monsoon	35
Papua New Guinea	30.0—36.0	April 1985	36
Indonesia	29.0—33.0	May 1985	36

Atlantic and Caribbean. However, in marginal seas in the Indo-Pacific where high evaporation and restricted circulation occur, and in enclosed lagoons or saline lakes of some coral atolls, salinities rise above 45‰. Lowest salinities on living coral reefs under normal conditions occur in Southeast Asian waters, where annual salinity minima routinely drop to around 25‰. The combined data suggest that corals and other reef organisms can live in normal salinities as low as 25‰ and as high as 45‰, even though most coral reefs occur in a more moderate salinity environment.

B. HYPERSALINE CORAL REEF ENVIRONMENTS

The landmark paper by Kinsman[17] describing reef corals growing in salinities of up to 48‰ in areas of high salinity greatly expanded the recognized upper tolerance limits of reef corals. Coral reefs can occur where high evaporation and low precipitation and freshwater input produce normal salinities much higher than the world average. Although hypersaline conditions may occur on coral atolls where lagoon waters become restricted from the open ocean,[19,24,25] the most extensive hypersaline areas occur mainly in the Red Sea and the Arabian Gulf. The normal salinity environment of coral reefs does not fall below 40‰ in the Arabian Gulf[17,28,33,34,37] and in the Gulfs of Aqaba and Suez in the Red Sea.[18,32,38]

Corals and coral reefs can flourish in these areas, although in the Arabian Gulf, relatively few species comprise the total coral coverage, and the numbers of coral species decrease with increasing salinity.[33,34] This reduced diversity may in part be due to isolation from the Indo-Pacific, but salinity elevation is a major factor restricting more sensitive coral species. Coral populations show systematic decreases in numbers of species with increasing salinity along the coasts of Abu Dhabi, Saudi Arabia, and Bahrain in the Arabian Gulf where salinities range from 42 to 50‰. In Abu Dhabi, only massive *Porites* survives salinities up to 48‰, and only a few *Acropora, Favia, Stylophora,* and *Pocillopora* live in 45 to 48‰.[17] No coral reefs have been found along the Saudi Arabian Gulf coastline south of a zone where salinities increase above 45‰, and fewer coral species occur at inshore reefs where salinities range up to 45‰ than on offshore reefs where salinities are 40 to 41‰.[33] However, Sheppard[34] reported eight coral species growing at 48‰ and three at 50‰ in a survey of the coral reefs of Bahrain. Sheppard[34] also found a decrease of approximately one coral species with each part per thousand increase across the 42 to 50‰ range on this survey, supporting the importance of salinity in restricting the diversity of the coral community.

Similar restriction of coral species with increasing salinity in atoll lagoons has been reported at Christmas and Canton Islands.[19,24,25] Numbers of coral species per quadrat sample decreased from up to 23 at the entrance to Canton Atoll lagoon where salinity is 35‰ to 0 to 3 species per sample 7.5 km from the lagoon pass where salinity was 39‰.[24,25] Coral coverage at Canton Island also decreased with increasing salinity. At Christmas Island, only a single coral species, *Acropora* cf. *grandis,* was reported to occur at 51 to 52‰.[19] This is the highest salinity environment recorded for a living coral. The report of salinities reaching 55‰ for corals at a reef in Tarut Bay, Saudi Arabia by Wells and Sheppard[39] is undocumented and perhaps a typographical error. No salinities above 43‰ occurred during 2 years of monitoring at this reef in 1985 to 1987,[40] nor have any been measured anywhere else in this vicinity of the western Arabian Gulf.[41,42]

Elevated salinity also restricts reef organisms other than corals. Clarke and Keij[43] reported the disappearance of several important groups, including perforate foraminifera, gastropods, lamellibranchs, and most echinoids, at the entrance to the Gulf of Salwah in the southwest Arabian Gulf where3 salinities increase above 45‰. At 48‰ all corals, alcyonarians, echinoids, melobesoid algae, and most arenaceous foraminifer disappear, leaving a restricted and less diverse reef fauna. Coles and McCain[41] also found increasing salinity to limit both numbers of species and numbers of individuals of benthic infauna along the Saudi Arabian Gulf coast.

Hypersaline reef environments also occur in the Florida and Caribbean area. The highest salinity ever reported at a coral reef was 70‰ at Turneffe Lagoon, British Honduras, in June 1939, during a period of very high air temperatures and low rainfall.[44] However, these measurements are suspect since they were taken by hydrometer 2 months before the author observed the reef. He found no indication of recent or unusual mortality to corals, gorgonians, calcareous algae, or "eel grass". Since a salinity of 70‰ would have devastated hard and soft corals (see below) and exceeds the known tolerance of most marine spermatophyes,[44-47] it is unlikely that this salinity report is accurate.

Robblee et al.,[48] in reviewing salinity records for Florida Bay, determined that salinities above 35‰ prevail throughout most of the Bay, with monthly means as high at 52‰. No information is given concerning how these high salinities affect the distributions of reef organisms in this area. However, the report suggests that hypersaline conditions in the vicinity of coral reefs commonly occur in areas of restricted circulation outside the Red Sea or Arabian Gulf.

C. SALINITY REDUCTION FROM RAINFALL AND STORM RUNOFF

Many reports describe salinity reductions causing massive coral and reef organism mortality. Usually these kills are due to extreme rainfall events, but other sources may produce freshwater intrusion with the same result. Wood-Jones[49] described the destruction of all living coral in the southwest portion of Cocos-Keeling atoll in 1876 by ''foul-water'' pouring from a volcanic vent in the southern side of the atoll. Groundwater discharge[50-53] may influence the distribution of corals and other organisms on the reef if discharge volumes are enough to alter ambient salinity.[54] Normal discharge from rivers and streams limits the growth and survival of corals at river mouths, forming channels in otherwise continuous fringing reefs.[2,55,58]

However, it is during major storm events that the effects of lowered salinity on coral reefs are most dramatic. Many such events have occurred in the Atlantic and the Pacific; however, only in few cases have salinities been measured synoptically that would rigorously define lower salinity tolerances.

One of the earliest such descriptions was of the effects of flooding in June 1926 on the corals and reefs of south Tahiti.[55] A rainfall of 32 in. in 6 d produced little effect on corals at Papeete, but 43 in. caused extensive mortality on the reef flat and edge at Moto Ini. At Faa mortality of *Porites* heads extended 6 in. below normal growing height, showing a sharp demarcation from live coral at the depth of fresh water penetration. Coral death began a few days after the flood and continued for 6 months, accompanied by mass mortality of other reef animals.

Orr and Moorhouse[57] reported salinity induced coral mortality at the Low Isles on the 1928 to 1929 Great Barrier Reef Expedition. The minimum salinity measured in an area of growing coral was 17.1‰ in a 12-cm-deep pool on the reef flat. This lowered salinity killed half the coral species present in 11.5 h, and nearly all died after 24 h of exposure. Slack-Smith[58] reported coral deaths in 1956 at Moreton Bay, 150 m south of the Great Barrier Reef, to be due to excessive summer rainfall and river outflow. Rainfall during that period was the highest in 27 years; however, salinities were not measured at the site during this event.

Goodbody[59] described three major rainfall events occurring in Jamaica in October 1956, May 1958, and October 1958 which produced various degrees of reef mortality. Monthly rainfalls during these periods exceeded 300 mm. Twenty-four hour maxima were more than 100 mm, with 156 mm falling during the most intense 24-h period in October 1958. The lowest salinities were 11.4‰ 3 d after the rain in May 1956, 13.3‰ shortly after the May 1958 rain, and 5.4‰ during the 5-d rain in October 1958. Limited mortality occurred following the first two events, but ''enormous mortality'' resulted from the October 1958 flooding. In Kingston Harbor a wide variety of invertebrates, including sponges, hydroids, ascidians, ophiuroids, holothurians, asteroids, and bivalves, died. Salinity returned to normal only after 2 to 3 months, during which recolonization and development of the benthic fauna occurred.

Similar levels of rainfall occurred in Puerto Rico during Hurricane Edith in September 1963, but with less devastating effects.[9,60] Over 200 mm of rain fell in one night at La Parquera, but salinity minima were only 27.2‰ in a mangrove canal nearshore and 33.8‰ at an offshore coral cay. All reef damage that occurred was due to mechanical breakage from hurricane-generated waves.

In contrast, salinity reductions and effects on corals were far more pronounced on Jamaican reefs from Hurricane Flora in October 1963.[61] Rainfall of 550 mm in 2 d reduced salinities in the Port Royal area to less than 3‰ on October 9. Salinities less than 30‰ persisted for more than 5 weeks. This resulted in zooxanthellae loss (bleaching) of all scleractinians, actinians, zooanthids, and *Millepora* to depths of 2.5 to 3 m in the Port Royal area. Similar effects occurred to 1-m depth at Sand Cay, 6 to 7 km from Port Royal. Most bleached corals showed a sharp demarcation between normally pigmented portions and bleached zones at the depth of low-salinity penetration. Many bleached corals survived the event and recovered slowly. Zooxanthellar pigmentation began returning as early as 13 d following the flood and required 14 weeks to complete.

Acevedo and Goenaga[62] observed similar bleaching of corals after flooding in southwestern Puerto Rico. A turbidity plume reached 30-f depth, and coral bleaching occurred up to 60 f and persisted at least 45 d. Salinities were not measured, but the authors attributed the bleaching to light restriction from turbidity rather than salinity changes.

In Fiji a hurricane in February 1965 produced 1270 mm of rainfall in the mountains and 355 mm at sea level in 5 d in the Suva area.[63] This caused unprecedented flooding of the Rewa River, which moved the river to a different channel. On Tomberua Island all invertebrates and fish died on reefs at the mouth of the new river channel. Five days after flooding began, massive deposits of putrefying organisms had accumulated on the reef flat and shoreline, and few live organisms could be found. Unfortunately, no measurements of salinity were made during this event.

Two high rainfall events in Kaneohe Bay, Hawaii provide records of the effects of lowered salinity on reefs which have been under long-term, continuous study. The first event occurred May 2 to 8, 1965 when nearly 700 mm of rain fell in 2 d in the mountains above the bay. This produced a thick layer of fresh, silt-laden water in the south bay where exchange with the open ocean is most restricted.[65,65] Ths first salinity measurements on May 7 showed a surface salinity of 8‰ nearshore in the south bay and 20 to 25‰ in areas of existing coral. Salinity profiles on May 10 showed salinities from surface to 1 m to have returned to above 30‰, but lower than normal salinities persisted to May 18.

The mortality that resulted from this salinity reduction was extensive and may show detrimental added or synergistic effects due to siltation, turbidity, and anoxia from dying organisms. Reef coral populations were devastated, and plankton, sponges, coelenterates, annelids, mollusks, crustaceans, and fish also died. Most shallow water corals bleached rapidly and died to depths of about 0.3 to 1.5 m below former live depths on nearshore and lagoon reefs in the central and south Bay.

By 1968 the first recolonization of reef corals had occurred,[65] but zoanthids occupied most of the exposed surfaces on the reef edges and flats. During this same period rapid increases in the discharge of treated sewage, streams, and siltation from poor land management radically altered the bay to favor filter-feeding organisms over reef corals, and no significant coral populations existed in south Kaneohe Bay.[65,66] Continued eutrophication of the south and midbay led to a decision to remove sewage discharge to an outfall outside the bay beginning in 1978. This resulted in rapid decreases in nutrient loading, particulate and plankton concentrations, and reduced populations of benthic filter feeders.[67] By 1983 coral populations in the south and mid bay were well on their way to recovery.[68-71] Coral coverage at one midbay station increased by over six times,[71] and average coverage throughout the bay doubled its 1971 to 1973 value.[68,70]

A second storm-related reef kill occurred in early January 1988 when heavy rains of more than 250 mm fell in 24 h on January 1 in some parts of the Kaneohe Bay watershed. Salinities on fringing reefs were 15 to 20‰ for several days. Massive mortality of corals, echinoderms, crustaceans, mollusks, and annelids similar to the 1965 kill resulted, and fishes moved into deeper water to escape the low-salinity surface layer.[72]

Observations made in August 1989 suggested that reef corals that had recruited to the south bay between 1979 and 1988 were all killed by low salinities to depths of up to 2 m on nearshore reef flats and slopes, but deeper coral colonies were unaffected.[73] Patch reefs in the central south bay, which had formerly been dredged to depths greater than 2 m, apparently had been unaffected by the storm. The flat sections of these deeper reefs, which had abundant coverage of corals that had colonized these reefs since 1979, showed no indication of the mortality that occurred on nearshore fringing reef flats.

On nearshore reef flats, corals up to 30 cm in diameter were all dead, but skeletons were still intact, suggesting death within the last 2 years. The demarcation between high live coral coverage and dead coral skeletons was distinct, and the depth of this transition zone decreased with distance from the main watersheds along the western shore of the bay. Very little recruitment of corals had occurred on reefs in the devastated areas by August 1989.

These case studies of mortality and sublethal effects on reef organisms resulting from lowered salinity show a wide range of responses. Stresses associated with high freshwater runoff such as increased siltation and turbidity undoubtedly influence mortalities. With these complicating variables and limited salinity data, these observations do not permit rigorous determination of specific salinity tolerance limits. Such information can be determined only from controlled experimental studies of the effects of lowered salinity, and these studies have been primarily limited to reef corals.

III. REEF CORAL SALINITY TOLERANCES

Although corals usually dominate coral reefs and provide much of the aesthetic attractiveness of these structures, corals often account for only a small portion of the total biomass or energy flow of the reef. However, reef corals are important because of the structural contribution their growth provides the reef, and because of their general sensitivity to physical stress. Death of the corals usually results in death or migration of other reef organisms,[74] and replacement by less productive or diverse communities. Therefore, salinity experiments have focused on the tolerances of reef corals. However, even for this group, surprisingly little experimental work on the effects of altered salinity has been done. Most of the available information comes from studies conducted more than 50 years ago in only two geographic areas. Considering the potential importance of salinity in determining the distribution, propagation, and survival of reef corals and coral reef ecosystems, such lack of information is a serious deficiency.

A. LETHAL TOLERANCES

Vaughan[75,76] made the first experimental measurements of the effects of reduced salinity on reef corals. Seventeen species of Florida corals were exposed to 18‰ for 6 to 24 h. All species survived 6 h, and most survived 24 h, with *Siderastrea* and *Porites* the most resistant genera (Table 2). Experiments by Edmondson[15] on Hawaiian corals showed similar levels of resistance, with most species surviving 23 h at 16‰ and 49 h at 22‰ (Table 2). Some genera proved to be highly resistant to reduced salinity. *Stephanaria* and *Favia* survived 6 d at 16‰ and, along with *Leptastrea* and *Fungia,* 1 to 4 months at 22‰. *Favia, Leptastrea,* and *Fungia* even survived 0.5 to 3 h of exposure to totally fresh water.

By contrast, eight species of corals tested by Mayer[77] at Murray Island on the Great Barrier Reef showed somewhat less resistance to lowered salinity. Two of eight species died after only 11.5 h at 50‰ seawater (about 18‰), and six of eight died after 24 h of exposure (Table 2). The two surviving species, *Porites* and *Coeloseris,* were half killed by this treatment.

Planulae of three species of Hawaiian corals were tested by Edmondson[78] for effects of brief exposures to lowered salinity. Planulae of all three survived 30 s in freshwater, and one species recovered when exposed for 2.5 min. Two of the three species tolerated 5 min

TABLE 2
Reef Coral Survival to Experimentally Lowered and Raised Salinities

Salinity	Time	No. species Lived	No. species Killed	Genera surviving
Florida[75,76]				
18	6 h	17	0	All
18	12 h	14	2	All but *Agaricia* & *Acropora*
18	24 h	12	4	*Oculina, Orbicella, Favia, Meandra, Siderastrea, Porites*
Florida[79]				
40	12 h	11	1	All but *Acropora*
43	12 h	11	2	All but *Acropora* & *Orbicella*
48	12 h	5	6	*Eusmilia, Favia, Manicina, Diploria, Siderastrea*
50	12 h	5	8	*Favia, Manicina, Diploria, Siderastrea*
55	12 h	5	8	*Favia, Manicina, Diploria, Siderastrea*
70	12 h	0	11	None
Great Barrier Reef, Australia[77]				
18	4.5 h	8	0	All
18	11.5 h	6	2	*Pocillopora, Goniastrea Coeloseris, Porites*
18	24 h	2	6	*Coeloseris, Porites*
Hawaii[15]				
0	0.5—3 h	3	18	*Favia, Leptastrea, Fungia*
16	12 h	23	0	All
16	23 h	13	10	All but *Cyphastrea*
16	39 h	4	19	*Stephanaria, Favia, Leptastrea*
16	6 d	2	21	*Stephanaria, Favia*
22	31 h	11	10	All
22	49 h	8	13	All but *Pocillopora*
22	1 week	5	16	*Montipora, Stephanaria, Favia, Leptastrea, Fungia*
22	1—4 mo.	4	17	*Stephanaria, Favia, Leptastrea, Fungia*
38	24 h	12	0	All
38	48 h	10	2	All but *Pocillopora*
38	4 d	9	2	All but *Pocillopora*
38	2 weeks	5	7	*Porites, Stephanaria, Cyphastrea, Favia, Leptastrea*
38	1—3 mo.	3	9	*Stephanaria, Cyphastrea, Favia, Leptastrea*
44	24 h	10	2	All but *Pocillopora*
44	45 h	7	5	All but *Pocillopora*
44	3—17 d	3	9	*Stephanaria, Favia, Leptastrea*
44	22 d	2	10	*Favia, Leptastrea*
52	24 h	2	10	*Stephanaria*

at one fourth normal salinity (about 9‰) and up to 2 h at one third normal salinity (about 12‰). These results suggest that the reduced salinities of storms and runoff in coral reef areas would not seriously damage the planula larval stages of corals, especially if they remained below the lowest salinities at the surface.

Edmondson[15] also exposed Hawaiian corals to elevated salinity and found them relatively sensitive compared with their responses to lowered salinity (Table 2). An increase to only 38.5‰ killed 7 of 12 species within 2 weeks. Exposure to 44‰ killed 5 of 12 species within 2 d. Only two species of *Stephanaria* were able to survive 24 h at 52‰. *Pocillopora* was the least tolerant genus and died within 48 h at 38.5‰. *Stephanaria, Favia,* and *Leptastrea*, the genera that were most tolerant to lowered salinity, also survived the highest salinities.

Wells[79] tested tolerances of Florida corals to elevated salinities, but these data are of limited usefulness because of the short exposure durations of only 12 h (Table 2). Little mortality occurred at salinities of 40 and 43‰, and 48, 50, and 55‰ exposures killed only about two thirds of the species tested. At the next highest salinity tested, all species died from 12 h of exposure to 70‰.

These four classic studies comprise most of the published experimental data on salinity tolerances of reef corals. The only recent study was that of Marcus and Thorhaug,[80] who compared salinity tolerances of Pacific *Porites compressa* in Hawaii with Atlantic *Porites porites* in Florida. Corals exposed to a salinity range of 5 to 45‰ for 20 d showed no stress symptoms in salinities from 25 to 37‰. Mortality occurred within 3 d in 5 and 45‰ and within 6 d in 10 and 15‰ for both species. Salinities of 40 and 20‰ caused release of mucus, bleaching, and mortalities of 30 to 40‰, showing these salinities to correspond to sublethal stress.

The two species from the two geographic areas responded in very similar manner to the salinity alterations. Florida *P. porites* appeared slightly more tolerant to lowered salinity and Hawaiian *P. compressa* slightly more tolerant to elevated salinities. Since the normal salinity range in Florida is somewhat higher than in Hawaii, these results contrast with what might be expected if salinity adaptation had occurred.

Recent experiments have compared the salinity response of these Hawaiian and Florida *Porites* with *Porites* sampled from a hypersaline environment in the Arabian Gulf.[81] *P. compressa,* believed to be the same species that occurs in Hawaii,[82] dominates inshore reefs in the northern area of the wetern Arabian Gulf where salinities average 40 to 41‰. Specimens of *P. compressa* were exposed to salinities from 19 to 53‰ for 20 d using the methods previously used on Hawaiian *P. compressa* and *P. porites.*

Mortality results with salinity for *Porites* from the three regions are summarized in Figure 1. The tolerance of Arabian Gulf *Porites compressa* is approximately 5‰ higher than both the 15 to 20‰ minimum tolerance and the 40 to 45‰ maximum tolerance of Hawaii and Florida *Porites*. Gulf *P. compressa* tolerated salinities down to 21‰ during the 20-d experiment, and those corals held at 23‰ died within 2 weeks following their return to 40‰. At the upper part of the range, specimens survived 20 d at up to 47‰ with little detrimental effect, with mortality occurring within 7 d at 49‰ or higher.

Salinity tolerances of Florida gorgonian soft corals were tested by Goldberg[83] with 24-h exposures. Gorgonians tolerated a range of 25 to 28.5‰ at the low end to highs of 41 to 43.5‰. This group, thus, appears to have an upper salinity tolerance similar to hard corals, but is less tolerant of salinity reduction.

The experimental results on salinity tolerances of corals substantiate the limited field data that suggest that 15‰ is about the lowest medium to long-term (days to weeks) salinity tolerable by most coral reef organisms. Although euryhaline species can survive lower salinities and longer exposures, most corals succumb if salinities remain below 15‰ for more than a day or two.

The experimental results for elevated salinity suggest the upper salinity tolerance of reef corals throughout most of the world to be around 40 to 41‰, or about 5‰ above normal ambient salinity. However, upper tolerance can apparently be raised at least 5‰ through adaptation by prolonged exposure to elevated salinities, as has apparently occurred for Arabian Gulf *P. compressa.* In this case adaptation to a higher salinity tolerance has accordingly raised the lower tolerance limit by an equivalent 5‰.

B. SUBLETHAL EFFECTS

Simple marine organisms such as coelenterates and sponges have few or no mechanisms for osmoregulation when salinities change abruptly. Therefore, deviation of salinity from a normal range imposes osmoregulatory stresses that may affect other physiological processes and an organisms's overall metabolism. In the short term this stressed metabolism is measurable as change in respiration from the organism's normal resting rate. In the long term, decreased potential for growth, reproduction, and survival may result from continued sublethal stress from altered salinity or any other important environmental factor.

In reef corals the symbiotic relationship with zooxanthellae provides a convenient mea-

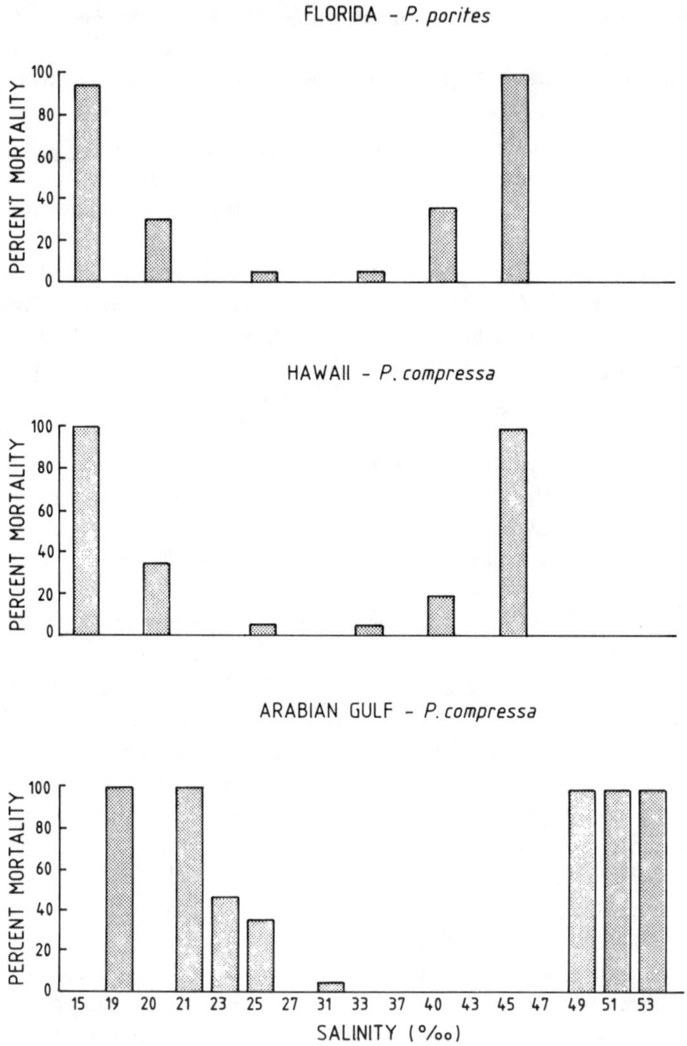

FIGURE 1. Mortality of Florida *Porites porites,* Hawaiian *P. compressa,* and Arabian Gulf *P. compressa* to salinities indicated. Vertical bars show the percentage mortality of ten specimens in each salinity after 20-d exposure. Florida and Hawaii data from Ref. 80, Arabian Gulf data from Ref. 81.

sure of the level of several environmental stresses, including salinity. The relationship between corals and their symbionts is tenuous. Ejection of the zooxanthellae is usually the first symptom of sublethal disturbance that, if continued, will lead to death of the host. The general condition of a coral under sublethal stress can therefore be determined by observing its pigmentation in relation to a normal well-pigmented coral. More sophisticated measurements such as areal zooxanthellar concentrations, cell counts, photosynthesis, respiration, and calcification can provide more complete diagnoses of the condition of a stress coral.

Such techniques have been utilized extensively to measure effects of other physical stresses, expecially temperature.[84-89] However, little information is available on the effects of sublethal salinity stress on reef corals or other coral reef organisms. Coffroth[90] found increased mucus formation for *P. furcata* exposed to 27‰, while this salinity did not significantly affect mucus production by *P. asteroides*. Hoegh-Guldberg and Smith[91] reported salinity reductions to 30‰ to have no effect on the biomass and physiological processes of *Stylophora pistillata* and *Seriatopora hystix*.

The only published study of sublethal salinity change on the metabolism of a reef coral is by Muthiga and Szmant[92] on *Siderastrea siderea*, one of the Florida species found most resistant to lowered salinity by Vaughan.[75] The normal salinity environment of this species is 28 to 30‰, but salinities can drop to as low as 5 to 10‰ during brief runoff periods. Muthiga and Szmant[92] measured the photosynthesis and respiration of corals held up to 6 d in salinities ranging from 16 to 42‰ and compared results with controls held at 30‰. Corals also were acclimated to 42‰ over a month's time, and their photosynthetic and respiration rates compared with controls.

Salinity reductions of 5 to 10‰ did not result in significant changes in photosynthesis or respiration, but a reduction of 14‰ from 30‰, to 16‰, produced significant decreases in both parameters. An increase of 10‰ above 32‰, to 42‰, caused no significant change in respiration but decreased photosynthesis significantly. A 14‰ increase above 28‰, to 42‰, caused significant decreases in both photosynthesis and respiration. Photosynthetic and respiration rates of corals acclimated to 42‰ were similar to controls, and these rates did not change significantly when salinity was restored to 35‰. A more pronounced salinity drop of 20‰ from 42‰, to 22‰, caused the most extreme reduction of photosynthesis and respiration of these experiments.

Since both photosynthesis and respiration decreased on either side of normal environmental salinity, the question remains of the net effect of salinity change on the coral-symbiont total metabolism. Both net photosynthesis (P_N) and respiration (R) are easily measured as oxygen changes in an aquatic medium in light and dark. Total or gross photosynthesis (P_G) during a light period can be estimated by summing P_N and R, assuming equivalent light and dark respiration. The P_G to R ratio is a dimensionless estimate of the autotrophic capability of an organism. P_G to R can be compared among test organisms and used as a short-term test to evaluate the effects of change in a physical parameter on a photosynthetic organism's net metabolism.[87]

Figure 2 shows P_G to R ratios calculated from the photosynthesis and respiration data of Muthiga and Szmant[92] plotted against salinity, along with P_G to R data for salinity experiments on *Porites compressa* done in Hawaii.[81] Salinity exposures for *Siderastrea* lasted from 17 h to 6 d, while photosynthesis and respiration were measured for *Porites* within 3 h of introduction to the experimental salinity. However, similar results were obtained, with P_G to R ratio decreasing above or below the normal salinity of 28 to 30‰ for *Siderastrea* or 35‰ for *Porites*.

These results imply that extended exposure to salinities outside the normal range reduce the net productivity of the coral-algal complex, and this may lead to decreased vitality of the coral. Assuming a 12-h light-dark cycle, a P_G to R ratio of 2.0 or more is necessary to maintain an autotrophic condition. In Figure 2, the P_G to R ratio falls below 2.0 only for *Porites* at 15‰, a salinity which produced mortality in 6 d.[80] However, the P_G to R ratio is still above 2.0 at 44‰, while 45‰ killed this species within 3 d.[80] Therefore, osmoregulatory problems may be more serious than reduced productivity at high salinities. The P_G to R ratio remains above 2.0 throughout the salinity range tested for *Siderastrea,* suggesting high autotrophic capability for this salinity tolerant species across a wide range of exposures.

An environmental stress such as altered salinity will seldom act independently of other physical factors which can have additive or synergistic detrimental effects. Despite the potential importance of such multiple effects, very little experimental data are available which analyze the interactive effects of salinity with other physical factors. The only such study was of the synergistic effects of temperature and salinity and light on a Hawaiian reef coral.[89] Mortality of temperatures above 30°C to *Montipora verrucosa* was aggravated by salinities below 30‰ and by holding corals in direct sunlight. By contrast, corals held in 40‰ survived slightly longer than those held at normal salinity. These results suggest that

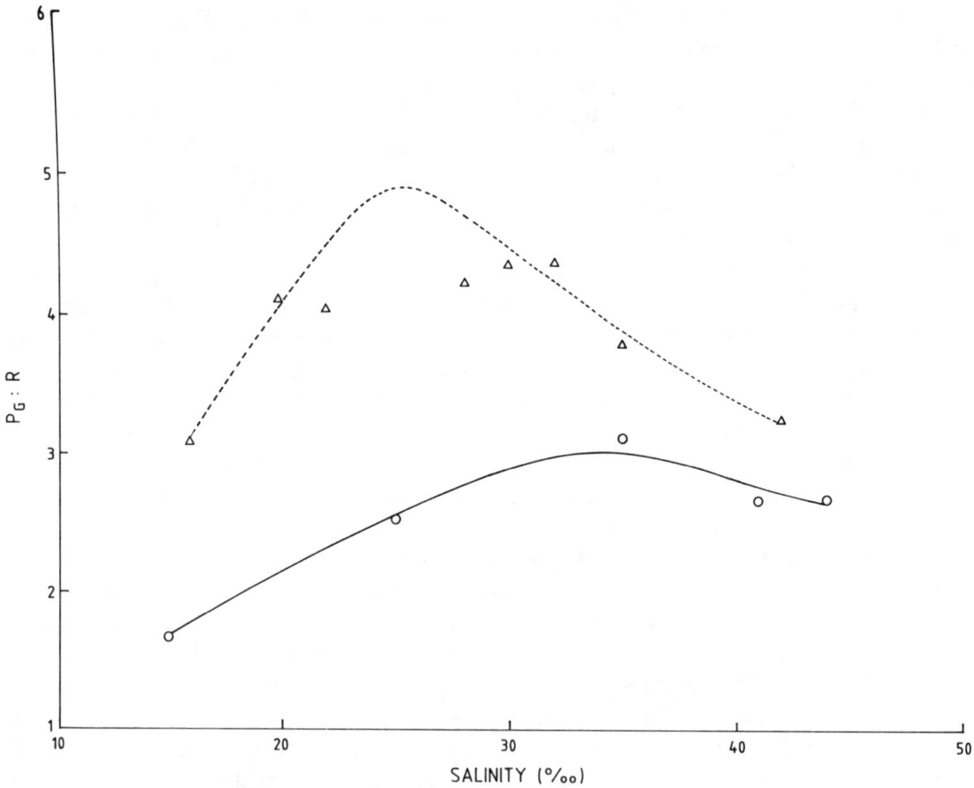

FIGURE 2. Gross photosynthesis to respiration (P_G:R) ratios with salinity for *Siderastrea siderea* in Florida[92] (---) and of *Porites compressa* in Hawaii[81] (——). Measurements of *S. Siderea* were made 17 h to 6 d after introduction to the indicated salinity; *P. compressa* measurements were made within 3 h of exposure.

slight elevations of salinity might in some cases impart increased resistance to other physical stresses, contrary to the detrimental effects usually expected.

IV. ECOSYSTEM EFFECTS

The general effects of salinity on coral reefs have been discussed in previous monographs on tropical marine pollution and ecology of coral reefs.[2,74,93] However, attention to salinity in these discussions has been brief, limited to acute tolerances of reef corals, or descriptions of specific instances of salinity-related disturbances of reef systems. Few studies have been made of the effects of salinity change on an entire coral reef community independent of other physical stresses. The most complete information was obtained by Kohout and Kolopinsky[54] on the effects of groundwater discharge on the benthic and fish communities in Biscayne Bay, Florida, where salinities ranged 5 to 30‰ at the shoreline to about 32‰ 16,000 ft from shore. Two distinct communities were found. Nearshore, comparatively few euryhaline or brackish water species occurred in high abundances to about 400 ft from the shoreline. Beyond this distance stenohaline species common to coral reefs occurred in moderate numbers. Two indicator organisms, the coral *Siderastrea siderea* and the sponge *Sphecospongia vesparia,* and two species of reef gastropods only occurred at distances greater than 700 ft from the shoreline. The general pattern showed salinity to be the most important physical factor affecting the distribution of the fauna, with decreasing numbers of species with reduced salinity and increased abundances of those forms adapted to low and fluctuating salinity conditions. Seasonal restriction of benthic organisms by annual fluctuations in salinity from >0 to 30‰ in Lagos Lagoon, Nigeria has also been shown.[94,95]

Studies of benthic communities in areas where salinities near or exceed the upper limits to coral reef development also show restriction of stenohaline coral reef species. In the Arabian Gulf where localized conditions may produce salinities well above 45‰,[96] many taxonomic groups of coral reef organisms found elsewhere in the region do not occur in hypersaline environments. Evans et al.[97] and Clarke and Keij[43] found restriction of major invertebrate groups, including hard and soft corals, melebesoid and other red algae, articulated brachiopods, and many species of mollusks and echinoderms from hypersaline areas in the Abu Dhabi and Gulf of Salwah locations in the Arabian Gulf. Quantitative studies of soft bottom benthos in sand and seagrass beds along the entire Gulf coast of Saudi Arabia by Coles and McCain,[41] showed systematic reductions in both numbers of species and numbers of individuals as mean salinity increased from 40 to 56‰. Echinoids, phoronids, penaeids, carideans, and halacarideans were absent from areas where salinity exceeded 45‰, and more than half of the species in many remaining groups did not occur above this salinity. Hypersalinity was found to be the most important environmental factor affecting the benthos, with salinity effects clearly exceeding any other natural trend or pollution related impact.

Under extreme circumstances of isolation of atoll lagoons from open ocean water, highly hypersaline conditions result[19,98,99] and the reef community becomes simplified to a few adapted organisms. In hypersaline lagoons on Laysan Atoll and Kauai, Hawaii where salinities reach up to 130‰, the biota are limited to *Dunaliella* mats supporting dense populations of *Artemia,* and dipteran insect larvae.[99] In saline lakes adjoining Christmas Island lagoon where salinities can reach 300‰, no marine macroorganisms occur above 53‰.[19]

These examples demonstrate a trend of exclusion of many coral reef species and simplication of the coral reef community under continuous conditions of either high or low salinity stress. However, on most coral reefs salinity stress occurs only infrequently, especially in the case of low salinity, and seldom exerts its effect alone. When drastic changes in salinity on coral reefs occur, ecosystem effects seldom if ever can be attributed solely to salinity changes. Other factors associated with intrusion of fresh water onto coral reefs may produce both short- and long-term changes in the physical environment which may be even more important than the effects of salinity change itself.

Most important of these factors are changes in sedimentation, turbidity, and nutrient loading that are likely to accompany salinity changes from land runoff, or increased nutrients associated with groundwater intrusion. Virtually all the cases of salinity reduction from high rainfall and storm runoff described above also produced temporary increases in turbidity and sedimentaiton of varying degree. The effects of these stresses on coral reefs have been reviewed elsewhere,[74] and it can be expected that substantially increased sediment loading and reduced light resulting from increased tubidity will aggravate osmoregulatory stresses that would result from salinity reduction alone. The combined stresses of these factors either kill coral reef organisms outright if the disturbance is severe enough or result in reduced viability of sensitive organisms manifested by such symptoms as coral bleaching. Motile organisms may leave the area until more optimal conditions of subsurface light, sediment load, and salinity return.

Increased nutrient concentrations are another secondary result of storm runoff or groundwater discharge which may correlate strongly with salinity decrease.[50,52,100,101] The effects of increased nutrients may act over a longer term and be more serious than the effects of salinity change alone. In the case of groundwater intrusion, salinity on coral reefs is altered without the complications of increased sedimentation and turbidity. However, inorganic nutrient concentrations in receiving water may be substantially enriched by the highly elevated concentrations of inorganic nutrients in groundwater. Johannes[51] reported nitrate from submarine groundwater discharge to a coastal lagoon in the Perth region to be several times as much as nitrate from land runoff and proposed the significance of groundwater sources of nitrate to marine ecological systems. Substantial groundwater intrusion has been reported in coral reefs or areas of coral growth in Guam,[102] Hawaii,[50] Jamaica,[52] Barbados,[53] and Reunion Island.[101] All of these areas showed significant elevations of nitrate levels in receiving waters.

An increased input of inorganic nutrients associated with decreased salinity into a coral reef represents a potential source of eutrophication and degradation of the system from its pristine state. However, most reports indicate uncontaminated submarine groundwater to be primarily increased in nitrate, with little increase in reactive phosphate.[52,53,100,102] Phytoplankton and benthic algae, therefore, appear still to be limited by phosphate in these circumstances. However, when phosphate is made available through terrestrial runoff[102] or contaminated groundwater,[101] algae blooms and severe change in the dominant organisms of the reef can occur. Cuet et al.[101] report a dramatic shift from a coral-dominated community to benthic and filamentous algae monopolizing the reef at Reunion Island as a result of intrusion of submarine groundwater high in both nitrate and phosphate from sewage contamination. Other significant changes in the reef ecosystem associated with eutrophication were a shift to detritus food webs as indicated by increased abundances of sponges and anemones and the regeneration and retention of nutrients on the coral reef from abundant decaying macroalgae.

These examples demonstrate that salinity alteration rarely acts alone in producing changes in a coral reef ecosystem. Especially in the case of salinity decreases, associated factors may act synergistically with the osmoregulatory stress of salinity to produce an end result which is usually some level of temporary or long-term degradation of the reef system. Much work remains to be done in identifying the factors which may interact with salinity and to determine levels of such stresses tolerable to coral reef organisms.

V. HYPER- AND HYPOSALINE DISCHARGES

Discharges into coral reef and coral growing areas that may substantially alter the salinity of receiving water and impact reef organisms are primarily of two types: sewage or industrial waste effluents, which are point sources of low salinity water independent of land-based freshwater runoff, and hypersaline effluent from the production of freshwater by desalination plants. However, as previously described for terrestrial runoff and submarine groundwater discharge, altered salinity of these effluents may be only a minor factor in producing the changes that can occur in coral reef communities when impacted by these sources of salinity change.

Sewage effluent, even when having received secondary treatment to reduce organic matter content, still contains elevated levels of reactive phosphate and nitrate that can promote eutrophication and degradation of reef communities when the effluent is disposed into an area with limited circulation with the open ocean. The best documented example of this process in the vicinity of coral reefs was in Kaneohe Bay, Hawaii. Disposal of secondary treated domestic sewage wastes into the southeast sector of the bay increased from about 2687 m^3/d (0.71 MGD [gal^6/d]) in 1963 to 11,540 m^3/d (3.05 MGD) in 1972 for a community of 30,000 people.[103] This discharge volume, although substantial for a point discharge into an embayment, was only about 20% of the net precipitation and runoff reaching the same sector of the bay.[103] Therefore, the relative effect of the sewage discharge on the salinity of the bay was minor compared with other freshwater sources. However, the treated sewage accounted for about 90% of the inorganic phosphorus and nitrogen input into the south sector of the bay by 1977.[103] This nutrient was responsible for eutrophication, increased concentrations of plankton, particulate matter and turbidity in the water, and replacement of corals and other reef organisms by suspension feeding benthic organisms and macroalgae.[65,66,74] The net result of this process was degradation of the coral reefs of the bay serious enough to warrant diversion of sewage from the bay in 1978, which reduced eutrophication and particulate levels[67] and permitted the reestablishment of the normal coral reef community within 6 years.[68-71]

Hypersaline discharges of diluted brine from desalination plants which produce freshwater from sea or brackish water are another source of salinity disturbance which have grown considerably in the last 20 years. Total worldwide capacity for freshwater production by desalination grew from 0.935×10^6 m^3/d (247 MGD) in 1970[104] to 9.92×10^6 m^3 (2621 MGD) in 1985[105] with most of the growth occurring in the Middle East and 60% of the total desalination capacity of the world located on the Arabian Peninsula. In Saudi Arabia alone freshwater production capacity is presently 2.31×10^6 m^3/d (611 MGD), with 61% of this capacity located on the Arabian Gulf and the remainder on the Red Sea.[106]

Most desalination plants produce freshwater by flash distillation, and effluent from these plants consists of high salinity blowdown diluted by cooling water by a 6 or 9.5 ratio, depending on the brine temperature. The total effluent produced by the 2.31×10^6 m^3/d (611 MGD). freshwater production in Saudi Arabia is, for example, around 24.4×10^6 m^3/d (6457 MGD). Total worldwide disposal of desalination effluent based on 1985 estimates[105] was, therefore, approximately 105×10^6 m^3/d (27,680 MGD). Brines produced by the distillation process range up to 70‰ and 45°C, but these temperatures and salinities are diluted with water used in once-through cooling of plant condensers.

Unfortunately, little published information is available on the marine impacts of effluents from specific desalination plants, especially in coral reef areas. However, the thorough study made by Chesher[104,107] of the environmental effects of a desalination plant at Key West, Florida provides substantial information on a variety of effects on a tropical marine community resulting from desalination effluent disposal. This facility discharged 8000 to 19,000 m^3/d (2.2 to 5.0 MGD) of effluent into a 9-m-deep man-made harbor connected with the ocean by an entrance canal. Studies of the effects of the effluents of the plant were made June 1970 to June 1971 and included a variety of physical, chemical, and biological properties that could be affected by the plant.

Effluent temperatures ranged 34 to 42°C and salinities 45 to 61‰, compared with ambient conditions of 22 to 30°C and 35 to 38‰. However, due to turbulent mixing at the discharge with receiving water, effluent rapidly formed a distinct 1.5-m-thick density layer with temperature and salinity increased only 0.3 to 0.5°C and 0.2 to 0.5‰. This layer persisted at a depth of about 5.5 m throughout the harbor and beyond the entrance canal. The direct effects of the minor increases in salinity and temperature of this stratified layer on biota were innocuous. However, the stratification inhibited further mixing with receiving water and continually exposed organisms in the harbor to elevated concentrations of heavy metals, particularly copper, that had been leached from the operating system of the plant. During plant operation concentrations of copper in effluent were often 5 to 10 times ambient levels, which elevated concentrations in sediments and exceeded bioassay-determined tolerances of local organisms.

Continuous exposure to this copper-rich effluent was toxic to most of the benthic organisms in the habor, which all vanished in 15 months of observations. Because of the effluent stratification, organisms living at shallow depths were not affected by effluent released during regular plant operations. However, plant startups after periodic shutdowns released effluents low in temperature and salinity but high in copper, nickel, and iron from corrosion that had built up during plant downtime. Copper concentrations during these releases reached as high as 6500 ppb. Since this effluent was low in density, it spread over shallow areas in the receiving zone, killing shallow water organisms that had been missed by operational effluent. The only organisms that were able to escape the toxic effects of the high metal concentrations from the plant were those such as barnacles and serpulid worms that could isolate themselves by closing their operculae, or fish which could migrate from the area during periods of high toxicity.

These case studies demonstrate that the salinity changes of hypo- or hypersaline discharges from man's activities are likely to be much less important in their impact on coral reefs than other factors that are associated with the discharged effluents. Industrial sewage wastes may contain elevated levels of toxic heavy metals in addition to the nutrient loads remaining from treatment of domestic wastes. The potential effects of metal contamination are, therefore, not limited to hypersaline discharges. Moreover, detrimental effects of metals are likely to be synergistically aggravated by decreased or increased salinity where an organism is living near its upper or lower tolerance limits. This effect has been clearly demonstrated by cadmium and copper with reduced salinity[108] and is likely to apply for increased salinities as well.

VI. CONCLUSIONS

Although reef corals and coral reefs are often considered to be restricted to a narrow salinity range, they exist in normal salinities ranging from 25 to 42‰. Vigorous reef development occurs thoughout this range of continuously occurring salinities. In hypersaline areas of the Red Sea and the Arabian Gulf, reefs flourish in up to 42‰, with a few species of reef corals surviving salinities up to 50‰ in the Gulf. Limited experimental data suggest that adaptation of reef corals to hypersaline conditions in the Arabian Gulf has elevated tolerances about 5‰ at both ends of their tolerance range. However, the numbers of species of corals and other reef organisms able to survive the most hypersaline conditions are relatively few and decrease with increasing salinity.

At the lower end of the salinity tolerance range, many examples exist of lethal and sublethal effects of lowered salinity on coral reefs which have occurred from storm events throughout the world. The minimum tolerable salinity suggested by these events is around 20‰ for exposures of no more than a day. However, few actual measurements of salinity during these events have been made, and results are complicated by other storm-related stresses such as sedimentation, turbidity, and anoxia. Therefore, our estimates of lower salinity tolerances of coral reef organisms rely heavily on the limited experimental data for reef corals. Lower tolerances of 15 to 20‰ are indicated for most corals, and these limits correspond with observations of reef organism mortality following storm-related events.

Salinity increases or decrease within lethal tolerances can produce sublethal changes in coral reef metabolism. However, available information is also principally limited to experiments on reef corals. Symptoms of salinity stress are mucus release, loss of zooxanthellae and pigment leading to energetic imbalance, and reduced growth and reproductive potential. The energetic indicators of photosynthesis and respiration suggest decreased net energy to be available above and below salinity optima which correspond to normal ambient values.

Simplication of reef populations and reduced diversity of coral reef communities have resulted from continuous reduction of salinities from groundwater intrusion. However, the ecosystem effects of altered salinity alone are usually difficult or impossible to measure under natural conditions, since large changes in the salinity environment on coral reefs seldom occur without similar levels of change in other important environmental variables. During massive storms, excessive land runoff elevates turbidity, sedimentation, and nutrient concentrations. All of these factors may be more responsible for reef organism mortality or changes in dominant reef organisms than the accompanying salinity change. Groundwater intrusion in some coral reef areas may alter the salinity environment substantially, but the input of high concentrations of inorganic nutrients from groundwater is a potentially important source of eutrophication that may dramatically alter the quality and composition of the reef biota.

The direct effects of salinity change resulting from man-related discharges into coral reef locations appear to be far less important than the stresses resulting from other factors associated with the discharges. Whether these are increased nutrients from hyposaline sewage effluent or heavy metals in hypersaline desalination plant blowdown and cooling water, the accompanying salinity changes are relatively innocuous by comparison.

ACKNOWLEDGMENTS

We wish to acknowledge the support of the Research Institute of the King Fahd University of Petroleum and Minerals and the Hawaii Institute of Marine Biology.

REFERENCES

1. **Wells, J. W.,** Corals, *Geol. Soc. Am. Mem.,* 67, 1087, 1957.
2. **Stoddart, D. F.,** Ecology and morphology of recent coral reefs, *Biol. Rev.,* 44, 433, 1969.
3. **Dietrich, G., Kalle, K., Krauss, W., and Siedler, G.,** *General Oceanography, an Introduction,* J. Wiley & Sons, New York, 1980, 222.
4. **Tchernia, P.,** *Descriptive Regional Oceanography,* Pergamon Press, New York, 1980, chaps. 5-7.
5. **Dole, R. B.,** Some Chemical Characteristics of Sea-water at Tortugas and around Biscayne Bay, Florida, Pap. Tortugas Lab. Vol. 5, Carnegie Inst. Wash., Publ. 182, 1914, 69.
6. **Dole, R. B. and Chambers, A. A.,** Salinity of Ocean Water at Fowey Rocks, Florida, Pap. Tortugas Lab. Vol. 9, Carnegie Inst. Wash., Publ. No. 231, 1918, 299.
7. **Vaughan, T. W.,** The Temperature of the Florida Coral-Reef Tract, Pap. Tortugas Lab. Vol. 9, Carnegie Inst. Wash., Publ. 31, 1918, 321.
8. **Jones, J. A.,** Ecological studies of the southeastern Florida patch reefs. I. Diurnal and seasonal changes in the environment, *Bull. Mar. Sci.,* 13, 282, 1963.
9. **Glynn, P. W.,** Ecology of a Caribbean coral reef. The *Porites* reef-flat biotope. I. Meteorology and hydrography, *Mar. Biol.,* 20, 297, 1973.
10. **Aller, R. C. and Dodge, R. E.,** Animal-sediment relations in a tropical lagoon Discovery Bay, Jamaica, *J. Mar. Res.,* 32, 209, 1974.
11. **Schroeder, W. W.,** Current and hydrographic characterization of the south central insular shelf of Grand Bahama Island, *Coral Reef Symp.,* 2, 518, 1977.
12. **Cortes, M. and Risk, M. J.,** A reef under siltation stress: Cahuita, Costa Rica, *Bull. Mar. Sci.,* 36, 339, 1985.
13. **Ferre-D'Amare, A. R.,** Coral reefs of the Mexican Atlantic: a review, *Proc. 5th Int. Coral Reef Congr.,* 6, 349, 1985.
14. **Leao, Z. M. A. N., Arujo, T. M. F., and Nolasco, M. C.,** The coral reefs of eastern Brazil, *Proc. 6th Int. Coral Reef Symp.,* 3, 339, 1988.
15. **Edmondson, C. H.,** The ecology of a Hawaiian coral reef, *Bull. Bernice P. Bishop Mus.,* 45, 1, 1928.
16. **Bathen, K. H.,** A Descriptive Study of the Physical Oceanography of Kaneohe Bay, Hawaii Inst. Mar. Biol. Tech. Rep. 25, Oahu, HI, 1968.
17. **Kinsman, D. J. J.,** Reef coral tolerance of high temperatures and salinities, *Nature (London),* 202, 1280, 1964.
18. **Fishelson, L.,** Ecological and biological phenomena influencing coral species at Eilat (Gulf of Aqaba, Red Sea), *Mar. Biol.,* 19, 183, 1973.
19. **Helfrich, P., et al.,** The Feasibility of Brine Shrimp Production on Christmas Island, Sea Grant Tech. Rep. UNIHI-SEAGRANT-TR-73-02, University of Hawaii, Honolulu, 1973.
20. **Dana, T. F.,** Reef-coral dispersion patterns and environmental variables on a Caribbean coral reef, *Bull. Mar. Sci.,* 26, 1, 1976.
21. **Chuang, S. H.,** Ecology of Singapore and Malayan coral reefs, *Proc. 3rd Int. Coral Reef Symp.,* 1, 56, 1977.
22. **Ricard, M.,** Phytoplankton contribution to primary productivity in two coral reef areas of Fiji Islands and French Polynesia, Proc. 3rd Int. Coral Reef Symp., 1, 344, 1977.
23. **Wolanski, E.,** Aspects of physical oceanography of the Great Barrier Reef lagoon, *Proc. 4th Int. Coral Reef Symp.,* 1, 375, 1981.
24. **Jokiel, P. L. and Maragos, J. E.,** Reef corals of Canton Island, *Atoll Res. Bull.,* 221, 71, 1978.
25. **Smith, S. V. and Jokiel, P. L.,** Water composition and biogeochemical gradients in the Canton Atoll lagoon, *Atoll Res. Bull.,* 221, 15, 1978.

26. **Yap, H. T. and Gomez, E. D.,** Growth of *Acropora pulchra* (Brook) in Bolinao, Pangasinan, Philippines, *Proc. 4th Int. Coral Reef Symp.*, 2, 207, 1981.
27. **Chansang, H., Boonyanate, P., and Charuchinda, M.,** Effect of sedimentation from coast mining on coral reefs on the northwest coast of Phuket Island, Thailand, *Proc. 4th Int. Coral Reef Symp.*, 1, 129, 1981.
28. **Downing, N.,** Coral reef communities in an extreme environment: the northwestern Arabian Gulf, *Proc. 5th Int. Coral Reef Congr.*, 6, 343, 1985.
29. **LeGall, J-Y. and Fesquet, J-M.,** Coral lagoon as aquaculture environment for the green marine turtle farm at Ile de Reunion (Indian Ocean), *Proc. 5th Int. Coral Reef Congr.*, 5, 481, 1985.
30. **Piyakarchana, T.,** Some ecological factors affecting the distribution of *Acanthaster planci* (L.) in the Gulf of Thailand, *Proc. 4th Int. Coral Reef Symp.*, 2, 613, 1981.
31. **Rougerie, F.,** The New Caledonian south-west lagoon: circulation, hydrological specificity and productivity, *Proc. 5th Int. Coral Reef Congr.*, 6, 17, 1985.
32. **Roberts, H. H.,** Carbonate platforms forming in a strong tidal current setting: Southern Gulf of Suez, *Proc. 5th Int. Coral Reef Congr.*, 6, 335, 1985.
33. **Coles, S. L.,** Limitations on coral reef development in the Arabian Gulf: Temperature or algal competition?, *Proc. 6th Int. Coral Reef Symp.*, 3, 211, 1988.
34. **Sheppard, C. R. C.,** Similar trends, different causes: responses of corals to stressed environments in Arabian seas, *Proc. 6th Int. Coral Reef Symp.*, 3, 297, 1988.
35. **McClanahan, T. R.,** Seasonality in East Africa's coastal waters, *Mar. Ecol. Prog. Ser.*, 44, 191, 1988.
36. **Scoffin, T. P., Tudhope, A. W., and Brown, B. E.,** Fluorescent and skeletal density banding in *Porites lutea* from Papua, New Guinea and Indonesia, *Coral Reefs,* 7, 169, 1989.
37. **Shinn, E. A.,** Coral reef recovery in Florida and the Persian Gulf, *Environ. Geol.*, 1, 241, 1976.
38. **Meshal, A. H.,** Hydrography of a hypersaline coastal lagoon in the Red Sea, *Estuar. Coast. Shelf Sci.*, 24, 167, 1987.
39. **Wells, S. M. and Sheppard, C.,** *Coral Reefs of the World. Vol. 2: Indian Ocean, Red Sea and Gulf,* IUCN Publ. Serv., Cambridge, 1988, 273.
40. KFUPM/RI, Marine Monitoring and Baseline Survey, Vol. III. Coral Reef Studies, unpublished Rep. prepared for the Saudi Arabian Meteorological Protection Administration by the King Fahd University of Petroleum and Minerals Research Institute, Dhahran, S. Arabia, 1988, 37-38.
41. **Coles, S. L. and McCain, J. C.,** Environmental factors affecting benthic infaunal communities of the western Arabian Gulf, *Mar. Environ. Res.*, 29, 289, 1990.
42. **John, V. C., Coles, S. L., and Abozed, A. I.,** Seasonal cycles of temperature, salinity and water masses of the western Arabian Gulf, *Oceanograph. Acta,* in press.
43. **Clarke, M. W. H. and Keij, A. J.,** Organisms as producers of carbonate sediment and indicators of environment in the southern Persian Gulf, in *The Persian Gulf,* Purser, B. H., Ed., Springer-Verlag, New York, 1973, 33.
44. **Smith, F. G. W.,** Sponge disease in British Honduras, and its transmission by water currents, *Ecology,* 22, 415, 1941.
45. **McMillan, C. and Moseley, F. N.,** Salinity tolerances of five marine spermatophytes of Redfish Bay, Texas, *Ecology,* 48, 503, 1967.
46. **Zieman, J. C.,** Seasonal variation of turtle grass, *Thalassia testudinum* Konig, with reference to temperature and salinity effects, *Aquat. Bot.*, 1, 107, 1975.
47. **Zieman, J. C.,** Tropical sea grass ecosystems and pollution, in *Tropical Marine Pollution,* Wood, E. J. F. and Johannes, R. E., Eds., Elsevier, Amsterdam, 1975, 68.
48. **Robblee, M. B., Tilmant, J. T., and Emerson, J.,** Quantitative observations on salinity, *Bull. Mar. Sci.*, 44, 523, 1989.
49. **Wood-Jones, F.,** *Corals and Atolls,* Lowell Reeve and Co., London, 1910, chap. 10.
50. **Bienfang, P.,** Water quality characteristics of Honokohau Harbor: a subtropical embayment affected by groundwater intrusion, *Pac. Sci.*, 34, 279, 1980.
51. **Johannes, R. E.,** The ecological significance of the submarine discharge of groundwater, *Mar. Ecol. Prog. Ser.*, 3, 365, 1980.
52. **D'Elia, C., Webb, K. L., and Porter, J. W.,** Nitrate-rich groundwater inputs to Discovery Bay, Jamaica: a significant source of N to local coral reefs?, *Bull. Mar. Sci.*, 31, 910, 1981.
53. **Lewis, J. B.,** Groundwater discharge onto coral reefs, Barbados (West Indies), *Proc. 5th Int. Coral Reef Congr.*, 6, 477, 1985.
54. **Kohout, F. A. and Kolipinsky, M. C.,** Biological zonation related to groundwater discharge along the shore of Biscayne Bay, in *Estuaries,* Lauff, G. H., Ed., Am. Assoc. Adv. Sci., Publ. 83, Washington, D.C., 1967, 488.
55. **Crossland, C.,** Notes on the ecology of reef-builders of Tahiti, *Proc. Zool. Soc. London,* 3, 717, 1928.
56. **Squires, D. F.,** Corals at the mouth of the Rewa River, Vitu Levu, Fiji, *Nature (London),* 195, 361, 1962.

57. **Orr, A. P. and Moorhouse, F. W.** Variations in physical and chemical conditions on and near Low Isles Reef, *Sci. Rep. Great Barrier Reef Exped.*, 2, 87, 1933.
58. **Slack-Smith, R. J.**, An investigation of coral deaths at Peel Island, Moreton Bay in early 1956, *Pap. Univ. Queensland Dept. Zool.*, 1, 211, 1960.
59. **Goodbody, I.**, Mass mortality of marine fauna following a tropical rain, *Ecology*, 42, 150, 1961.
60. **Glynn, P. W., Almodovar, L. R., and Gonzalez, J. G.**, Effects of Hurricane Edith on marine life in La Parguera, Puerto Rico, *Caribb. J. Sci.*, 4, 335, 1964.
61. **Goreau, T. F.**, Mass expulsion of zooxanthellae from Jamaican reef communities after Hurricane Flora, *Science*, 145, 383, 1964.
62. **Acevedo, R. and Goenaga, C.**, Note on coral bleaching after a chronic flooding in southwestern Puerto Rico, *Caribb. J. Sci.*, 22, 225, 1986.
63. **Cooper, M. J.**, Destruction of marine flora and fauna in Fiji caused by the hurricane of February, 1965, *Pac. Sci.*, 20, 137, 1966.
64. **Banner, A.**, A fresh-water "kill" on the coral reefs of Hawaii, Tech. Rep. No. 15, Hawaiian Inst. Mar. Biol. Honolulu, 1968.
65. **Banner, A. H.**, Kaneohe Bay, Hawaii: urban pollution and a coral reef ecosystem, *Proc. 2nd Int. Coral Reef Symp.*, 2, 685, 1974.
66. **Maragos, J. E.**, A Study of the Ecology of Hawaiian Reef Corals, Ph.D. thesis, University of Hawaii, Honolulu, 1972.
67. **Smith, S. V., Kimmerer, W. J., Laws, E. A., Brock, R. E., and Walsh, T. W.**, Kaneohe Bay sewage diversion experiment: perspectives on ecosystem responses to nutritional perturbation, *Pac. Sci.*, 35, 279, 1981.
68. **Maragos, J. E., Evans, C., and Holthus, P.**, Reef corals in Kaneohe Bay six years before and after termination of sewage discharges (Oahu, Hawaiian Archipelago), *Proc. 5th Int. Coral Reef Congr.*, 4, 189, 1985.
69. **Alino, P. M.**, A comparison of coral community structure on reef flats in Kaneohe Bay, Hawaii, in Coral Reef Population Biology, Jokiel, P. L., Richmond, R. H., and Rogers, R. A, Eds., Haw. Inst. Mar. Biol. Tech. Rep. No. 37, Honolulu, 1986, 91.
70. **Evans, C. W., Maragos, J. E., and Holthus, P. F.**, Reef corals in Kaneohe Bay. Six years before and after termination of sewage discharges (Oahu, Hawaiian Archipelago), in Coral Reef Population Biology, Jokiel, P. L., Richmond, R. H., and Rogers, R. A., Eds., Haw. Inst. Mar. Biol. Tech. Rep. No. 37, Honolulu, 1986, 76.
71. **Holthus, P. F., Evans, C. W., and Maragos, J. E.**, Coral reef recovery subsequent to the fresh water kill of 1965, in Coral Reef Population Biology, Jokiel, P. L., Richmond, R. H., and Rogers, R. A., Eds., Haw. Inst. Mar. Biol. Tech. Rept. No. 37, Honolulu, 1986, 66.
72. **Jokiel, P. L.**, unpublished data.
73. **Coles, S. L.**, unpublished observations.
74. **Johannes, R. E.**, Pollution and degradation of coral reef communities, in *Tropical Marine Pollution*, Wood, E. J. F. and Johannes, R. E., Eds., Elsevier, Amsterdam, 1975, chap. 2.
75. **Vaughan, T. W.**, The corals of the Bahamas and of South Florida, Carnegie Inst. Wash. Yearb. No. 13, 1914, 222.
76. **Vaughan, T. W.**, The results of investigations of the ecology of the Floridian and Bahamian shoal-water corals, *Proc. Natl. Acad. Sci., U.S.A.*, 2, 95, 1916.
77. **Mayer, A. G.**, Ecology of the Murray Island coral reef, Pap. Tortugas Lab., Vol 9., Carnegie Inst. Wash. Publ. 213, 1918, 1.
78. **Edmondson, C. H.**, Behavior of coral planulae under altered saline and thermal conditions, *B. P. Bishop Mus. Occ. Pap.*, 18, 283, 1946.
79. **Wells, J. W.**, Study of the reef corals of the Tortugas, Carnegie Inst. Wash. Yearb., 31, 290, 1932.
80. **Marcus, J. and Thorhaug, A.**, Pacific versus Atlantic responses of the subtropical hermatypic coral *Porites* spp. to temperature and salinity effects, *Proc. 4th Int. Coral Reef Symp.*, 2, 15, 1981.
81. **Coles, S. L.**, unpublished data.
82. **Coles, S. L. and Fadlallah, Y. H.**, Coral survival and mortality at low temperatures in the Arabian Gulf: new species-specific lower temperature limits, *Coral Reefs*, 9, 231, 1990.
83. **Goldberg, W.**, Ecological aspects of salinity and temperature tolerances of some reef dwelling gorgonians from Florida, *Caribb. J. Sci..*, 13, 173, 1973.
84. **Jokiel, P. L. and Coles, S. L.**, Effects of heated effluent on hermatypic corals at Kahe Point, Oahu, *Pac. Sci.*, 28, 1, 1974.
85. **Coles, S. L.**, A comparison of effects of elevated temperature versus temperature fluctuations on reef corals at Kahe Point, Oahu, *Pac. Sci.*, 29, 15, 1975.
86. **Coles, S. L., Jokiel, P. L., and Lewis, C. R.**, Thermal tolerance in tropical versus subtropical Pacific reef corals, *Pac. Sci.*, 30, 159, 1976.

87. **Coles, S. L. and Jokiel, P. L.,** Effects of temperature on photosynthesis and respiration on hermatypic corals, *Mar. Biol.*, 43, 209, 1977.
88. **Jokiel, P. L. and Coles, S. L.,** Effects of temperature on the mortality and growth of Hawaiian reef corals, *Mar. Biol.*, 43, 201, 1977.
89. **Coles, S. L., and Jokiel, P. L.,** Synergistic effects of temperature, salinity and light on the hermatypic coral *Montipora verrucosa, Mar. Biol.*, 49, 195, 1978.
90. **Coffroth, M. A.,** Mucous sheet formation on poritid corals: effects of altered salinity and sedimentation, *Proc. 5th Int. Coral Reef Congr.*, 4, 165, 1985.
91. **Hoegh-Guldberg., O. and Smith, J. S.,** The effect of sudden changes in temperature, light and salinity on the population density and export of zooxanthellae from the reef corals *Stylophora pistillata* (Esper) and *Seriatopora hystix* (Dana), *J. Exp. Mar. Biol. Ecol.*, 129, 279, 1989.
92. **Muthiga, N. A. and Szmant, A. M.,** The effects of salinity stress on the rates of aerobic respiration and photosynthesis in the hermatypic coral *Siderastrea siderea, Biol. Bull.*, 173, 539, 1987.
93. **Moore, H. B.,** Aspects of stress in the tropical marine environment, *Adv. Mar. Biol.*, 10, 217, 1972.
94. **Hill, M. B. and Webb, J. E.,** The ecology of Lagos lagoon. II. Topography and physical features of Lagos harbour and Lagos lagoon, *Phil. Trans. R. Soc. London B*, 241, 319, 1958.
95. **Webb, J. E.,** The ecology of Lagos lagoon. III. Topography and physical features of Lagos harbour and Lagos lagoon, *Phil. Trans. R. Soc. London B*, 241, 335, 1958.
96. **Basson, P. W., Burchard, J. E., Jr., Hardy, J. T., and Price, A. R. G.,** Biotopes of the western Arabian Gulf, Aramco Dept. Env. Affairs, Dhahran, 1977, chap. 10.
97. **Evans, G., Murray, J. W., Biggs, H. E. J., and Bush, P. R.,** The oceanography, ecology, sedimentology and geomorphology of parts of the Trucial Coast barrier island complex, Persian Gulf, in *The Persian Gulf*, Purser, B. H., Ed., Springer-Verlag, New York, 1973, 233.
98. **Schoonmaker, J., Tribble, G. W., Smith, S. V., and Mackenzie, F. T.,** Geochemistry of saline ponds, Kirimati (Republic of Kirimati), *Proc. 5th Int. Coral Reef Congr.*, 3, 439, 1985.
99. **Caspers, H.,** On the ecology of hypersaline lagoons on Laysan Atoll and Kauai Island, Hawaii, with special reference to the Laysan duck, *Anas laysanensis* Rothschild, *Hydrobiologia*, 82, 261, 1981.
100. **Johannes, R. E. and Hearn, C. J.,** The effect of submarine groundwater discharge on nutrient and salinity regimes in a coastal lagoon off Perth, Western Australia, *Estuar. Coast. Shelf Sci.*, 21, 739, 1985.
101. **Cuet, P., Naim, O., Faure, G., and Conan, J. Y.,** Nutrient-rich groundwater impact on benthic communities of La Saline fringing reef (Reunion Island, Indian Ocean): preliminary results, *Proc. 6th Int. Coral Reef Symp.*, 2, 207, 1988.
102. **Marsh, J. A.,** Terrestrial inputs of nitrogen and phosphorus on fringing reefs of Guam, *Proc. 3rd Int. Coral Reef Symp.*, 1, 331, 1977.
103. **Smith, S. V.,** Kaneohe Bay. A preliminary report on the responses of a coral reef ecosystem to relaxation of sewage stress, *Proc. 3rd Int. Coral Reef Symp.*, 2, 577, 1977.
104. **Chesher, R. H.,** Biological impact of a large-scale desalination plant at Key West, Florida, in *Tropical Marine Pollution*, Wood, E. J. F. and Johannes, R. E., Eds., Elsevier, Amsterdam, 1975, chap. 6.
105. **Ukayli, M. A.,** Comparative evaluation of surface water availability, water reuse and desalination in Saudi Arabia, *Water Int.*, 13, 218, 1988.
106. **Butt, F. M.,** King Fahd University of Petroleum and Minerals Research Institute, personal communication.
107. **Chesher, R. H.,** Biological Impact of a Large-Scale Desalination Plant at Key West, Rep. for U. S. EPA Proj. No. 18080GBX, U. S. Government Printing Office, Washington, D.C., 1971.
108. **McClusky, D. S., Bryant, V., and Campbell, R.,** The effects of temperature and salinity on the toxicity of heavy metals to marine and estuarine invertebrates, *Oceangr. Mar. Biol. Annu. Rev.*, 24, 481, 1986.

Section IV: Water Quality Management in Tropical Ecosystems

Chapter 7

STANDARDS AND CRITERIA FOR POLLUTION CONTROL IN CORAL REEF AREAS

Darryl W. Hawker and Des W. Connell

TABLE OF CONTENTS

I. INTRODUCTION

Coral reefs are found largely in warm water regions of the world, i.e., in the tropical belt and areas such as Southern Florida and Japan that lie in the path of warm-water currents. The greatest species diversity occurs in the central Caribbean and the region from the Philippine archipelago to Northern Australia.[1] The stony or scleractinian corals that are the major constructors of coral reefs require specific environmental conditions for optimal growth. For example, living coral reefs are rarely found in cold-water regions of the world, or even in tropical areas that are subject to cold currents, e.g., the west coast of Africa. Furthermore, corals appear to be sensitive to salinity variations. Coral reefs are not found in areas subject to freshwater runoff. The large freshwater flow from the Amazon River prevents the reefs of the Caribbean area from extending southward along the coast of South America. Fringing reefs surrounding islands often have breaks in them opposite river or stream mouths. Scleractinian corals also require adequate light for survival and growth. In general, coral growth is inversely related to the turbidity of the surrounding water. Most reef-forming corals do not grow below a depth of approximately 50 m due to attenuation of light, and they flourish best in water shallower than 30 m. For this reason, they are typically found in shallow inshore waters.

A key factor in the strict environmental requirements of corals is the symbiotic relationship between scleractinians and zooxanthellae. Zooxanthellae are tiny unicellular algae of the species *Gymnodinium microadriaticum*. They occur in large numbers embedded in the tissues of hermatypic corals which include those of the order Scleractinia.

The benefits of this symbiotic relationship to the zooxanthellae are the provisions of living space, protection, and a plentiful supply of basic nutrients in excretory materials and CO_2 from the host, which are converted into proteins, carbohydrates, and other useful substances. The corals themselves benefit by the most efficient removal of waste products and the substantial transferral of photosynthetically fixed carbon (carbohydrates) from the alga for use as a food source. Perhaps the most significant benefit is the enhancement of the calcification process and skeleton formation. Coral growth is correlated with the light requirements of contained zooxanthellae. It has been found that $CaCO_3$ secretion is much faster during daylight, and the calcium uptake from seawater is fastest on a clear, sunny day but is reduced by 50% on a cloudy day and 90% in total darkness.

Since growth and survival of coral colonies is very dependent on this symbiosis, sensitivity to environmental change, whether natural or anthropogenic, is enhanced. Disappearance of one of these organisms would result in disappearance of the other. If lack of light, due to suspended sediment, for example, caused mortality of the zooxanthellae, the coral would effectively cease to grow and ultimately die. Conversely, if coral dies, the entrained zooxanthellae would not be able to survive.[2]

The sensitivity to environmental change of reef-building corals suggests that coral reef areas would be quite susceptible to the effects of pollution. The relationship between pollutant level and effect on aquatic ecosystems is complex but has received increasing attention over recent years. Evaluation of such relationships for freshwater areas has resulted in the development of standards and criteria which can be used as guides to potential pollutant impacts. There have been few comparable investigations in marine areas and, in particular, fewer in coral reef systems.[3] With tourism and development increasing worldwide, such areas are likely to be subject to increasing pollution stresses. Preliminary standards and criteria for pollution control in coral reef areas may be developed by collating and evaluating the limited data available concerning the direct and indirect responses of corals to this type of stress.

II. DISCHARGES IN CORAL REEF AREAS

Reef-building corals are sessile organisms, adapted to the environment in which they exist. There are a number of types of discharges which may impact on and alter the ambient water characteristics and quality, ultimately affecting the survival of coral species.

Sewage discharges typically containing relatively high levels of nitrogen and phosphorus nutrient forms may result from outfalls in coral reef areas. Sewage and similar waste discharges also contain organic matter and suspended solids. The levels of these components discharged into the reefal environment depend primarily on the extent of treatment of the sewage.

Surface water runoff from rivers and streams, comprising freshwater suspended sediment and nutrients, represents a second type of discharge. For isolated oceanic atolls, runoff discharges will arise from local sources. Coral reefs are most often found in tropical areas characterized by episodic high rainfall and consequent high volume runoff events. The impact of the components of runoff discharge from relatively remote sources may become important where reefs are suitably located in the vicinity of large mainland rivers. Anthropogenic activities such as land clearing, construction work, and road and track building produces extensive erosion in coral areas of the world, contributing to the sediment load of runoff.[2]

Because of the episodic nature of rainfall and increasing resident and tourist populations in coral reef areas, desalination plants are commonly used to supplement local reservoirs. Desalination plant effluent consists of hypersaline brine together with heavy metals (Cu, Fe, Ni), chlorine, and various biocides added to control scaling and fouling of the plant and associated structures.[4] Therefore desalination plant effluent is a common type of discharge in coral reef areas.

Power station- and industrial plant-cooling water (and desalination plant effluent) is generally of much higher temperature than the receiving seawater. This heated, or hyperthermal, discharge is a pollutant when introduced into a tropical marine environment such as a coral reef where organisms generally live only a few degrees below their lethal upper temperature limit.[4]

A further type of discharge in coral reef areas is that of petroleum hydrocarbons. Regions such as the Caribbean, Southeast Asia, and the Middle East possess high levels of coral diversity, together with significant oil recovery, refining, and transport activities.[5] Discharge of petroleum hydrocarbons results from any one of these sources. On a smaller scale, spillage of petroleum hydrocarbons at fuel-handling facilities of offshore tourist development together with heavy use of power boats with outboard motors in confined areas are additional discharge sources.

Leachate from antifouling agents and paints applied to the hulls of boats and immersed marine structures and biological waste such as offal from fish cleaning may also be considered as types of discharge into coral reef areas.[6]

III. EFFECTS OF WASTE DISCHARGES IN CORAL REEF AREAS

A. SEWAGE

The detrimental effects of sewage are related to the capacity of the receiving waters to accept, dilute, and disperse the effluent. To some extent, its noxious quality depends on the degree and extent of prior treatment, but treatment can give rise to sewage sludge, which has disposal problems of its own.

There are a number of reports on the effects of sewage on coral and coral reef ecosystems from various parts of the world. Possibly the most extensive study on the effect of sewage on a coral reef community is that of Kaneohe Bay, Hawaii. It is a partially enclosed embayment, 12.7 km long and 4.3 km broad, with ocean frontage of 8.8 km on the northeastern side of the island of Oahu. Since 1939, the surrounding population has increased tenfold, and sewage discharges into the bay increased, culminating in the construction of

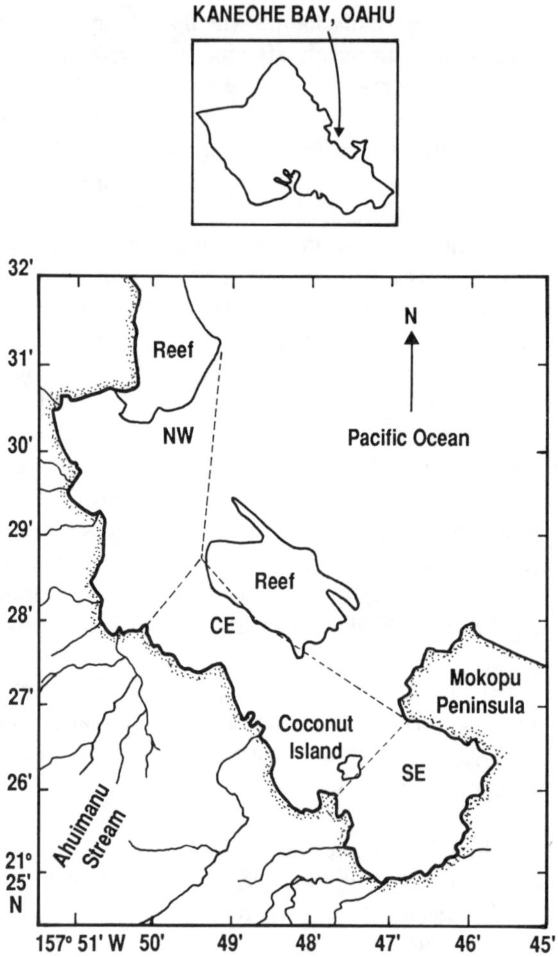

FIGURE 1. The location of Kaneohe Bay, Oahu, Hawaii. (Adapted from Kinsey, D. W., Great Barrier Reef Marine Park Authority, Townsville, Australia, 1988).

large sewage outfalls in the southeast section of the bay in 1963.[7-9] By 1977, the total sewage effluent volume totaled over 20,000 m^3 d^{-1}, with 95% being discharged into the southern section of the bay (see Figure 1). An analysis of the composition of Kaneohe Bay sewage effluent is found in Table 1.

Just after the World War II, the most abundant coral species were reported to be *Porites compressa* and *Montipora verrucosa*. Other corals of less importance included *Pocillopora damicornis, Fungia scutaria, Cyphastrea ocellina, Leptastrea bottae, Pavona varians,* and the ahermatypic *Tubastrea aurea*. Available information for Kaneohe Bay suggests that the most conspicuous effects of the sewage were in terms of increased biomass and productivity, together with altered community structure.[10] The sewage discharge was reported not to have markedly affected the pH, dissolved oxygen, or BOD in the bay away from the immediate areas of discharge. The most obvious changes to the once-living reef areas were the loss of almost all living corals from the southeast sector of the bay, and the replacement of living corals by the alga *Dictyosphaeria cavernosa* in the central section. The benthic fauna of the southeast section of the bay saw a dominance of filter and detrital feeders such as sponges, sea cucumbers, oysters, and clams depending on suspended organic material in the water.

TABLE 1
Composition of Kaneohe Bay Sewage Effluent

	mol m^{-3}	g m^{-3}
Nitrogen		
Ammonium	0.90	—
Dissolved organic	0.76	—
Particulate	0.20	—
Phosphorus		
Phosphate	0.17	—
Dissolved organic	0.04	—
Particulate	0.01	—
Silicon	0.5	—
Suspended solids	—	38
Particulate carbon	1.3	—
Biological oxygen demand	—	32
Residual chlorine	—	2.0
Trace metals		
Aluminum	—	N.D.—0.25
Cadmium	—	N.D.—0.05
Chromium	—	N.D.—0.006
Copper	—	N.D.—0.02
Iron	—	0.6—1.05
Mercury		N.D.
Nickel		N.D.—0.025
Lead		N.D.—0.03
Zinc		N.D.—0.38

From Smith, S. V., Kimmerer, W. J., Laws, E. A., Brock, R. E., and Walsh, T. W., *Pac. Sci.*, 35, 279, 1981. With permission.

In addition, the bottom sediments were fine and black, with indications of anaerobic decomposition taking place near the surface. The reef flats had massive growths of algae such as *Acanthophora, Graciliara,* and *Hydroclathrus* (but no *D. cavernosa*). In many places, the old dead heads of *Porites compressa* were still present, although covered with fine sediment, together with sponges and sea cucumbers. Sporadic outbreaks of phytoplankton concentrations knows as "red tides" also became a feature of the bay. Kinsey[2] has attributed these changes to chronic stress from nutrients and sediments, together with periodic acute stress from freshwater runoff.

In 1977, 95% of the sewage that had previously been discharged into the poorly flushed southeast sector (residence time of the order of weeks)[8] was diverted to an ocean outfall. This event presented investigators with the opportunity to study the role of sewage in the coral mortality evident and perhaps observe the recovery of the marine ecosystem. In the southern bay, water clarity had improved considerably by 1979, but changes in community structure were not marked although some of the filter feeders, particularly the sponges and barnacles, had died as a result of decreasing particulate plankton. By 1982, the previously heterotrophic reef flats had lost most of their filter feeders, and the dead substrate of the early reefs had become largely covered with an algal population dominated by red algae. By 1985, algal populations had declined, and new coral growth was evident all over the reef flats.[2]

Benthic biological composition has not yet completely returned to presewage conditions, however, partly because some key organisms are relatively long-lived and partly because the bay substratum has been perturbed by sewage input and acts as a reservoir of nutrients and organic detritus.[10]

It is noteworthy that diversion of sewage to a deep ocean outfall has had no noticeable adverse impact on the reef communities adjacent to the outfall. The ocean outfall site is exposed to strong currents, waves, and water circulation, with residence times measured on the order of hours. This rapid turnover of water prevents a buildup of nutrients, plankton biomass, and the eutrophic waters that characterized Kaneohe Bay during the period of sewage discharge.[8]

Since one of the principal stresses on a reef community as a result of sewage discharges appear to result from elevated concentrations of plankton in the water, Laws and Redalje[11] have proposed that various measures of plankton concentration such as chlorophyll-a or adenosine triphosphate (ATP) level may be more relevant toward judging the ecological impact of sewage than are inorganic nutrient concentrations.

In areas of the Red Sea near Aqaba where coral mortality due to the effects of sewage and phosphate pollution have been recorded, phosphate levels (0.96 $\mu g\ l^{-1}$) were found to be over three times greater than in control levels (0.26 $\mu g\ l^{-1}$).

Nitrate and nitrite levels were found not to be elevated. However, ammonium ion concentrations, generally larger than the former in areas of sewage influence, were not measured. The increased growth of the sea lettuce *Ulva* sp. was taken as an indication of increased NH_4^+ levels. Again, increases of nutrient concentrations and phosphate — in this case by a factor of three over background levels — resulted in severe coral mortality.[12]

On the Great Barrier Reef, fertilization (i.e., addition of nitrogen and phosphorus) of a microatoll within the lagoon of One Tree Island was carried out with a view to eventual cropping of particular trophic levels, e.g., fish. For 8 months, urea and monoammonium phosphate were added to the study area at each daytime low tide so that the concentrations of phosphorus and nitrogen would be 2 $\mu g\ l^{-1}$ (2 μM) and 20 $\mu g\ l^{-1}$ (20 μM), respectively. The microatoll was completely flushed twice a day by high tides, so the effect on the planktonic community should have been minimal. Substantial enhancement of photosynthesis was achieved, although there was negligible increase in planktonic activity. The authors speculated that this may not be the case with nutrient enrichment of the lagoon as a whole at these levels. Because replacement of this water at high tides averages less than 25%, buildup of nutrients, plankton blooms, and, hence, coral stress was thought likely.[13]

The effect of increasing phosphate concentrations in seawater on calcification has been examined.[14] Experiments combining the dye alizarin (1,2-dihydroxyanthroquinone) and ^{45}Ca indicated a high correlation, meaning that incorporation of alizarin was an accurate measure of calcification. Varying the phosphate concentration from 0.2 $\mu g\ l^{-1}$, which was considered the average level for relatively unpolluted Kaneohe Bay water, to 20 $\mu g\ l^{-1}$ caused a very significant drop in calcification rate. At 2×10^3 $\mu g\ l^{-1}$, no uptake was observed at all.

The coral skeleton is composed largely of calcium carbonate in two crystalline forms: aragonite and calcite.[2] Studies have shown that various forms of phosphorus inhibit calcium carbonate crystal formation, with increasing numbers of distorted crystals formed as phosphorus concentration increases.[15]

Based on the above investigations, Table 2 contains a summary of the physicochemical and biological effects of sewage discharges on coral reef systems.

The sequence of interactions is illustrated diagrammatically in Figure 2. The initial discharge results in an increase in the nutrient concentrations present in the ambient water. This results in increased primary production, in the form of phytoplankton, benthic algae, and blooms, and can be observed as increases in chlorophyll-a and turbidity. This material is consumed by filter and detritus feeders stimulating their population growth. The phytoplankton population causes shading of the coral, and through increased turbidity together with competition from benthic algae and toxicity due to phosphate, the coral numbers decline. In addition, sediment organic matter increases, due to deposition of phytoplankton on death, with the lowering of redox potential.[3]

TABLE 2
Summary of Physicochemical and Biological Effects of Sewage Discharges on Coral Systems

Characteristic	Change with increasing enrichment
Biomass	Increase
Primary production	Increase
Coral numbers	Decrease with *Porites compressa* among the most sensitive: numbers effective zero in extreme situations
Chlorophyll-a	Large increase
Filter and detritus feeders	Large increase
Benthic algae	Large increase
Sediments	Medium size low in organic matter to fine and high in organic matter
Sediment redox potential	High, with high dissolved oxygen in the interstitial water, to patches with low oxygen and some anaerobic areas
Water characteristics	pH, DO, BOD altered near source
Turbidity	Increase
Occurrence of blooms	Large increase

From Hawker, D. W. and Connell, D. W., *Int. J. Environ. Stud.*, 34, 179, 1989. With permission.

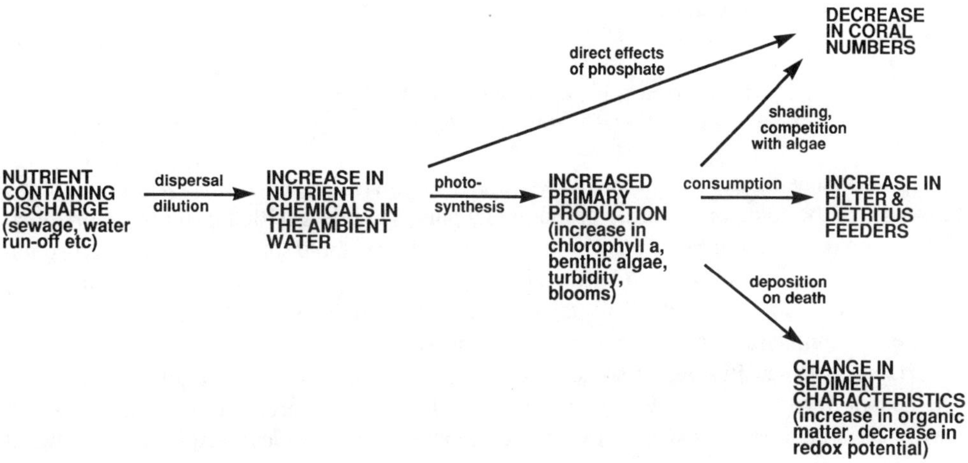

FIGURE 2. A diagrammatic illustration of the general effects of nutrient containing discharge on a coral reef system. (From Hawker, D. W. and Connell, D. W., *Int. J. Environ. Stud.*, 34, 179, 1989. Wtih permission.)

B. SEDIMENTATION

The process of sedimentation may also constitute a significant stress on coral reef ecosystems. Sediment can arise from sources such as terrestrial runoff, dredging, and sewage. Suspended solids in receiving waters for sewage discharges originate from three sources: particles contained in effluents, particulate organic matter produced by nutrient enrichment, and natural seston.

Corals reject sediment deposition on the surface of the colony by five mechanisms: passive rejection, polyp distension by ingestion of water, tentacular movements, ciliary action, and mucous production. The ability of corals to reject sediment is limited by several factors. For those species unable to coordinate transport of sediment off the colony by the shortest possible route, the sediment rejection process is more efficient in small colonies than in large ones. Second, silt or fine particulate matter (<62 μm) is the largest particle size effectively removed by many coral species. Larger size fractions, which are removed by some species but not others, are transported by polyp distension or tentacular movement rather than by relatively weak ciliary action.

TABLE 3
Relative Sensitivity of Some Coral Species to Sedimentation

Species	Sensitivity: Low	Moderate	High
Acropora cervicornis		X	
A. corymbosa			X
A. hyacinthus			X
A. palmata			X
Other *Acropora* spp.		X	
Agaricia agaricites		X	
Diploria strigosa	X		
Fungia spp.	X		
Madricis marabilis		X	
Manicinia areolata	X		
Montastrea annularis		X	
M. cavernosa	X		
Pocillopora spp.	X	X	
Porites astreoides			X
Other *Porites* spp.		X	
Siderastrea radians	X		
S. siderea	X		

From Pastorok, R. A. and Bilyard, G. R., *Mar. Ecol. Prog. Ser.*, 21, 175, 1985. With permission.

Species sensitivities to sedimentation are determined largely by the particle-trapping properties of the colony and ability of individual polyps to reject settled materials. Horizontal platelike colonies and massive growth forms present large stable surfaces for the interception and retention of settling solids. Conversely, vertical plates and upright branching forms are less likely to retain sediments. Tall polyps and convex colonies are also less susceptible to sediment accumulation than are other growth forms.

The relative sensitivities of some coral species, to sedimentation are given in Table 3. The genera *Acropora, Porites,* and *Pocillopora* illustrate the variation in sensitivity which may occur among species within a genus. In general, coral species inhabiting the seaward margins of a reef are less tolerant of high sediment loads than species found in nearshore areas.

Growth inhibition by sublethal sediment loads may be caused by decreased light availability, abrasion, and increased energy expenditure for sediment rejection. High turbidity interferes with light penetration, thereby limiting photosynthesis of zooxanthellae and, indirectly, coral growth. Rejection of particulates by corals is also energetically expensive. As energy required for sediment removal is diverted from other metabolic functions, growth and reproductive output may be reduced. Moreover, if coral polyps are occupied with sediment rejection activities, they may be unable to capture zooplankton effectively. At the population and community levels, increased sedimentation may inhibit population recruitment, cause changes in the relative abundances of coral species, reduce substrate cover by living corals, and reduce species diversity. Species richness, percent cover, and mean colony size of corals are each inversely related to sedimentation rate.[16]

Much of the particulate and dissolved organic matter discharged in sewage effluent is organic and readily decomposed by microbial activity in the water column or sediments. Because most tropical organisms, including corals, are living near their critical tolerance levels for dissolved oxygen, depressed oxygen levels caused by decomposition of sewage particulate matter may also constitute a significant stress. Depression is most critical at night when oxygen levels are usually at their daily minima.[4,17]

C. HYPO- AND HYPERSALINE DISCHARGES

A hyposaline environment can be created by discharges of relatively large volumes of nonsaline water, such as those from domestic effluent and from rainwater runoff. Hypersaline discharges are commonly associated with desalination plants.

Initial responses of corals to altered salinity states appear to be increased mucous production and/or localized bleaching due to expulsion of zooxanthellae. Reduction of salinity to 27‰ for specimens of *Porites astreoides* and *P. furcata* from the Caribbean coast of Panama caused elevated mucous production in both species. Thus, mucous sheet formation would appear to be an attempt by the coral polyps to alleviate the stress of reduced salinity.[18]

The impact of dramatic lowering of salinity has been observed in regions subject to massive terrestrial runoff following flood rains. Such an event in Jamaica in 1963 induced widespread bleaching of hermatypic corals. Salinity levels in reef areas were below 10‰ for approximately 48 h. Although fully or partially bleached, polyps continued to expand and feed normally. Some colonies were killed, however, and they became covered in a thick green slime which disintegrated into mucous strings, then sloughed off. Subsequently, the skeleton became overgrown by green filamentous algae. It was observed that gorgonians were even more susceptible to this stress than the least-tolerant hard corals, *Millepora complanata, Palythoa caribbaea,* and *Montastrea annularis.* This is consistent with studies of salinity tolerances of reef-dwelling gorgonians from Florida by Goldberg.[19] It has been suggested that expulsion of zooxanthellae is induced by contact with water of lowered osmotic pressure.[20] A similar mass expulsion of zooxanthellae from corals *(Pocillopora danae, P. damicornis,* and *Porites lobata)* on Easter Island in the southeast Pacific following torrential rain has been recorded.[21] Both here and in Jamaica, regeneration of the depleted zooxanthellae populations was very slow, with total recovery taking many months.

Overall studies show that members of the Acroporidae, Pocilloporidae, and Poritidae families may be more susceptible to osmotic (saline) stress than corals with larger polyps, such as those from Faviidae and Fungiidae families.[22]

It should be noted, however, that investigations of saline effluent effects are greatly complicated by different stress factors that can act synergistically on corals and other reefal organisms.

D. HYPO- AND HYPERTHERMAL DISCHARGES

Heated effluents released by power plants, some desalination plants, and other installations are a cause of serious stress in shallow tropical waters since it is well known that unspoiled tropical waters routinely reach summer temperatures only a few degrees below the upper thermal limits of many tropical marine organisms. Corals are among the most sensitive species involved, being killed by heated effluents at greater distances from the outfall than most other biota.[4,23]

Apart from coral mortality, it has been found that fishes, crustaceans, and echinoderms characteristic of reef flat areas in Guam disappeared when heated water from a power plant was introduced. In addition, there was considerable destruction of marine coralline red and calcareous green algal species. The areas where the natural algal communities were damaged by hyperthermal effluent became colonized by blue-green species. Benthic dwellers may also be affected since, although sediments are poorer conductors of heat than seawater, once sediments become heated, they retain it much longer. Therefore, any biological effects will be due to the duration of heating as well as the temperature attained, and such effects may be cumulative if the sediment acts as an integrator of fluctuating temperature.[24]

As for corals themselves, Coles and Jokiel[25] found that extrapolation of decreasing photosynthesis to respiration (P:R) ratios to a 24-h value of one (indicating net consumption of organic materials) coincided with the upper lethal temperature limit indicated for some subtropical Hawaiian corals (33°C). Similar extrapolation for tropical specimens from Eniwetok in the Marshall Islands suggested lethal temperatures 2 to 5°C higher. These results indicate that tropical and subtropical corals may be rigorously adapted to ambient water temperature conditions.

In contrast to the high temperature sensitivity of recently settled corals, coral planulae are relatively resistant to temperature elevation, and short-term exposure to elevated temperature may enhance planular settlement and subsequent reef development, provided subsequent thermal impingement on juvenile corals does not occur. On this basis, therefore, coral reef growth may cooccur with and even be promoted by discharge of thermal effluent when reasonable and proper design and operating conditions are met.

E. OTHER DISCHARGES

1. Hydrocarbons

Depending on the particular species and exposure time, oil directly contacting live coral tissue can cause localized tissue necrosis and death (sometimes death of the entire colony) or be lifted clear of tissue by increased mucous secretion. Hydrocarbons are soluble in seawater to a limited extent, and the action of currents and waves causes the formation of emulsions. This means that floating oil can, in fact, deliver relatively large amounts of hydrocarbons into contact with corals. Chemical dispersants exacerbate this problem, and mixtures of hydrocarbons and dispersants are often more toxic than petroleum hydrocarbons alone.[26,27] Exposure of corals to water-soluble or accommodated hydrocarbons has also produced a range of responses. Following brief exposure to high levels, corals have died[28] or been killed indirectly by the hydrocarbon pollutant stimulating mucous secretion which promoted microbial activity resulting in tissue destruction.[29,30] Short exposures have produced sublethal responses including severe planulae mortality and abnormal feeding responses.[31,32]

Where corals have been exposed to low concentrations of soluble and/or aggregated hydrocarbons for short periods, widely varying responses have been observed including short-term reduction in photosynthesis and altered calcification rates.[33] Chronic exposure to low levels has resulted in tissue necrosis, reduction in zooxanthellae concentration, reduced fertility, and, in some cases, even coral mortality.[34,35]

2. Metals

Our knowledge of the effect of heavy metals on reef corals is relatively limited. Effluents from power plants and desalination units are a source of metals in ionic form which may be up to 30 or 40 times the metal concentrations of receiving waters. Sewage discharges are another major source of metals in many chemical forms, but assessment of their influence on coral reefs is difficult in view of the blanketing effects of sedimentation and nutrient enhancement which promotes algal overgrowth of corals.

Several pathways by which heavy metals may be incorporated into corals have been described.[36,37] Soluble metals in seawater probably represent the most obvious and direct route of metal uptake available to corals, although this pathway may not be the primary contributing factor to their metal status. The feeding activities of corals may, in fact, play a major role in accumulation of metals.

Zooxanthellae may also be involved in the direct uptake of metals in situations where, for example, potentially hazardous elements are metabolically substituted for essential ones such as phosphorus. In addition, as well as being implicated in metal uptake, the zooxanthellae may themselves be adversely affected by exposure to metals, leading to coral growth inhibition.

Data concerning the effects of metals on corals are strictly limited. Available evidence suggests that chronic exposure to elevated levels of metals may result in mucous production, polyp retraction, partial bleaching, and, in extreme situation, in death. Other residents of coral communities, such as mollusks, also accumulate heavy metals. Such organisms have great potential as indicators of heavy metal pollution because they can concentrate metals to very high levels (sometimes of the order of parts per thousand in various organs) with no apparent deleterious effect, and depurate only very slowly.[38]

3. Surfactants

Surfactants (surface active agents) are present in detergents and dispersants and can have deleterious effects on marine systems. The most obvious is the formation of foam or scum around outfalls discharging waste containing surfactants. These compounds have also been shown to be directly toxic toward marine biota. Most relevant information is derived from work on oil-spill treatment by dispersion. Dispersants are generally of two types, water-based or oil-based. The components of water-based dispersants are, besides water, a solvent such as ethylene glycol or isopropyl alcohol and a surfactant. Oil-based dispersants consist basically of a petroleum fraction which is virtually water-insoluble, together with a surfactant.[39]

Surfactants probably act on biological membranes, disrupting them at the cellular and subcellular level, and also disturbing the function of some important enzymes. Importantly, toxicity has been found to increase with temperature, making tropical organisms particularly vulnerable.[39]

A study of the effects of both water and oil-based dispersants on fish, crustaceans, and bivalve mollusks indicated the importance of the chemical and physical nature of the outer layer of the body surface. Corexit 7664, a water-based dispersant, was most toxic toward fish and least toxic toward crustaceans, while this toxicity order was reversed for oil-based surfactants. The high resistance to water-based dispersants of the crustaceans in comparison with fish and mollusks has been attributed to their hydrophobic cuticle which is usually covered with a lipophilic waxy layer. Hydra, which is a coelenterate and, therefore, of the same phylum as corals, are particularly sensitive toward surfactants. They have no protective outer layers of any consequence, and nonionic, anionic, and amphoteric surfactant varieties all cause serious cellular disruption. It might be anticipated, then, that reef corals, with an exposed epidermal layer of tissue, would be similarly vulnerable.

Laboratory experiments have shown progressive bleaching and gradual deterioration of coral specimens on exposure to surfactants. A delayed stress response has also been noted as some coral polyps have died several months after exposure, while the remaining tissue gradually recovered.[22]

4. Pesticides and Herbicides

Detectable levels of herbicide residues, most notably 2,4-D (2,4-dichlorophenoxyacetic acid) and 2,4,5-T (2,4,5-trichlorophenoxyacetic acid) have been found in the tissues of stressed and dying coral colonies on the Pacific coast of Panama. Laboratory-based studies on the scleractinian coral *Pocillopora damicornis* from Hawaii have shown that exposure to nominal levels of 2,4-D of 10 ppm or greater produced 100% mortality within 48 h following extensive tissue loss. The effects of lower nominal concentrations were variable, but planulae were released by all specimens exposed to herbicide.

The variable response of *P. damicornis* was suggested to be due not only to variable herbicide concentrations, but also to the reproductive state of the coral. *P. damicornis* has a monthly planulation cycle, and changes associated with that cycle may influence coral response to the herbicide mixture.[40]

5. Chlorine

Chlorine is commonly used as a disinfectant for sewage water and an antifouling agent for power-generating and desalination plant-cooling water systems. The chemistry of chlorine in seawater is complex and still poorly understood, and the effect of free residual chlorine on many marine organisms is unclear.

Static, controlled temperature bioassays with NaOCl as the chlorine source have shown toxic responses from larvae of the sea urchin *Echinometra mathael* and the gastropod *Stylocheilus longicauda,* together with two species of tropical phytoplankton *(Dunaliella tertiolecta* and *Chaetoceros gracilis),* with LC_{50}'s in the part per million range. Preliminary experiments with planulae of Hawaiian corals have shown that exposure to 0.49 ppm chlorine for several hours was not lethal.[16]

6. Antifouling Agents

Although chlorine is often used as an antifouling agent in some desalination and power plants, antifouling paints are also used on these facilities, as well as jetties, pylons, and boat hulls. Copper and organotin compounds are the major active ingredients in antifoulant formulations. Organotin paints are generally more toxic to marine life than copper-based ones.

Bis(tri-*n*-butyl)tin oxide or TBTO was found to have a 96-h LC_{50} of 1 ppm with some marine copepods. Sublethal effects were apparent at levels as low as 0.3 ppb. As a comparison, copper was found to be ten times less toxic with this organism.[41]

Little comparable data are available for tropical marine ecosystems, and reefal areas in particular, but it is probable that antifouling agents are also hazardous to biota of these systems.

IV. TOLERANCE LEVELS OF POLLUTANTS IN CORAL REEF AREAS

A. NUTRIENTS

The range of nutrient levels within which coral reefs occur is quite broad, and, therefore, they are not oases in the marine deserts of the world as they are often represented.[2]

The responses of coral communities to natural variations in nutrient loading provide an interesting comparison with enrichment effects from anthropogenic sources. Periodic upwelling of deep oceanic water, rich in nutrients, enables fouling organisms such as filamentous algae, bryozoans, and tunicates to colonize open areas rapidly and overgrow most coral recruits. As nutrient levels decrease, the light levels increase, the rate of biomass accumulation on benthic substrates declines, and hermatypic corals have a better chance of reaching a size large enough to avoid being completely overgrown. Thus, moderate anthropogenic inputs may mimic nutrient enrichment from natural upwelling by promoting rapid growth of benthic organisms, high biomass of filamentous algae, low diversity of corals, and domination by benthic filter-feeders. The relative importance of direct and indirect effects of nutrient enrichment on coral reefs probably varies with reef type and reef trophic status. In nutrient-poor regions, anthropogenic nutrient inputs may cause profound shifts in community structure by reducing the importance of corals and increasing the importance of selected filter feeders. In areas influenced by upwelling, such as outer-shelf and midshelf reefs, moderate anthropogenic inputs may be less likely to cause dramatic changes, since reef biota are already adapted to nutrient perturbations.[16]

Based on studies from Kaneohe Bay, the Gulf of Aqaba, and the Great Barrier Reef, it is likely that total phosphorus levels elevated to two or three times the normal ambient levels can cause increased primary production and biomass in both phytoplankton and benthic algal populations, affecting coral nutrition, growth, and, ultimately, survival. Enhancement of nutrient levels by a factor of three or more would appear to constitute a major stress on coral reef communities. Therefore, elevated nutrient (particularly total phosphorus) concentrations can be used as an indication of potential detrimental effects of a discharge, and as a preliminary estimate, levels should not exceed three times the normal ambient levels.

To assess quantitatively the effects of nutrient enrichment on hermatypic reef-building

TABLE 4
Linear Regression Equations between Transformed Average Coral Growth Rates ($Y\ cm\ yr^{-1}$) and Environmental Variables Where All the Raw Data Are Transformed by (x + 1)

Suspended particulate matter (mg/l)	
$\log Y = -0.638 \log SPM + 0.760$	$r^2 = 0.79$
Volatile particulate matter (mg/l)	
$\log Y = -0.340 \log VPM + 1.670$	$r^2 = 0.79$
Chlorophyll-a ($mg\ m^{-3}$)	
$\log Y = -0.863 \log CHL + 0.452$	$r^2 = 0.75$
BOD (mg/l)	
$\log Y = -1.368 \log BOD + 0.611$	$r^2 = 0.72$
Sediment organic content (%)	
$\log Y = -0.169 \log ORG + 0.367$	$r^2 = 0.63$
Surface illumination (%)	
$\log Y = 0.619 \log ILL - 0.701$	$r^2 = 0.56$
Reactive phosphate (μg at/l)	
$\log Y = -1.940 \log PO_4 + 0.335$	$r^2 = 0.51$

Note: Ammonium, nitrate, and nitrite, temperature, salinity, and current velocity all had $r^2 < 0.48$.

From Tomascik, T. and Sander, F., *Mar. Biol.*, 87, 143, 1985. With permission.

corals, 14 environmental variables were monitored along a transect of seven locations off the west coast of Barbados.[42] The physicochemical and biological data indicate that a gradient of effects exists related to distance away from the primary discharge source. Growth rates of *Montastrea annularis,* a principal reef-builder measured along this gradient, exhibited a high correlation with a number of water-quality variables. Among the inorganic nutrients, phosphate showed the strongest negative relationship with growth, followed by ammonium, and nitrate plus nitrite concentrations. Table 4 contains linear regressions between the logarithms of transformed average coral growth rates and environment variables, where all the raw data are transformed by (x + 1).[3]

It is somewhat arbitrary as to what decrease in growth rate constitutes an unacceptable stress, but a 20% decrease is sometimes taken as a threshold level with environmental variables. When this growth decrease is combined with the relevant equations for phosphorus (as phosphate) and nitrogen (as NH_4^+ and $NO_2^- + NO_3^-$) from Table 4, percentage increases over ambient concentrations of 23, 285 and 285%, respectively, may be derived. It is emphasized that water quality parameters, including nutrient levels, may vary naturally by larger amounts over specific time spans. The calculated percentage increases represent increases over *long-term average background* levels. The factor increases in nutrient concentration for 90, 50, and 10% decreases in coral growth may be calculated by substitution into the appropriate equations from Table 4 and are found in Table 5.

Of the nutrients, coral is clearly most sensitive to phosphate and a 90% decrease in growth rate, probably effective death, results from a two- to threefold increase in reactive phosphate concentration. This calculated increase is in very reasonable agreement with experimental observations from both Hawaii and the Red Sea, as reported previously.

In order to develop a quantitative estimate of tolerance levels for a particular coral reef area, local long-term average background levels are required. Using the Great Barrier Reef as a general example, for inorganic nutrients, average phosphate concentrations in local reef areas appear to be approximately 0.2 $\mu g\ l^{-1}$,[43,44] while NH_4^+ and $NO_3^- + NO_2^-$ con-

TABLE 5
Factor Increases over Ambient for Various Proportions of Growth Inhibition with Some Water Quality Parameters[a]

Water quality parameter	% growth decrease 90	50	10
Suspended particulate matter	× 4.23	× 1.94	× 1.13
Chlorophyll-a	× 4.88	× 2.48	× 1.22
Reactive phosphate concentration	× 2.25	× 1.61	× 1.11

[a] Derived from the equations in Table 4.

From Hawker, D. W. and Connell, D. W., *Int. J. Environ. Stud.*, 34, 179, 1989. With permission.

centrations average 0.17 and 0.34 μg l^{-1}, respectively. Tolerance limits, then, are 0.25, 0.65, and 1.31 μg l^{-1}. From these data, nitrogen levels are relatively poor indicators of eutrophication, as also found by Laws and Redalje.[11]

As a caveat, it should be remembered that these calculated tolerance limits are derived using data from one Barbadian coral genus, an arbitrary stress limit, and limited local water quality information. Improved limit definition could be obtained by more extensive biological, chemical, and physical investigations of local waters. The calculated tolerance levels must be regarded as preliminary since they are based on relationships established in one particular reefal area, where similar but different environmental and biological factors exist, to other reefal areas.

B. SUSPENDED SEDIMENT

Studies have revealed that gorgonians (soft corals, sea whips, sea fans) are among the most tolerant of the reef macrofauna to sediment loading and turbidity because the flexible branches and erect growth forms prevent sediment accumulation. Scleractinian (hard) corals tolerate short-term (few days' duration) sediment loading, but prolonged exposure to siltation and high turbidity results in loss of zooxanthellae, polyp swelling, or abnormal rates of mucous secretion. Observations suggest that these corals may be more tolerant to sediment loading events of relatively short duration than to sustained high turbidity. Sponges are tolerant to short-term sediment loading, with some appearing to slough off adhering particles by the secretion of a mucoid material.

From the equation relating suspended particulate matter to coral growth rate in Table 4, a 20% reduction in annual growth rate corresponds to a 28% increase in average long-term background suspended particulate matter or suspended sediment levels. These background levels tend to vary from region to region and are site specific. Wolanski et al.,[45] for example, found that total suspended particulate concentration along a cross-shelf transect from Cape Ferguson to Keeper Reef on the Great Barrier Reef varied from 15 mg l^{-1} inshore to 3 mg l^{-1} at the midshelf Keeper Reef, assuming a particulate density of 1.5 g cm^{-3}. These levels were determined in mid-July, however, and it is likely that there is considerable variability during the year, with large increases inshore during the rainy season. The coral communities that exist across the shelf (i.e., outer shelf, midshelf, fringing, and inner shelf reefs) are all adapted to existence in environments of differing water qualities. For example, inner shelf reefs would be dominated by species capable of survival in turbid waters and efficient at sediment removal.[46] Since the offshore suspended sediment levels are subject to less variability, nomination of a long-term average background value is most relevant for these areas, i.e., for offshore or outer-shelf reefs. It is suggested that suspended matter

TABLE 6
Degree of Impact on Coral Communities by Various Levels of Sedimentation

Sedimentation rate ($mg\ cm^{-2}\ d^{-1}$)	Degree of impact
1—10	Slight to moderate
	Decreased abundance
	Altered growth forms
	Decreased growth rates
	Possible reductions in recruitment
	Possible reductions in numbers of species
10—50	Moderate to severe
	Greatly reduced abundance
	Greatly decreased growth rates
	Predominance of altered growth forms
	Reduced recruitment
	Decreased numbers of species
	Possible invasions of opportunistic species
>50	Severe to catastrophic
	Severely decreased abundance
	Severe degradation of communities
	Most species excluded
	Many colonies die
	Recruitment severely reduced
	Regeneration slowed or stopped
	Invasion by opportunistic species

From Pastorok, R. A. and Bilyard, G. R., *Mar. Ecol. Prog. Ser.*, 21, 175, 1985. With permission.

concentrations, not to be exceeded for any extended periods on midshelf and outer shelf locations, be 3×1.28 or 3.85 mg l^{-1} (3.85 ppm). It should be noted that these values are based on limited data and do not take into account natural temporal variation. More realistic levels might be set by consideration of mean annual suspended solids concentration.

Average sedimentation rates measured over extended periods in natural coral reef habitats in the Caribbean range from 0.3 to 37 mg $cm^{-2}\ d^{-1}$.[47,48] In the Indo-Pacific region, corresponding sedimentation values range from 0.1 to 228 mg $cm^{-2}\ d^{-1}$.[49] In shallow waters of both regions, higher sedimentation rates beyond the upper ends of the quoted ranges may be found. Based on studies from Guam, observations have suggested that where average sediment loads are about 160 to 200 mg $cm^{-2}\ d^{-1}$, a poor coral community of fewer than 10 species covering less than 2% of the available solid substrate could be expected. Conversely, a rich community of over 100 species covering over 12% of the solid substrate could be expected where average sedimentation rates were about 5 to 32 mg $cm^{-2}\ d^{-1}$.

From a number of investigations, Pastorok and Bilyard[16] developed a tentative impact scale for various levels of sediment deposition, with consequent effects on coral reef communities. This scale is shown in Table 6. Based on these data, a tentative tolerance limit of 30 mg $cm^{-2}\ d^{-1}$ is proposed, although in some inshore reef areas, levels greater than this may be a common natural occurrence.

C. BOD AND CHLOROPHYLL-a

Based on the relevant equations in Table 4, a 20% decrease in annual growth rate corresponds to 19 and 48% increases, respectively, over long-term average background BOD and chlorophyll-a levels. Chlorophyll-a surface concentration in Great Barrier Reef waters seems to be about 0.4 mg m^{-3}, compared with 0.13 mg m^{-3} in the adjacent Coral Sea.[44,50,51]

Again, values vary throughout the year, with a minimum in winter and generally higher concentrations inshore.[45] Assuming a mean level of 0.4 mg m^{-3}, however, the maximum tolerable long-term concentration would be 1.48×0.4, or 0.59 mg m^{-3}. It is stressed, however, that local variations of chlorophyll need to be determined before this parameter can be used and that it may not be appropriate in all cases. There is little information available on background BOD levels in tropical waters, since it is a water quality parameter primarily associated with sewage effluent and is often only measured after discharge has commenced. Both chlorophyll-a and BOD would appear to be sensitive indications of eutrophication, particularly where receiving waters are partially enclosed and circulation relatively poor.

A 90% reduction is growth rate, or effective mortality, corresponds to 128 and 388% increases in BOD and chlorophyll-a levels, respectively.

D. TEMPERATURE

Heated effluent discharged into waters off Oahu, Hawaii has been found to cause almost total destruction of all corals in areas where water temperatures were elevated 4 to 5°C above ambient. Bleaching and high mortality rates occurred in areas characterized by temperature increases of 2 to 4°C.[52] Although it was suggested that the absolute temperature was critical, rather than the degree of enhancement over ambient levels, damage was most severe in late summer, when seawater temperatures are maximal. Therefore, it can be argued that the extent of enhancement of summer ambient water temperature is the critical factor in hyperthermal stress.

In confirming this hypothesis, Marcus and Thorhaug[53] found sublethal hyperthermal limits of 32°C for *Porites compressa* from Hawaii and 33°C for *P. porites* from Florida. Summer ambient temperatures are 27 and 29°C, respectively. Overall results showed that *Porites* species exhibited stress symptoms at temperatures 4 to 5°C above summer ambient, while elevations of 5 to 6°C for periods of up to 5 d were lethal. Low temperature tolerances of 16 to 20°C for *P. porites* compared with a winter ambient of 22°C indicated that corals may live as close to their hypothermal sublethal limit in winter as they do to their hyperthermal sublethal limit in summer. Coles et al.[54] and Coles and Jokiel[55] found a sublethal upper temperature limit of 32°C for *P. compressa* and three other reef corals, but a limit of 34°C for these species at Eniwetok Atoll where the summer ambient is 2°C higher. Temperature tolerances of gorgonians from Florida appeared much the same as those for coral from this area in comparison with summer and winter ambient temperatures of 22 and 29°C.[19] Hyperthermal limits (ambient + 2°C) were roughly comparable with that for *P. porites*. Based on these data the maximum recommended temperature elevation over summer ambient levels is +3 to 4°C, with a similar reduction below winter ambient water temperatures.

Endean[56] has observed that water temperatures in the Great Barrier Reef region vary with latitude. Maximum sea temperatures in the northern sector are approximately 30°C, while in the southern, approximately 28°C. The ambient summer temperature found in open waters near Lizard Island is 29°C.[57] This compares with levels of 29.6°C at Davies Reef in the central Great Barrier Reef area[58] and 29°C for Heron Island reef flats. It is clearly important to measure annual seawater temperature fluctuations and, in particular, average summer maxima, in areas into which effluent discharge will occur. Assuming, however, an average summer maximum seawater temperature of 29°C, the hyperthermal tolerance level is 32 to 33°C for Great Barrier Reef waters. Extended exposure of hermatypic corals in this area to temperatures of 33°C or greater is likely to result in significant stress and mortality.

E. SALINITY

The upper and lower salinity tolerances of reef organisms are taken to be their sublethal limits. Outside these limits, organisms are capable of marginal survival. However, with other synergistic natural and anthropogenic stresses, mortality would almost certainly ensue.

P. porites and *P. compressa* from the Atlantic (Florida Keys) and Pacific Oceans (Oahu, Hawaii), respectively, exhibit a similar and fairly limited tolerance to salinity variation. No stress from salinity occurred from seawater at 25 to 37‰ within 20 d of exposure. Sublethal symptoms occurred in both species at 40‰, although slightly more severe in *P. porites,* i.e., partial bleaching and production of a light layer of mucous. The lower sublethal limit was determined to be 20‰, while lethal limits (5 d) were 45 and 15‰.[53]

As a comparison, salinity tolerances of some reef-dwelling gorgonians from Florida have been measured by Goldberg.[19] Gorgonians are also known as sea fans or sea whips. The upper sublethal salinities varied from 39 to 42.5‰ and lethal limits from 43 to 44.5‰. The upper optimal salinities for gorgonians are, thus, approximately the same as for the scleractinian species *P. porites*. Sublethal hyposaline levels are, however, much higher, ranging from 29.5 to 30.5‰, with lethal concentrations 25 to 28.5‰. Scleractinian corals then are most vulnerable to hypersaline conditions such as those generated by desalination plant effluent. Conversely, gorgonians are most susceptible to hyposaline conditions. Average salinity of Great Barrier Reef waters is 34 to 36‰,[44,45,51] and based on the preceding investigations, salinity tolerances are approximately 30 to 40‰. Normal saline conditions are, therefore, intermediate between tolerance limits, although extreme natural variations may periodically exceed these limits for short periods of time.[45]

F. OTHER POLLUTANTS

1. Hydrocarbons

Estimation of tolerance levels for hydrocarbons is complicated by the extremely low aqueous solubilities of these compounds. Concentrations calculated from the addition of a nominal amount of oil to water may bear little relation to the actual concentration of soluble and aggregated hydrocarbons actually in contact with a marine organism. Therefore, reports of toxicity of hydrocarbons toward coral community members must be interpreted carefully.

Lewis[28] found a sublethal limit of 50 ppm hydrocarbon in experiments involving four Caribbean corals (*P. porites, Agaricia agaricites, Favia fragum,* and *Madracis asperula*). After 24-h exposure at 50 ppm, the percentage of colonies with expanded tentacles and normal feeding behavior compared with controls decreased.

With staghorn coral (*Acropora formosa*) from the Great Barrier Reef, exposure to 5 ppm of oil hydrocarbon for periods longer than 24 h had an eventual lethal effect (within 5 d) on colonies. For 6- to 12-h exposures, most corals eventually recovered after initial mucous production and bleaching.[22] Based on this work, and considering petroleum hydrocarbons to be toxic but noncumulative (although this latter assumption is not always strictly correct),[59] an application factor of 0.05[60] affords a tolerance level of 0.25 ppm. An application factor is the ratio between a "safe" level and a toxic concentration of a contaminant. An arbitrary factor of 0.01 is often used if the contaminant is persistent and cumulative, and 0.05 if it is not. Other coral species may well have different susceptibilities, and longer-term exposure at sublethal levels may also have a deleterious effect.[61]

The sensitivity of other tropical marine organisms has been measured in static bioassay tests.[62] The most vulnerable of the test species were goatfish, hermit crabs, and sea urchins, with LC_{50}'s of 6 ppm and a consequent tolerance limit above which harmful effects may occur of 0.3 ppm. Soft corals appeared to be intermediate in sensitivity, with an LC_{50} of 18 ppm. The derived tolerances from this investigation are of the same order of magnitude as that found by Harrison et al.[22] for coral from the Great Barrier Reef, and these results suggest that scleractinian corals are the most vulnerable members of coral reef communities to effluent containing hydrocarbons. Based on limited data, the tolerance level for corals is proposed to be 0.25 ppm.

2. Metals

In this section, tolerance levels are derived for six heavy metals, viz., zinc, copper, nickel, cadmium, mercury, and lead. Some members of this series are among the most toxic to marine organisms, and their presence at low levels has been detected in both biotic and abiotic components of coral reef areas.

Little information is available on zinc toxicity to marine organisms, and virtually none with regard to numbers of coral reef communities. The U.S. Environmental Protection Agency (EPA)[63] has recommended an application factor of 0.01 be used with 96-h LC_{50} data for the most sensitive organisms. On the basis of all available evidence, it was suggested that aqueous concentrations of zinc greater than 100 $\mu g\ l^{-1}$ constitute a hazard to the marine environment, and those less than 20 $\mu g\ l^{-1}$ present minimal risk of deleterious effects. In view of the lack of specific toxicity data and present low ambient levels, a tolerance level of 20 $\mu g\ l^{-1}$ or 20 ppb for zinc is proposed.

Copper has been used for eliminating marine algae, and its salts have bactericidal properties. The element is toxic to many invertebrates and has been used extensively in marine antifouling paints. Exposure of two Hawaiian reef-building corals to copper concentrations of 0.01 mg l^{-1} resulted in symptoms of severe stress and moribundity after 48 h. At exposures of 0.1 mg l^{-1} or greater, all corals died within 24 h.[64] With the recommended application factor of 0.01,[63] the tolerance level for copper is calculated as 0.1×0.01 mg l^{-1} or 1 $\mu g\ l^{-1}$ (ppb). This level compares with current ambient seawater concentrations of 0.22 to 0.32 $\mu g\ l^{-1}$. Copper concentrations, therefore, have to be monitored relatively closely, since increases of three to five times current levels would approach derived tolerance levels.

Short-term nickel toxicity tests on marine organisms show 96-h LC_{50} values varying between 6 mg l^{-1} for the copepod *Nitocra spinipes* to 350 mg l^{-1} for the fish *Fundulus heteroclitus*.[65] Given an application factor of 0.01, and based on these data, concentrations greater than 6×0.01 or 60 $\mu g\ l^{-1}$ would be expected to be hazardous. The EPA recommends a limit of 2 $\mu g\ l^{-1}$, which should pose minimal risk.[63] Therefore, a suggested tolerance level for nickel in tropical marine water is 2 $\mu g\ l^{-1}$.

Cadmium has marked acute and chronic effects on aquatic organisms. It can also act synergistically with other elements. Seawater concentrations are at present generally below 0.01 $\mu g\ l^{-1}$ (ppb).[66] The current sublethal limit for marine organisms is 0.2 $\mu g\ l^{-1}$.[63] In the absence of any data for tropical reef-dwelling biota and the extremely low ambient levels, it is suggested that this limit (0.2 $\mu g\ l^{-1}$) be adopted as the tolerance level for cadmium.

For inorganic mercury, on the basis of data available, a concentration of greater than or equal to 0.10 $\mu g\ l^{-1}$ has been suggested as harmful in the marine environment.[63]

A 12-week LC_{50} of 0.5 mg l^{-1} lead and an 18-week LC_{50} of 0.3 mg l^{-1} lead have been observed with oysters, while sublethal effects were noted as 0.1 mg l^{-1} lead following 12-week exposure.[63] Natural levels in seawater are around 0.02 $\mu g\ l^{-1}$, and levels in Great Barrier Reef surface water have been shown to be consistently below analytical detection limits of 0.06 $\mu g\ l^{-1}$.[66] It is suggested, based on existing evidence, that levels of less than 0.01 mg l^{-1} or 10 $\mu g\ l^{-1}$ present minimal risk.[63] Although this tolerance limit is higher than for most other metals and is considerably higher than current ambient lead levels, it is recommended in the absence of more specific data and criteria.

3. Surfactants

It has been suggested that anionic surfactants could exert sublethal effects on marine life at levels of 20 to 50 ppb.[67] Bode et al.[68] found that concentrations of nonionic and anionic surfactants of 15 mg l^{-1} (ppm) or more were toxic to hydra. Assuming the surfactants are degradable and using the appropriate application factor of 0.05, a tolerance level based on this work is 750 ppb. The LD_{50} value of the dispersant Shell LTX mixed with seawater for the coral *Madracis mirabillis* was found to be 700 ppb, resulting in a tolerance level of 35 ppb.[69] For other tropical marine organisms, exposure to the nonionic surfactant ST5 resulted in 96-h LC_{50}'s ranging from 6 ppm (goatfish, hermit crab, and sea urchin) to 64 ppm (drill), and, hence, tolerance levels of 300 ppb and 3.2 ppm, respectively.[62] From the data considered, it appears that hermatypic corals are the most sensitive marine reef-dwelling fauna toward surfactants, although some algae may be slightly more susceptible. A maximum tolerance level of 35 ppb is proposed, based on existing evidence.

4. Pesticides and Herbicides

Studies with the hermatypic corals *Montastrea annularis, Acropora cervicornis,* and *Madracis mirabilis* revealed no changes in feeding, behavior, or polyp extension on exposure to 10 ppb aqueous mixtures of p,p'-DDT, dieldrin, and Aroclor 1254. A decrease in photosynthetic rate and, therefore, possibly a long-term reduction in growth rate was observed, however.[16] In another investigation, a commercial weed-killing formulation caused sublethal effects at 2,4-D (Na salt) concentrations of 60 ppb.[40] On the basis of this evidence, a tolerance level of 10 ppb is proposed. It must be emphasized, though, that data are obviously scarce and that tolerance levels for other reefal biota may be somewhat lower. For example, herbicides and defoliants may be hazardous to phytoplankton and algae at concentrations much less than the nominated tolerance level of 10 ppb ($\mu g\ l^{-1}$).

5. Chlorine

The toxicity of chlorine on marine organisms, particularly larvae, is well known.[70] The effects on some coral reef phytoplankton and invertebrate larvae were reduced growth rate (50%) of phytoplankton at levels as low as 90 ppb and LC_{50}'s of 460 to 840 ppb for urchin larvae and greater than 1950 ppb for mollusk larvae.[71] In addition, chlorine has been found to be a potent fertilization inhibitor of marine invertebrates such as sea urchins and worms at concentrations of 50 ppb.[72]

It is recommended that an application factor of 0.1 be used with LC_{50} data from seawater bioassays of the most sensitive species involved.[63] From this criterion, and in conjunction with other data considered, a tolerance level of 50 ppb ($\mu g\ l^{-1}$) is proposed.

6. Antifouling Agents

The toxicity of antifouling agents is primarily due to either copper or organotin compounds. Tolerance levels for copper have been derived in the section on heavy metals, and, therefore, this section focuses on organotin antifouling agents.

The main organotin compound used is tributyltin oxide (TBTO) and some long-term studies on fish and invertebrates have indicated that the tolerance level is below 1 ppb.[73] The 96-h LC_{50} for the copepod *Acartia tonsa* with TBTO was found to be 1.0 $\mu g\ l^{-1}$ (ppb), while oysters showed sublethal effects at concentrations of 0.15 $\mu g\ l^{-1}$.[41,74] Often, the larval stage of many marine species are extremely sensitive to TBTO. Larvae of the common mussel *Mytilus edulis* showed a 15-d LC_{50} value of 0.1 $\mu g\ l^{-1}$, indicating that long-term tolerance levels are much less than this concentration.[75] Assuming that organotin compounds are relatively resistant to degradation, use of an application factor of 0.01 affords a tolerance level of 1×0.01 or 0.01 $\mu g\ l^{-1}$ (ppb), based on available evidence.

G. SUMMARY OF TOLERANCE LEVELS

The derived long-term tolerances for various pollutants below which minimal disruption to coral reef communities should occur are summarized in Table 7. Quantitative estimates of these tolerance levels for Great Barrier Reef waters are also presented. Tolerance levels have been derived on the basis of concentrations causing a 20% decrease in growth rate or, alternatively, sublethal limits determined by direct observation or application factors rec-

TABLE 7
Summary of Derived Coral Maximum Tolerance Levels[a] in Ambient Water

Sewage	Increase over ambient levels (%)	Quantitative estimates of tolerance levels for Great Barrier Reef, Australia
Suspended material	28	3.85 mg l^{-1}
Sedimentation rate		30 mg cm^{-2} d^{-1}
BOD	19	0.84 mg l^{-1}
Chlorophyll-a	48	0.59 mg m^{-3}
PO_4^{3-}	23	0.25 μg l^{-1}
NH_4^+	285	0.65 μg l^{-1}
NO_2^- and NO_3^-	285	1.31 μg l^{-1}
Petroleum hydrocarbons		0.25 ppm
Salinity		30—40 ‰
Temperature		Mean summer ambient + 3°C (32°C)
Pesticides		10 ppb
Surfactants		35 ppb
Chlorine		50 ppb
Heavy metals		Zn 20 ppb Cu 1 ppb Ni 2 ppb Cd 0.2 ppb Hg 0.1 ppb Pb 10 ppb
Antifouling agents		0.01 ppb

[a] Derived on the basis of a tolerance of 20% growth decrease.

Compiled from Hawker, D. W. and Connell, D. W., *Int. J. Environ. Stud.*, 34, 179, 1989.

ommended by the EPA for marine waters, based on reef organism LC_{50}'s where possible. For many years, 0.1 of the 48-h LC_{50} was used as an indication of safe levels. More recent studies tend to use an arbitrary application factor of 0.01, if the pollutant is considered persistent or cumulative, and an application factor of 0.05 for noncumulative pollutants.[60]

The derived tolerance levels must be regarded as preliminary and tentative. For example, those developed for nutrients are based on growth rate and nutrient concentration data for one species of Barbadian coral. Results should be extrapolated with caution. Improved tolerance definition would be achieved by more extensive biological, chemical, and physical investigations in coral reef areas.

The tolerance levels found in Table 7 should also be regarded as conservative, since synergistic or additive deleterious effects are possible, but difficult to quantitate. It is also difficult to gauge the effects of natural stresses such as turbidity, temperature, salinity, borers, and *Acanthaster planci* to coral ecosystems already stressed by waste discharges.

REFERENCES

1. **Sheppard, C. R.C.,** *A Natural History of the Coral Reef,* Blandford Press, Poole, Dorset, 1983, chap. 1.
2. **Kinsey, D. W.,** Responses of coral reef systems to elevated nutrient levels, in *Proc. Workshop on Nutrients in the Great Barrier Reef Region,* Great Barrier Reef Marine Park Authority, Townsville, Australia, 1985, 55.
3. **Hawker, D. W. and Connell, D. W.,** An evaluation of the tolerance of corals to nutrients and related water quality characteristics, *Int. J. Environ. Stud.*, 34, 179, 1989.

4. **Johannes, R. E.,** Pollution and degradation of coral reef communities, in *Topical Marine Pollution,* Ferguson-Wood, E. J. and Johannes, R. E., Eds., Elsevier, Amsterdam, 1975, chap. 2.
5. **Rodrigues, A.,** Marine coastal environmental stresses in the wider Caribbean region, *Ambio,* 10, 283, 1981.
6. **Dutton, I. M.,** Management of waste discharge into the Great Barrier Reef Marine Park, presented to 47th Annu. Conf. Australian Institute of Health Surveyors (Qld), Townsville, Australia, 1981.
7. **Smith, S. V.,** Kaneohe Bay: a preliminary report on the responses of a coral reef/estuary ecosystem to relocation of sewage stress, in *Proc. 3rd Int. Coral Reef Symp.,* Miami, 1977, 578.
8. **Maragos, J. E., Evans, C., and Holthus, P.,** Reef corals in Kaneohe Bay six years before and after termination of sewage discharges (Oahu, Hawaiian Archipelago), in *Proc. 5th Int. Coral Reef Congr.,* Vol. 4, Tahiti, 1985, 189.
9. **Banner, A. H.,** Kaneohe Bay, Hawaii: urban pollution and a coral reef ecosystem, in *Proc. 2nd Int. Coral Reef Symp.,* Brisbane, 1974, 685.
10. **Smith, S. V., Kimmerer, W. J., Laws, E. A., Brock, R. E., and Walsh, T. W.,** Kaneohe Bay sewage diversion experiment: perspectives on ecosystem responses to nutritional perturbation, *Pac. Sci.,* 35, 279, 1981.
11. **Laws, E. A. and Redalje, D. G.,** Effect of sewage enrichment on the phytoplankton population of a subtropical estuary, *Pac. Sci.,* 33, 129, 1979.
12. **Walker, D. I. and Ormond, R. F. G.,** Coral death and phosphate pollution at Aqaba, Red Sea, *Mar. Pollut. Bull.,* 13, 21, 1982.
13. **Kinsey, D. W. and Domma, A.,** Effects of fertilization on a coral reef environment — primary production studies, in *Proc. 2nd Int. Coral Reef Symp.,* Brisbane, 1974, 49.
14. **Lamberts, A. E.,** Measurement of alizarin deposited by coral, in *Proc. 2nd Int. Coral Reef Symp.,* Brisbane, 1974, 241.
15. **Rasmussen, C.,** Effects of nutrients carried by mainland run-off on reefs of the Cairns area: A research plan and preliminary results, in *Proc. Workshop on Nutrients in the Great Barrier Reef Region,* Great Barrier Reef Marine Park Authority, Townsville, Australia, 1988, 66.
16. **Pastorok, R. A. and Bilyard, G. R.,** Effects of sewage pollution on coral reef communities, *Mar. Ecol. Prog. Ser.,* 21, 175, 1985.
17. **Kinsey, D. W.,** Small scale experiments to determine the effects of crude oil films on gas exchange over the coral back-reef at Heron Island, *Environ. Pollut.,* 4, 167, 1973.
18. **Coffroth, M. A.,** Mucous sheet formation in Poritid corals: effects of altered salinity and sedimentation, in *Proc. 5th Int. Coral Reef Cong.,* Vol. 4, Tahiti, 1985, 165.
19. **Goldberg, W.,** Ecological aspects of salinity and temperature tolerances of some reef dwelling Gorgonians from Florida, *Caribb. J. Sci.,* 13, 173, 1973.
20. **Goreau, T. F.,** Mass expulsion of zooxanthellae from Jamaican reef communities after Hurricane Flora, *Science,* 145, 383, 1964.
21. **Egana, A. C. and Di Salvo, L. H.,** Mass expulsion of zooxanthellae by Easter Island corals, *Pac. Sci.,* 36, 61, 1982.
22. **Harrison, P. L., Collins, J. D., Alexander, C. G., and Harrison, B. A.,** The effects of lowered salinity and oil pollution upon the tissue of the staghorn coral *Acropora formosa,* Great Barrier Reef Marine Park Authority, 1986.
23. **Johannes, R. E.,** How to kill a coral reef. I, *Mar. Pollut. Bull.,* 1, 186, 1970.
24. **Zieman, J. C. and Ferguson-Wood, E. J.,** Thermal pollution and tropical-type estuaries, in *Tropical Marine Pollution,* Ferguson-Wood, E. J. and Johannes, R. E., Eds., Elsevier, Amsterdam, 1975, chap. 5.
25. **Coles, S. L. and Jokiel, P. L.,** Effects of temperature on photosynthesis and respiration in hermatypic corals, *Mar. Biol.,* 43, 209, 1977.
26. **Coates, M. L., Connell, D. W., and Barron, D. M.,** Aqueous solubility and octan-1-olto water partition coefficients of aliphatic hydrocarbons, *Environ. Sci. Technol.,* 19, 628, 1985.
27. **Mackay, D., Chang, S., and Wells, P. G.,** Calculation of oil concentrations under chemically dispersed slicks, *Mar. Pollut. Bull.,* 13, 278, 1982.
28. **Lewis, J. B.,** Effect of crude oil and an oil spill dispersant on reef corals, *Mar. Pollut. Bull.,* 2, 59, 1971.
29. **Mitchell, R. and Chet, I.,** Bacterial attack on corals in polluted seawater, *Microb. Ecol.,* 2, 227, 1975.
30. **Ducklow, H. W. and Mitchell, R.,** Composition of mucous released by coral reef coelenterates, *Limnol. Oceanogr.,* 24, 706, 1979.
31. **Loya, Y. and Rinkevich, B.,** Effects of oil pollution on coral reef communities, *Mar. Ecol. Prog. Ser.,* 3, 167, 1980.
32. **Reiner, A. A.,** Effects of crude oil on corals, *Mar. Pollut. Bull.,* 6, 39, 1975.
33. **Cook, C. B. and Knap, A. H.,** Effects of crude oil and chemical dispersant on photosynthesis in the brain coral *Diploria strigosa, Mar. Biol.,* 78, 21, 1983.
34. **Peters, E. C., Meyers, P. A., Yevich, P. P., and Blake, N. J.,** Bioaccumulation and histopathological effects of oil on a stony coral, *Mar. Pollut. Bull.,* 12, 333, 1981.

35. **Rinkevich, B. and Loya, Y.,** Harmful effects of chronic oil pollution on a Red Sea scleractinian coral population, in *Proc. 3rd Int. Coral Reef Symp.*, Miami, 1977, 586.
36. **St. John, B. E.,** Heavy metals in the skeletal carbonate of scleractinian corals, in *Proc. 2nd Int. Coral Reef Symp.*, Vol. 2, Brisbane, 1974, 461.
37. **Livingstone, H. D., and Thompson, G.,** Trace element concentrations in some modern corals, *Limnol. Oceanogr.*, 16, 786, 1971.
38. **Klumpp, D. W. and Burdon-Jones, C.,** Investigations of the potential of bivalve molluscs as indicators of heavy metal levels of tropical marine waters, *Aust. J. Mar. Freshwater Res.*, 33, 285, 1982.
39. **Nagell, B., Notini, M., and Grahn, O.,** Toxicity of four oil dispersants to some animals from the Baltic Sea, *Mar. Biol.*, 28, 237, 1974.
40. **Glynn, P. W., Howard, L. S., Corcoran, E., and Freay, A. D.,** The occurrence and toxicity of herbicides in reef building corals, *Mar. Pollut. Bull.*, 15, 370, 1984.
41. **U'ren, S. C.,** Acute toxicity of bis (tributyltin) oxide to a marine copepod, *Mar. Pollut. Bull.*, 14, 303, 1983.
42. **Tomascik, T. and Sander, F.,** Effects of eutrophication on reef building coral: growth rate of the reef building coral *Montastrea annularis, Mar. Biol.*, 87, 143, 1985.
43. **Kinsey, D. W., and Davies, P. J.,** Effects of elevated nitrogen and phosphorus on coral reef growth, *Limnol. Oceanogr.* 24, 935, 1979.
44. **Andrews, J. C.,** Water masses, nutrient levels and seasonal drift on the outer central Queensland Shelf (Great Barrier Reef), *Aust. J. Mar. Freshwater Res.*, 34, 821, 1983.
45. **Wolanski, E., Jones, M., and Williams, W. T.,** Physical properties of Great Barrier Reef lagoon water near Townsville. II. Seasonal variations, *Aust. J. Mar. Freshwater Res.*, 32, 321, 1981.
46. **Done, T. J.,** Patterns in the distribution of coral communities across the central Great Barrier Reef, *Coral Reefs,* 1, 95, 1982.
47. **Loya, Y.,** Effects of water turbidity and sedimentation on the community structure of Puerto Rican corals, *Bull. Mar. Sci.*, 26, 450, 1976.
48. **Rodgers, C. S.,** Sublethal and lethal effects of sediments applied to common Caribbean reef corals in the field, *Mar. Pollut. Bull.*, 14, 378, 1983.
49. **Schumacher, H.,** Ability of fungid corals to overcome sedimentation, in *Proc. 3rd Int. Coral Reef Symp.*, Vol. 2, Miami, 1981, 504.
50. **Andrews, J. C. and Gentien, P.,** Upwelling as a source of nutrients for the Great Barrier Reef ecosystems: a solution to Darwin's question?, *Mar. Ecol. Prog. Ser.*, 8, 257, 1982.
51. **Ikeda, T., Gilmartin, M., Revelante, N., Mitchell, A. W., Carleton, J. H., Dixon, P., Hutchinson, S. M., Hing Fay, E., Boto, G. M., and Iseki, K.,** Biological, Chemical and Physical Observations in Inshore Waters of the Great Barrier Reef, North Queensland, 1975-1978, Tech. Bull., Oceanography Ser. No. 1, Australian Institute of Marine Science, Townsville, Australia, 1979.
52. **Jokiel, P. L. and Coles, S. L.,** Effects of heated effluent on hermatypic corals at Kahe Point, Oahu, *Pac. Sci.*, 28, 1, 1974.
53. **Marcus, J. and Thorhaug, A.,** Pacific versus Atlantic responses of the subtropical hermatypic coral *Porites* spp. to temperature and salinity effects, in *Proc. 4th Int. Coral Reef Symp.*, Vol. 2, Manila, 1981, 15.
54. **Coles, S. L., Jokiel, P. L., and Lewis, C. R.,** Thermal tolerance in tropical versus subtropical Pacific reef corals, *Pac. Sci.*, 30, 159, 1976.
55. **Coles, S. L. and Jokiel, P. L.,** Effects of temperature on photosynthesis and respiration in hermatypic corals, *Mar. Biol.*, 43, 209, 1977.
56. **Endean, R.,** *Australia's Great Barrier Reef,* University of Queensland Press, Brisbane, 1982, chap. 2.
57. **Harriot, V.,** Mortality rates of scleractinian corals before and during a mass bleaching event, *Mar. Ecol. Prog. Ser.*, 21, 18, 1985.
58. **Andrews, J. C., Dunlop, W. C., and Bellamy, N. F.,** Stratification in a small lagoon in the Great Barrier Reef, *Aust. J. Mar. Freshwater Res.*, 35, 273, 1984.
59. **Solbakken, J. E., Knap, A. H., Sleeter, T. D., Searle, C. E., and Palmork, K. H.,** Investigation into the fate of ^{14}C labelled xenobiotics (naphthalene, phenanthrene, 2,4,5,2′,4′,5′-hexachlorobiphenyl, octachlorostyrene) in Bermudian corals, *Mar. Ecol. Prog. Ser.*, 16, 149, 1984.
60. **Hart, B. T.,** A Compilation of Australian Water Quality Criteria, Australian Water Resources Council, Tech. Pap. No. 7, 1974.
61. **Peters, E. C., Meyers, P. A., Yevich, P. P., and Blake, N. J.,** Bioaccumulation and histopathological effects of oil on a stony coral, *Mar. Pollut. Bull.*, 6, 39, 1981.
62. **Eisler, R.,** Acute toxicities of crude oils and oil-dispersant mixtures to Red Sea fishes and invertebrates, *Isr. J. Zool.*, 24, 16, 1975.
63. Water Quality Criteria, A Report of the Committee on Water Quality Criteria, National Academy of Sciences, Washington, D.C., 1972.
64. **Howard, L. S. and Brown, B. E.,** Heavy metals and reef corals, *Oceanogr. Mar. Biol. Annu. Rev.*, 22, 195, 1984.

65. **Carey, J.,** Nickel mining and refinery wastes in coral reef environs, in *Proc. 4th Int. Coral Reef Symp.*, Vol. 1, Manila, 1981, 137.
66. **Denton, G. R. W. and Burdon-Jones, C.,** Trace metals in surface waters from the Great Barrier Reef, *Mar. Pollut. Bull.*, 17, 96, 1986.
67. **Fraizier, A., Franck, D., Benente, P., Jown, R., and Debiard, J. P.,** Observations on the various forms of pollution of the Tahiti Lagoon, in *Proc. 5th Int. Coral Reef Congr.*, Vol. 6, Tahiti, 1985, 445.
68. **Bode, H., Ernst, R., and Arditte, J.,** Biological effects of surfactants — the hydra as a highly sensitive assay animal, *Environ. Pollut.*, 17, 175, 1978.
69. **Elgerhuizen, J. H. B. W. and de Kruijf, H. A. M.,** Toxicity of crude oils and a dispersant to the stony coral, *Madracis mirabilis, Mar. Pollut. Bull.*, 7, 22, 1976.
70. **Campbell, D. G.,** Bahamian chlorine bleach fishing: a survey, in *Proc. 3rd Int. Coral Reef Symp.*, Miami, 1981, 593.
71. **Best, B. R., Braley, R. D., Marsh, J. A., and Matlock, D. B.,** Effect of chlorine on some coral reef phytoplankters and invertebrate larvae, in *Proc. 4th Int. Coral Reef Symp.*, Vol. 1, Manila, 1981, 169.
72. **Muchmore, D. and Epel, D.,** The effects of chlorination of wastewater on fertilization in some marine invertebrates, *Mar. Biol.*, 19, 93, 1973.
73. **Laughlin, R. B. and Linden, O.,** Fate and effects of organotin compounds, *Ambio*, 14, 88, 1985.
74. **Waldock, M. J. and Thain, J. E.,** Shell thickening in *Crassostrea gigas:* organotin antifouling or sediment induced, *Mar. Pollut. Bull.*, 14, 411, 1983.
75. **Beaumont, A. R. and Budd, M. D.,** High mortality of the larvae of the common mussel at low concentrations of tributyl tin, *Mar. Pollut. Bull.*, 15, 402, 1984.

Chapter 8

RECREATIONAL WATER QUALITY

Anne R. McNeill

TABLE OF CONTENTS

I. INTRODUCTION

Recreational activities which involve contact with water are very popular, especially in the warmer tropical waters. A wide range of leisure activities are pursued at coastal areas or inland, at lakes and rivers and in more recent times in swimming pools. Activities take place in both fresh and marine aquatic environments.

There is often some degree of risk to human health while pursuing recreational activities; consider that for skiing, motorbike riding, football. Likewise, there are risks when undertaking swimming or other water-contact recreation activities; only those related to the contraction of a communicable disease will be considered here.

The risks vary with the type of activity, and those where there is prolonged intimate contact with the water, and thereby greater likelihood of ingesting water, pose greater risk. Furthermore, greater risk occurs when the recreational water has been contaminated by infectious agents. These agents could originate from: untreated or ineffectively treated domestic and industrial effluent or sludge, sanitary wastes from adjacent residence and amenity facilities, fecal wastes from boats and aircraft, drainage from sanitary tips, storm water runoff, and excretions of lower animals (pasture runoff, saleyards, abattoirs, animal feedlots, etc.). In addition, the aquatic environment itself may be the source of the infectious agents. The health risks associated with each of these sources are not equal, and untreated human fecal wastes pose the greatest hazard. However, such discharges are amenable to corrective action, and it is common practice for treatment plants to conform to criteria and guidelines at impacted targets such as recreational areas.

When establishing microbiological criteria for recreational waters, it is also necessary to consider the effect of applying stringent limits. Such limits may lead to the possible closure of these areas and the consequent deprivation of beneficial and enjoyable activities to the community.

If no limits or formalized policy on recreational water quality are adopted, the public may be exposed to an unnecessary risk. Also, deterioration of the water quality may occur in existing or potential recreational waters. The time and cost to regenerate such waters would be substantial.

From the aesthetic viewpoint it could be said that all water should be free from the following: material that will settle to form objectionable deposits; floating debris, oil, scum, and other matter; substances producing objectionable color, odor, taste, or turbidity; substances and conditions or combinations thereof in concentrations which produce undesirable aquatic life.

While water should be appealing and nonobjectionable to all the senses of the user, the individual's aesthetic preferences are often subjective or dependent on conditioning and, therefore, vary from individual to individual and community to community.

There is still no international consensus among public health scientists as to which bacteriological indicator systems are best for judging the quality and acceptability of fresh and marine recreational waters.

When measuring the microbiological quality of marine and freshwater, a number of questions arise with respect to water guidelines: (1) For which activities are water quality guidelines needed? (2) Are the same indicators of health hazards appropriate for both fresh and marine waters? (3) Are these indicators the same for temperate and tropical waters? (4) Are these indicators appropriate to show increased risk of disease arising from both the presence of pathogenic organisms of fecal origin as well as those from the environment itself?

Not only do the public health problems associated with infectious agents transmitted by water need to be considered, but also the risks of disease caused by organisms which occur naturally in the aquatic environment (especially those in the tropical environment).

It is increasingly evident that the indicators of sanitary quality are not appropriate markers for the presence of pathogens that are autochthonous (native) to the aquatic or marine environment. Studies of the occurrence and survival of pathogenic and indicator microorganisms in temperate aquatic environments have directed much of the water quality criteria adopted today; however, recent investigations in tropical environments have exposed problems with the application of traditional indicators.

While all the issues raised are unable to be addressed in detail in this chapter, attention will be given to waterborne and water-contact infections, indicator and pathogen occurrence, and survival in the tropics; alternative indicators for the tropics; history and application of microbiological guidelines for recreational waters. Readers are directed to the major review documents cited for more comprehensive information.

One such major review of the pre-1982 literature has been compiled in an Australian monograph,[1] outlining primarily the temperate aquatic situation and including waterborne infections arising from use of recreational waters, occurrence and survival of pathogens and indicators in aquatic environments, methodologies for detecting pathogens and indicators, and recreational microbiological criteria.

This chapter, therefore, concentrates on recent material and, more importantly, on that relating to tropical studies, of which many have been undertaken by Hazen and co-workers in Puerto Rico;[2-16] otherwise the literature is sparse.

II. WATERBORNE AND WATER-CONTACT INFECTIONS

Primary contact recreation activities are characterized by bodily immersion or submersion where there is direct contact with water, and includes swimming, diving, water skiing, and surfing. The upper body orifices are exposed to water, and so both waterborne infections, through the ingestion of fecally contaminated water, as well as water-contact infections, through direct contact with organisms of the aquatic environment, are possible.

The first group of infections relate to enteric illnesses that may result from the ingestion of water while swimming in fresh or marine waters which have been contaminated with effluents from sewage treatment plants or nonpoint sources of human and animal wastes. These illnesses have a tendency to affect young rather than old individuals, and the risk of illness is associated with the number of diseased individuals in the population discharging the wastes. Risk from this type of source can be eliminated using appropriate sewage treatment technology.

The second group of infections relate to direct contact with contaminated water rather than ingestion. Although some of the causative agents may be excreted in feces, infection is not established by the fecal-oral route, and some of the causative agents may be natives to the aquatic environment; hence, their presence cannot be controlled like organisms originating from domestic sewage plants. Infection may occur when skin is cut, abraided, or macerated.

A wide variety of bacterial, viral, protozoan, and helminthic pathogens excreted in feces are capable of initiating waterborne infections (Table 1). Furthermore, the blue-green algae *Cylindrospermopsis raciborskii* has recently been reported[17] to have caused an outbreak of hepatoenteritis following copper sulfate treatment of a drinking water source dam.

Waterborne spread of infection by pathogenic agents depends on factors such as pathogen survival in water and the dose required for establishing infection in susceptible individuals. In addition to pathogen survival, latency (the period between pathogen excretion and host infectivity) and a pathogen's ability for multiplication in the environment are factors influencing the infective dose. The minimum infective dose has been determined only for some of the bacteria, viruses, protozoa, and helminths which are fecally excreted and, therefore, potentially water transmitted.

TABLE 1
Human Pathogenic Microorganisms Potentially Waterborne

Pathogen	Clinical syndrome
Bacteria	
Aeromonas hydrophila	Acute diarrhea
Campylobacter spp.	Acute enteritis
Enterotox. *Clostridium perfringens*	Diarrhea
Enterotox. *E. coli*	Diarrhea
Francisella tularensis	Mild or influenzal, febrile, typhoidal illness
Klebsiella pneumoniae	Enteritis (occas.)
Plesiomonas shigelloides	Diarrhea
Pseudomonas aeruginosa	Gastroenteritis (occas.)
Salmonella typhi	Typhoid fever
Other salmonellae	Gastroenteritis
Shigella spp.	Shigellosis (''bacillary dysentry'')
Vibrio cholerae	Cholera dysentery (01 serovars) or choleralike infection (non-01)
V. fluvialis	Gastroenteritis
Lactose-positive *Vibrio*	Pneumonia and septicemia
V. parahaemolyticus	Gastroenteritis
Yersinia enterocolitica	Enteritis, ileitis
Cyanobacteria	
Cylindrospermopsis spp.	Hepatoenteritis
Viruses	
Enteroviruses	Aseptic meningitis, respiratory infection, rash, fever
Poliovirus	Paralysis, encephalitis
Coxsackie virus A	Herpangina, paralysis
Coxsackie virus B	Myocarditis, pericarditis, encephalitis, epidemic pleurodynia, transient paralysis
Echovirus	Meningitis, enteritis
Types 68—71	Encephalitis, acute hemorrhagic conjunctivitis
Hepatitis A	Infectious hepatitis type A
Hepatitis non-A, non-B	Hepatitis type non-A, non-B
Influenza A	Influenza
Norwalk and other parvoviruslike agents	Epidemic, acute nonbacterial gastroenteritis
Rotavirus	Nonbacterial, endemic, infantile gastroenteritis; epidemic vomiting and diarrhea
Protozoa	
Balantidium coli	Balantidiasis (balantidial dysentery)
Cryptosporidium	Cryptosporidiosis
Entamoeba histolytica	Amoebiasis (amoebic dysentery)
Giardia lamblia	Giardiasis — mild, acute, or chronic diarrhea
Helminths	
Ascaris	Ascariasis (roundworm infection)
Ancylostoma	Hookworm infection
Clonorchis	Clonorchiasis (Chinese liver fluke infection)
Diphyllobothrium	Diphyllobothriasis (broadfish tapeworm infection)
Dracunculus mediensis	Dracontiasis (Guinea worm infection)
Fasciola	Fascioliasis (sheep liver fluke infection)
Fasciolopsis	Fasciolopsiasis (giant intestinal fluke infection)
Paragonimus	Paragonimiasis (Oriental lung fluke infection)
Spirometra mansoni	Sparganosis (plerocercoid tapeworm larvae infection)
Taenia	Taeniasis (tapeworm infection)
Trichostrongylus	Trichostrongyliasis
Trichuris	Trichuriasis (whipworm infection)

Compiled from McNeill, A. R., Australian Water Resources Tech. Pap. No. 85, Australian Government Publishing Service, Canberra, 1985, 561 and Hawkins, P. R., Runnegar, M. T. C., Jackson, A. R. B., and Falconer, I. R., *Appl. Environ. Microbiol.*, 50, 1292, 1985.

TABLE 2
Human Pathogenic Microorganisms Potentially Water-Contact Transmitted

Pathogen	Clinical syndrome
Bacteria	
Aeromonas hydrophila	Wound and ear infections, septicemia, meningitis, endocarditis, corneal ulcers
A. sobria	Wound and ear infections
Chromobacterium violaceum	Septicemia
Clostridium perfringens	Wound infection — gas gangrene
Klebsiella pneumoniae	Pneumonia, bacteremia
Legionella spp.	Legionellosis (Legionnaires' disease)
Leptospira spp.	Leptospirosis (Weil's disease — jaundice, hemorrhages, aseptic meningitis)
Mycobacterium marinum	Skin infection (''swimming pool granuloma'')
M. ulcerans	Skin infection (progressive subcutaneous ulceration)
Pseudomonas aeruginosa	Otitis externa and media; follicular dermatitis (pruritic pustular rash)
P. pseudomallei	Meliodosis (glanderslike infection)
Staphylococcus aureus	Wound and skin infections
Halophilic vibrios (incl. *Vibrio parahaemolyticus, Vibrio alginolyticus,* lactose positive *Vibrio)*	Wound and ear infections, conjunctivitis, salpingitis, pneumonia, septicemia
Viruses	
Adenovirus	Pharyngoconjunctivitis (swimmning pool conjunctivitis), respiratory infection
Adenosatellovirus	Associated with adenovirus type 3 conjunctivitis and respiratory infection in children but etiology not clearly established
Protozoa	
Naegleria fowleri	Primary amoebic meningoencephalitis (PAME)
Helminths	
Schistosoma spp.	Schistosomiasis (bilharzia)
Avian schistosomes *(Trichobilharzia, Austrobilharzia)*	Schistosome dermatitis (swimmer's itch)
Ancylostoma duodenale	Hookworm infection
Necator americanus	Hookwork infection

Compiled from McNeill, A. R., Australian Water Resources Tech. Pap. No. 85, Australian Government Publishing Service, Canberra, 1985, 561.

A single cyst for protozoa, a single cercariae for helminths, and 1 to 10 PFU (plaque forming units) for viruses are generally said to constitute an infective dose, while bacterial infective doses are generally much higher, ranging from 10^2 organisms for species of *Salmonella* and *Shigella* to 10^{10} for some strains of enteropathogenic *Escherichia coli,* although *Vibrio cholerae* and *Salmonella typhi* may also be infective at much lower doses of less than ten organisms, depending on the health status of the individual. Several reviews[1,2,18] cite minimum infective doses for bacterial, viral, protozoan, and helminthic pathogens.

Water-contact causative agents encompass bacterial, viral, protozoan, and helminthic organisms (Table 2). Water-contact infections from pathogenic species of Acanthamoeba may also be possible, although this has not been established epidemiologically. Also, the yeast-like fungus *Candida albicans* could be a potential health hazard; however, there are no epidemiological studies of its role in water-contact infections except the observation that increases in vaginal infections due to this organism were associated with bathing in polluted coastal waters.[19] It has also been implicated in a skin infection referred to as ''surfer's foot'' in Australia.[20]

For incidences of recreational waterborne and water-contact outbreaks consult other reviews.[1,18,21,22]

III. OCCURRENCE, SURVIVAL, AND USE OF INDICATORS AND PATHOGENS IN THE TROPICS

The routine monitoring for all pathogens potentially present in recreational waters (Tables 1 and 2) has until recently been thought to be impractical due to pathogen diversity and the fact that detection methods for some pathogens are nonexistent while others are often time-consuming, complex, specialized, and generally nonquantitative. However, the development of direct enumeration of pathogens by fluorescent staining and nucleic acid analysis may provide a basis for radical changes to old traditions.

The application of "indicator" organisms to herald the presence of pathogens has been historically used in the water potability and food safety areas. Indicator organisms that demonstrate the presence of organisms from fecal material have been most commonly used, e.g., total and fecal coliforms, *E. coli,* fecal streptococci, and *Clostridium perfringens.*

To indicate the sanitary quality of water and assess the health risks associated with water usage, the ideal indicator(s) should fulfill the following criteria:

1. Pathogen source specificity (including fecal specificity)
2. Simultaneous presence with bacterial, viral, protozoan, and helminthic pathogens and in sufficient numbers to provide an accurate density estimate when pathogen levels associated with unacceptable health risks occur
3. Survival in the environment comparable to that of pathogens
4. Resistance to treatment processes and disinfection similar to that of pathogens
5. No regrowth in the environment
6. Applicability to all water types
7. Applicability to all geographic areas
8. Accurate, precise, sensitive, cost-effective, facile, detection methodology providing quantitative results in a short time period

Currently there is no absolute indicator system which complies with all the above criteria for all water types in all geographic regions. If all the criteria were met, then the "indicator" would in effect be replaced by the "pathogen" itself because only the pathogen can fulfill all the criteria. An indicator system, if used to obviate the need for pathogen monitoring is, therefore, that which best fits the criteria.

Unfortunately, the selection of suitable water quality indicators to index the potential health hazards associated with the use of different water types in different geographic regions remains an unresolved enigma, as evidenced from major symposia held to this end.[23,24] The traditional coliform group indicator system, developed in temperate climate countries, has been recently reported to be deficient in tropical regions.[2-13,25,26]

While it is not possible to discuss here the distribution and survival of viral, bacterial, protozoan, and helminthic pathogens in fresh and marine recreational waters, the temperate data up to 1980 have been reviewed[1] as have the indicator data.[27] The data pertaining to tropical regions will be briefly presented, and some have already been generally reviewed.[2,3]

Physical dilution and microbial inactivation processes are the major reasons for the disappearance of the pollutant microorganisms in any aquatic environment. Studies have shown the survival of indicator bacteria and bacterial pathogens in the aquatic environment is influenced by physical, chemical, and biological factors and the interactions of these factors. As with virus survival, water temperature is one of the most important factors controlling bacterial survival which is negatively correlated with increasing temperatures. Solar radiation is significant in reducing naturally occurring indicator bacteria in wastewater and seawater and has been observed to cause sublethal injury to *E. coli.* The presence or absence of natural microbiota and associated predation by protozoa and bacteria, as well as bacteriophage activity, also appears to play a role in bacterial survival. Sedimentation of solid-associated bacteria provides protection against predators and bacteriophage, thus, prolonging survival. Bacterial die-off has been attributed more to solar radiation, sedimentation, and nutrient-related effects than to predation, bacteriophage activity, algal and bacterial toxins, and physicochemical factors such as pH and osmosis. The ability of certain bacteria, including *Klebsiella* and *Enterobacter* spp., to utilize internal endogenous reserves may aid their survival over other bacteria.

Studies comparing survival of fecal coliforms and pathogenic bacteria in marine waters have frequently yielded conflicting results, and differences in experimental methods probably contribute to disparities. The preparation and handling of test suspensions can affect cell viability, and recovery methods can be too harsh to recover stressed cells after exposure to the marine environment, all leading to misinterpretation of the fate of cells and an underestimation of the associated health risk. The viable, but nonculturable phenomenon, whereby organisms retain viability but are unable to be cultured by routine microbiological methods, significantly affects findings. Finally, the whole effect of the dynamic environment is unable to be experienced either in *in vitro* or *in situ* studies, and so the complex interacting physical, chemical, and biological factors, that vary from site to site, are generally unable to be adequately tested.

Reviewed[1] survival times for various bacteria in fresh and marine waters at 18 to 25°C (marine shown in parenthesis) are: coliforms 35 to 44 d (<8 h), fecal coliforms <4 to 8 d (<8 h), *Escherichia coli* 8 d, fecal streptococci 1 to >20 d, *Salmonella* spp. 2.5 to 28 d, *V. cholerae* <18 to 48 h, *Campylobacter* spp. 2 d. Survival times at lower temperatures were longer. Total and fecal coliform survival in groundwater is reported to be even longer, with times of 87 to 126 d.[1]

Recent temperate studies[28] in which the survival characteristics of a range of indicator bacteria, pathogenic bacteria, and bacteriophages were compared in fresh and marine waters, in the light, and at a range of temperatures (2 to 25°C), nutrient concentrations, and salinities in the dark, showed *E. coli* survival to be very poor compared with Salmonellae in the light and best in freshwater, but in the dark, survival in marine water was better. A linear relationship was demonstrated for T_{90} and light intensity, and log T_{90} and temperature, but nonlinear responses occurred for a range of salinities and nutrient concentrations. Results support the use of fecal streptococci (or enterococci) as an indicator of marine bathing water quality, and light again was shown to be an important factor affecting the survival of bacteria.

Brazilian tropical and subtropical studies[29] of indicator and pathogenic bacteria in marine waters (around 35°C) found fecal and total coliforms or fecal streptococci to be adequate indicators of recent fecal pollution and suggest yeast or heterotrophic bacteria counts as complementary indicators. Waters with fecal coliform counts over 1000/100 ml showed increased incidence of presumptive pathogenic yeasts, *Pseudomonas aeruginosa,* and *Salmonella.*

In a temperate and tropical study[30] salinity did not appear to play a great role in the survival of *Klebsiella pneumoniae,* and the die-off rates were primarily found to relate to sunlight. Sewage effluent from Honolulu discharged into the warm sea was rapidly (within a few 100) dispersed and the bacteria killed by UV light which penetrates into the clear water. In the colder waters of Morecombe Bay, U.K., bacteria were found much further out (20 mi) to sea, and they survive for much longer periods because sunshine levels were relatively low and light penetration was inhibited by high turbidity from suspended solids.

Although in temperate climate waters, total and fecal coliforms generally appear to persist for as long as or longer than pathogenic *Salmonella* spp. and *V. cholerae,* it may not be the case in the warmer waters of tropical regions.

Tropical data from Puerto Rico *in situ* survival studies for *Y. enterocolitica,*[15] *K. pneumoniae,*[7] *A. hydrophila,*[3] *Bifidobacteria,*[4] *C. albicans,*[10] *S. typhimurium,*[11] *V. cholerae*[12] were different from those of *E. coli*; however, discussions relate only to survival figures over 2 to 5 d.

A review[22] of the current significance of human rotavirus, Norwalk virus, and hepatitis A virus in water and the methods which can be used for their study discusses findings in terms of temperate and tropical climates (see below).

No information currently exists on the survival of human rotavirus in the environment, but studies with simian rotavirus SA-11 indicate that it can survive for prolonged periods of time in both fresh and marine waters. Laboratory studies, using freshwater at 23°C, showed a 3-log reduction of rotavirus SA-11 to vary from 3 to >14 d. Enteric virus inactivation in marine waters is always more rapid. Rotavirus is reported[22] present in wastewater, lakes, rivers, groundwater, marine water, tap water, and soil. Furthermore, it is widely distributed in the tropical fresh and marine waters of Mexico.[31]

The Norwalk virus has been associated with outbreaks of gastroenteritis involving recreational waters;[32-34] however, current methods do not allow its detection in environmental samples.[22]

In temperate climates such as North America, England, and Australia, peak prevalence of rotaviral diseases occurs during the colder months. Whether this seasonal pattern occurs in developing countries with tropical climates is unclear. Some studies have failed to observe seasonal variations, while others (Costa Rica, India) suggest the dry and cool seasons favor rotavirus infection, or, to the contrary, in regions where temperatures vary little (Indonesia), increase in rotavirus gastroenteritis coincides with seasonal change from dry to wet.

In temperate climates, hepatitis A occurs in yearly seasonal epidemic waves with peaks in late autumn and early winter. In many tropical countries the peak of reported disease occurs during the rainy season with low incidence in dry periods. A small number of isolations from the environment have been made[35-37] and studies[38] of hepatitis A virus added to estuarine water at 24°C showed a loss of 1 log of infectivity after 1 d and 3 logs after 5 d, with poliovirus 1 behaving similarly.

Other studies reviewed[1] showed virus maximum survival times (shown in parenthesis) in marine waters to be polio (130 d), coxsackie (90 d), Echo (120 d), Reo (4 d), Simian rota (14 d), Coliphage (64 d).

In Spain, coliphage survival times in marine waters were stated to be extraordinarily low, surviving less than 15 min after their discharge into the sea,[39] while for tropical regions it was thought that they will survive like their test host, *E. coli*.[2]

South African studies[40] of polluted seawater, analyzed for human enteric viruses and indicators of fecal pollution, revealed that the ratio of counts of viruses and indicators varied extensively in samples. Viruses, or their antigens, were detected in a number of samples which contained negative results in conventional tests for at least one indicator, again showing the shortcomings of coliforms as water quality criteria with regard to viruses. New proposed criteria include coliphages, fecal coliforms, and fecal streptococci because their presence correlates well with viruses, the latter showing best correlation.

A subtropical Brazilian marine beach study[41] found that neither fecal coliforms nor coliphage counts were always predictive of the presence of Salmonella and enteroviruses and that where pathogens are found, coliphage-pathogen ratios are smaller than fecal coliform-pathogen ratios.

Up until now criteria set in temperate areas have been accepted without question by tropical nations. Despite this trend, there is a growing body of evidence that the underlying assumptions of assays being used are not valid in tropical climates. While the reviewed tropical studies (Tables 3 and 4) are presented in some detail, they are, to date, the majority of works undertaken and highlight the paucity of specific tropical studies.

Many authors (Table 3) from tropical regions report high densities of *E. coli* found in the absence of any known fecal source. The general conclusions made by authors of the

TABLE 3
Tropical Water Studies Other than Those in Puerto Rico

Tropical country/observations	Conclusions and comments	Ref.
Ceylon, Egypt, India, Singapore TC & EC survival long Coliform growth possible	Doubt specificity of lactose 44°C test to EC; Kenya temperatures (18°C) accelerated death of indicators. Only minor differences in distribution of intestinal bacteria between UK and Kenya. EC thought to be in lower numbers in tropical dwellers but not substantiated by Wright.[45] EC concluded to be less satisfactory indicator of fecal pollution (than in temperate regions)	42
India Natural waters: coliforms	No reduction in first 4—5 weeks of water storage. After hot weather, organisms still isolated after 4 months	43
New Guinea Rural area waters FC 0—10,000, $\bar{x}$ 100 FS 0—6,000, $\bar{x}$ 100 FC/FS ratio 0.20 to 1.31	All sites grossly fecally polluted; FC and FS densities correlated more with domestic animals (pigs) than humans; positive isolates not completely identified; EC survival could inflate FC/FS ratio	44
Sierra Leone 29 Settlement source waters FC 40—240,000 EC 30—120,000 FS 7—64,000 CP 40—1,500	No correlation between indicators and *Salmonella* spp.; CP densities lower than other indicators in human feces; *S. faecalis* 14—100% of FS confirmed positives; FC should not be synonomous with EC	45
28 source waters	Repeated no correlation findings above; methods gave lowest confirmation rates at dry-wet season; dry-wet transition period possibly one of greatest risk	46
Nigeria Variety of sources FC 760—18,000 FS 680—17,500	No consistent FC/FS ratio found; highest counts at dry-wet transition; higher proportion *K. pneumoniae* to EC in Imo State but reverse in Zimbabwe	47
Botswana Drinking water wells FC & EC, O, TC low/high, but *Salmonella* spp. present	No correlation of *Salmonella* spp. presence with TC, FC, EC	48
Ivory Coast West Africa Well water sources FC 12 $\bar{x}$; TC 38 $\bar{x}$	EC 55% of TC population and FC 66%; local human feces showed EC 92% of TC and FC 89%; average water temp. 30°C; FC count greater than TC in 24% of water analyses showing noncoliform inhibition effects for TC determination	49

TABLE 3 (continued)
Tropical Water Studies Other than Those in Puerto Rico

Tropical country/observations	Conclusions and comments	Ref.
Brazil		
Marine waters TC 21—32,000,000 FC 11—35,000,000 FS 20—240,000	TC/FC/FS values are ranges of yearly means; EC 70% of FC population; FC levels >1000 showed increased incidence of *P. aeruginosa, Salmonella* spp., and presumptive pathogenic yeasts. Use of coliforms or FS supported as indicators of recent fecal pollution in tropical marine waters	29
Hawaii		
Streams FC 100—10,000 FS 100—10,000 CP 0—46 56—210 (discharges)	Indicators present in streams, not known to have fecal source; CP increased in uncontaminated site after rainfall. FC suitability as indicator questioned and CP evaluated as alternative	25
Area A, 2 sites FC 1—4, 11—95 FS 0—16, 48—217 Area B, 2 d FC 710—1020; 130—200 FS 360—850, 70—140	Area A: Ahuimanu stream source and 1.3 km downstream Area B: Waikele-Kipapa stream at remote site on rainy and sunny days Stream temperatures generally 18—24°C. In both study areas source/remote sites in uninhabitated forest areas, therefore, fecal source unknown; suspect indicator bacteria multiplying in warm and humid soil/water environment but original source unknown. Appropriateness of FC as recreational hygenic marker questioned.	26

Note: EC: *E. coli,* FC: fecal coliforms, TC: total coliforms, FS: fecal streptococci, CP: *Clostridium perfringens;* all bacteria counts organisms/100 ml; $\bar{x}$: mean.

TABLE 4
Marine and Freshwater Studies in Tropical Puerto Rico

Organism(s)	Study conclusions/comments	Ref.
Bifidobacterium spp. *Eschericia coli*, fecal coliforms, total aerobic bacteria	Rainforest watershed (water 21—27°C): *In situ* study over 70 h, 6 sites; at 2 sites found: EC to survive "indefinitely," remain physiologically active, regrow at sites dependent on ambient nutrient levels. *B. adolescentis* did not survive, decreasing 50% in 48 h. Compare: 80% in 24 h in temperate study.[50] YN-6 medium lacks specificity and resolution when high background growth. *Bifido* bacteria suggested[42,51] as alternative indicator of fecal pollution since: found in gut of humans: density $>10^{10}$ cells/g feces; obligate anaerobe and therefore incapable of surviving in oxygenated extraenteral environment	4
Fecal coliforms, *E. coli*	Epiphyte water (21—24°C): 3 sites located up to 30 ft above ground level. TC 1.5×10^6, FC 4—400, EC 72% of FC pop. Random confirmation of FC pop. showed EC in "unpolluted" sites. EC perhaps part of phyllosphere microflora and so validity as indicator questioned. Every single bromeliad sampled on different days/times of day/month/year was positive for EC, thus, supposedly ruling out "recent" fecal contamination by birds/mammals; however, what of insect vectors?	5
	Nucleic acid analyses on DNA of 9 environmental FC strains from pristine sites in rainforest, including epiphyte waters, identified all phenotypically as EC. The use of DNA probes to detect pathogens directly in water suggested as alternative procedure since EC likely to be a natural inhabitant	6
E. coli, K. pneumoniae	Rainforest watershed (water 20—25°C): *in situ* study, 11 sites found: EC: survival, 36 h, 20% AODC; 108 h, 69% CC; activity, 6 h, 40%; 96 h, 30%; respiring, 96 h, 10% *K. pn.*: survival, 36 h, 33% AODC; 108 h, 34% CC; activity, 109 h, 90%; respiring, 108 h, 25%. Site profiles ranged: FC 5—2196; *K. pn.* 2—52; Hazen[2] cites EC surviving 294 h	7
	Marine waters (27—33°C): Sites receiving untreated sewage (US): FC 0—300, *K. pn.* 10; sites receiving rum distillery effluent (RD) FC 800 source, 0 upcurrent, *K. pn.* 100—2000 source, <80 upcurrent. After 24 h of *in situ* exposure near RD 50% EC cells transformed to micrococci. *In situ* study, 4 sites found: US: EC: survived 84 h, 3 h decreased 2 orders CC; activity, 3 h, 40%; respiring, 3 h, 20%; *K. pn.*: survived 84 h; activity, 3 h, 20%; respiring, 3 h, 10%; RD: EC: survived 84 h; activity, 3 h, 95%; respiring, 3 h, 99% *K. pn.*: survived 83 h; activity, 3 h, 40%; respiring, 3 h, 90%. Both organisms survived 84 h study in marine tropical water and EC remained physiologically active under high nutrient loading	8
A. hydrophila	Marine waters (27—33°C): *In situ* study, 6 sites found: *Aeromonas hydrophila* survived and increased in rum effluent plume and declined upcurrent. *V. cholerae* and *K. pneumoniae* diversities higher in rum effluent waters	9

TABLE 4 (continued)
Marine and Freshwater Studies in Tropical Puerto Rico

Organism(s)	Study conclusions/comments	Ref.
E. coli, C. albicans	Marine/freshwaters (21—27°C): *in situ* study, 7 sites found: EC and *Candida albicans* (CA) survived 108 h. CA survived equally well both in fresh and marine waters and grew in marine waters receiving high organic effluents (from rum distillery). CA levels did not correlate with FC densities in watershed and neither considered a good indicator of recent fecal pollution in tropical waters. CA levels may under certain circumstances reach densities which may represent a health hazard. Hazen[2] cites EC surviving 206 h in rainforest streams	10
E. coli, S. typhimurium	Rainforest watershed (water 18—21°C): *In situ* studies using CC and AODC counts found: EC and *Salmonella typhimurium* active in surviving >5 d and numbers declining by 1 log after 105 h; after 24 h EC more active. Authors condemn FC (EC) as enforceable criteria and suggest no indicator detection but direct search for pathogens by DNA probe and immunofluoresence	11
E. coli, V. cholerae	Coral reef waters (21—26°C): EC and *Vibrio cholerae* (VC) densities by CC declined within 24 h and over 108 h by 2 orders; with AODC less decline and just cell morphology changes. EC and VC activity <60% within 24 h. Coral reef stands, sediments, and grass environment, less stressful on VC than EC; thus, EC unable to be used as indicator of fecal-borne pathogens or VC. VC concluded to be a potential bacterial contaminant of fish and shellfish	12
	Rainforest watershed (water 21—26°C): *In situ* study, 12 sites, over 108 h found: EC and VC densities did not change significantly, both remaining active. VC and EC densities did not correlate and VC concluded to be indigenous to tropical fresh waters. Concluded FC or EC inappropriate to assess public health risks in tropical waters	13
Legionella spp.	Marine/freshwaters (26—30°C): study of 5 sites found: several *Legionella* spp. widely distributed, densities of *L. pneumophila* ave. 10^4/ml and presence correlated with sulfate and phosphate concentrates and pH. *L.* spp. found in epiphytes of rainforest trees. Sewage-contaminated coastal waters recorded highest *L. pn* densities. 15% of Puerto Rican hospital patients with atypical pneumonia had *L. pn* infections.	14
E. coli, Y. enterocolitica	Rainforest waters (20—27°C): *In situ*, 5 sites found: EC: density increased over time, doubling after 2 d; activity dropped then stabilized at 85%; respiring cells decreased to 10%; *Yersinia enterocolitica* (YE): density decreased 1 order over 6 h then remained constant; activity increased over 6 h then declined to 50%; respiring cells remained constant at 25%. Both survive in tropical freshwaters. Natural densities of YE fluorescent antibody cells <10/ml and no culturable isolations from 1—3 liter samples. Low incidence of Yersiniosis in tropics[52]	15

S. faecalis	Hazen[2] reports long survival periods from studies undertaken by Muniz et al. and Santo Domingo et al., unpublished to date	
Fecal coliform enumeration, comparison using membrane filtration	Marine/freshwaters: Study on tropical waters, similar to temperate Canadian study,[53] involving 4 membrane methods; compared accuracy, specificity and recovery. All methods tested were unacceptable as significantly higher false positive/negative errors. Only 40% of FC isolates EC compared with other temperate studies where level is 90%. Spanish[54] membrane filtration evaluation of 14 selective media for the enumeration of coliforms from seawater found m-Endo[55] the most selective in a comprehensive evaluation	16

Note: EC = *E. coli*/100 ml; FC = fecal coliforms/100 ml; TC = total coliforms/100 ml; CC = Coulter Counter; AODC = acridine orange direct count; d = day.

many Puerto Rico studies (Table 4) of pristine rainforest streams, epiphyte waters, and marine waters are that pathogens could be present in the absence of *E. coli* and that the poor correlations between pathogens and traditional indicators render the indicator system useless. Direct detection of pathogens is being recommended.

Finding *E. coli* in pristine environments is extremely unusual because this bacterium inhabits the intestine of warm-blooded animals, and its presence is expected only in environments that have been exposed to recent fecal contamination. Furthermore, it seems that this bacterium is capable of surviving indefinitely in tropical environments,[2-4,7,10,43] suggesting it could be a natural inhabitant in these environments. Given the doubts concerning the appropriateness of fecal coliforms (including *E. coli*) as indicators and their traditional use in international criteria for judging water quality, there is clearly a need for further research into suitable criteria.

The search for suitable indicators in both temperate and tropical regions is ongoing, although methodology problems associated with recovery and selectivity continue to thwart researchers. Obligate anaerobes are currently[2] being seen as the best candidate for tropical-source waters since they will not survive extraenterally.

The recently exposed problem of "viable but nonculturable" bacteria has led researchers to different methodologies involving direct viable counts and uptake of radiolabeled nutrients.[56]

As an alternative to indicator organism enumeration and utilizing recent scientific advances, the direct detection of pathogens has become feasible. Detection of pathogenic microorganisms has been significantly improved using the new tools of biotechnology which include DNA probes and monoclonal antibodies.[56-59]

DNA probes are small pieces of DNA that recognize specific genes, and they make it possible to identify and isolate genetic information of any organism. Already probes are available to detect *V. cholerae* and enterotoxigenic *E. coli* toxin-producing genes, as well as *Legionella* spp., *Salmonella* spp., *Bacteroides* spp., *Campylobacter* spp., enteric viruses, and protozoa. Current applications in the enumeration and identification of bacteria from environmental samples using this technique has been reviewed.[58]

While DNA probes have advantages such as (1) being highly specific and sensitive (similar to cell culture for viruses), (2) no culturing medium needed, (3) species and strain able to be detected from mixed population, (4) quick test allowing handling of large numbers of samples, the present major limitations[56] are (1) their relative insensitivity with a detection limit for environmental samples of 10^5 cells and possible cross-reactivity among family members (e.g., *Enterobacteriacae*), (2) difficulties in quantifying probes and rate of DNA recovery not always maximal, (3) method is still expensive[56] relative to other bacteria detection methods yet considered inexpensive[59] for virus detection.

Monoclonal antibody ligated to a fluorescent molecule may permit direct detection of coliforms and *Salmonella* spp. when numbers are 1 to 10 cells per liter.[57] Specificity of *Escherichia coli* monoclonal antibodies on medical, fecal, and water samples was evaluated,[60] and the inability to distinguish between live and dead *E. coli* cells highlighted with researchers recommending method sensitivity to be compared with classic culturable methods or other techniques involving immunoenzymatic reactions. Similarly with viruses, a major disadvantage of gene probes is that they may not differentiate between infectious and noninfectious viruses.[59]

The rapidity of these new techniques, less than 3 h, clearly will have a major impact on water microbiology and virology of the future.

IV. HISTORY AND APPLICATION OF MICROBIOLOGICAL GUIDELINES FOR RECREATIONAL WATERS

A recreational water quality criterion is defined as a quantifiable exposure-effect relationship based on scientific research and experimentation between the level of some indicators of the quality of the water concerned and the potential human health risks associated with the recreational use of the water. A water quality guideline derived from such a criterion is a suggested maximum density of the indicator in the water which is associated with unacceptable health risks. The concept of acceptability implies that social, cultural, economic, political as well as medical factors are involved. A water quality standard obtained from the criterion is a guideline fixed by law.

These factors of acceptability are invariably considered when criteria are established for local application, and amendments are also likely to take into account technological feasibility, risk-benefit and cost-benefit analyses as well as political and administrative decisions.

Effective water quality management involves a systematic program of sampling and analyses, and both the process control and the surveillance program involve monitoring. These requirements presuppose that the criteria values established in objectives, goals, guidelines, and standards are amenable to monitoring. The specified limits expressed in criteria necessitate an appreciation of the statistical limitations associated with such criteria. Both the sampling and analytical measurements need to be standardized so that reliable data-generation procedures are achieved.

The values of many international, national, and local guidelines and standards are presented (Table 5), and U.S. primary contact recreation standards, as of 1978 (Table 6), show diversity of application.

The history of existing recreational microbiological criteria has been reviewed[1,61] and is not presented in any detail primarily because existing criteria, derived in temperate regions, probably have little or no application in the tropics.[2] However, the following discussion gives a brief historical background on the use of fecal coliforms as indicators of risk for the occurrence of gastroenteritis associated with swimming.

Studies conducted by the U.S. Public Health Service during the early 1950s on the Ohio River showed that a count of 2300 to 2700 total coliforms per 100 ml was the threshold for finding significant effects of illness from water pollution. A study at the same location in the 1960s showed that a fecal coliform count of 400/100 ml of river water was approximately equivalent to the earlier total coliform counts.[74] After application of a safety factor, 200 fecal coliforms per 100 ml was recommended as the limit.[62]

In the mid-1970s, a major series of prospective epidemiological and microbiological studies were conducted for the EPA by Cabelli and co-workers[75] at public marine beaches in New York City. The aim of the studies was to determine measurable health effects associated with swimming in sewage-polluted waters. Results showed a direct linear relationship between the frequency of gastrointestinal illnesses in swimmers and the microbiological quality of bathing water as measured by enterococci. Other indicators evaluated included total and fecal coliforms *(E. coli, Klebsiella, Enterobacter, Citrobacter), Pseudomonas aeruginosa, Clostridium perfringens, Aeromonas hydrophila, V. parahaemolyticus,* and Staphylococci. The studies led to the development of a mathematical model that related mean enterococci densities with rates of gastrointestinal illness, especially severe symptoms, in swimmers. There was virtually no relationship between densities of fecal coliforms, *P. aeruginosa* and *C. perfringens* and swimming-associated gastrointestinal symptoms. Enterococci directly correlated with the risk of gastrointestinal illness among swimmers as compared with nonswimmers, and this association occurred with relatively low densities of enterococci (approximately 10/100 ml of water).

A related series of prospective studies was performed by this group[76] in freshwater bathing beaches, and the water quality indicators employed were *Escherichia coli*, enterococci, and fecal coliforms. A strong correlation was observed between swimming-associated gastrointestinal symptoms and with *E. coli* and enterococci densities in water, but not with fecal coliforms.

TABLE 5
Summary of Microbiological Guidelines for Recreational Waters[1]

Water type	Conditions	Values	Comments
US (FWPCA, 1968)[62]			
Water designated for rec. use			
Primary contact	Log $\overline{X}$ ($\not<$5 samples in $\not>$30 days)	200 FC	Primary contact — intimate contact with water involving considerable risk of ingesting water in quantities sufficient to pose signif. health hazard
	Max. in 10% samples (30 days)	400 FC	
Other than primary contact	log $\overline{X}$	1,000 FC	For enhancement of rec. value of waters for rec. uses other than primary contact
	max, in 10% samples	2,000 FC	
Waters not designated for rec. use			
General use (secondary contact)	Ave.	$\not>$2,000 FC	
	Max.	4,000 FC	
US (EPA, 1986)[63,64]			
Waters designated for rec. use			
Primary contact	Log $\overline{X}$ ($\not<$5 samples in $\not>$30 days)		Update of 1976 EPA "Quality Criteria for Water" based on FWPCA 1968. EPA expects gradual transition from FC criteria to new indicator bacteria by all States. Different confidence levels are recommended for four levels of swimming use: no sample should exceed a one-sided confidence limit (C.L.) calculated using the following as guidelines: Designated bathing beach (75% C.L.) Moderate use for bathing (82% C.L.) Light use for bathing (90% C.L.) Infrequent use for bathing (95% C.L.) based on a site-specific log standard deviation, or if site data are insufficient to establish a log standard deviation, then using 0.4 as the log standard deviation for both indicators for freshwater; or for seawaters, then using 0.7 as the log standard deviation
	Freshwater	$\not>$126 EC	
		$\not>$ 33 Ent.	
	Marine water	$\not>$ 35 Ent.	
Canada (CCREM, 1987)[65]			
Swimming	Log $\overline{X}$ ($\not<$5 samples in $\not>$30 days	200 FC	Others: pathogens, coliphage, enterococci, *E. coli*, *P. aeruginosa*, *G. lamblia* — No limits set but current U.S. proposals given and other advice re acceptable limits provided
	Resampled if greater than	400 FC	
		Others	

European Economic Community (Council Directive 76/160/EEC) (CEC, 1976)[66]			
Bathing water (both fresh & sea)	80% of TC & FC, 90% of FS samples to meet guide values. 95% of samples to meet mandatory values	Guide & Mandatory 100 2,000 FC 500 10,000 TC 100 — FS — 0 Salm. — 0 Virus	Includes areas where: bathing explicitly authorized or not prohibited & traditionally practiced by many. Min. sampling freq. for FC & TC, fortnightly reduce to monthly if prev. years show results better than values given & unlikely to worsen. Commence sampling 2 weeks prior to bathing season. Monitor for FS, Salm. (absent/l), Enterovirus (0 PFU/10 l) whenever area suspected (from inspection) to have these or water quality deteriorates.
USSR: All-Union State Standard 17.1.5.02—80 (USSR, 1980)[67]			
Primary contact	Max.	1,000 TC	TC–"lac.–pos. intest. bacilli (LIB)". If primary contact value exceeded, test for *S. typhi, S. paratyphim Shigella,* staphylococci, enteroviruses. If positive, use of water may continue if TC $\not>$10,000. Min. sampling freq. 4 times/mth during swim. season. Analyze for EC, enterococci & bact. phage to define contamin. source
General (boating, sailing)	Max.	10,000 TC	
South Africa (Grabow, 1977)[68]			
Primary contact	Max. Log $\bar{X}$	1,000 TC; 200 EC gp I	Or absence of undisintegrated feces
Not primary contact	Log $\bar{X}$	1,000 EC gp I	
South Africa (Grabow, 1989)[40]			
Seawater — direct contact	50% of samples	100 FC 40 FS 50 Coliphages 0 Virus	FC/FS/Coliphages/100 ml and human viruses/10 l
WHO: (WHO, 1975)[69]			Water classification
Bathing areas	Consistently	<100 EC	Highly satisfactory (bathing area waters)
	Consistently	≤1,000 EC	Acceptable
WHO: (WHO/UNEP, 1977)[70]			
Bathing areas	Max. in 10% samples (≥10 consec. samples in bathing areas)	≤1,000 EC	Refinement of WHO 1975 criterion above & 100 EC/100 ml criterion to be basis for design of treatment & disposal systems
NZ: Water Pollution Regs. (NZ, 1969)[71]			
Inland waters	Consistently	≤1,000 TC	Accessible & regularly used waters for public bathing
Coastal waters, Class SB	Median	≤200 FC	

TABLE 5 (continued)
Summary of Microbiological Guidelines for Recreational Waters[1]

Water type	Conditions	Values	Comments
Australia (NH&MRC, 1990)[72]			
Primary	Max. in 20% samples	600 FC	Max. interval between samples 1 month. In freshwaters >24°C zero protozoa. Swimming near sewage outfalls and drains to be avoided
	Median of 5 samples	150 FC	
		0 Protozoa	
Secondary	Max. in 20% samples	4,000 FC	
	Median of 5 samples	1,000 FC	

Note: $\overline{X}$ = mean; ≮ = not less than; ≯ = not more than; TC = total coliforms/100 ml; FC = fecal coliforms/100 ml; EC = *E. coli*/100 ml; FS = fecal streptococci/100 ml; Ent. = Enterococci/100 ml.

TABLE 6
Some U.S. Applied Standards for Primary Contact Recreational Marine Waters

Location	Water class[a]	Total coliform value/100 ml					Fecal coliform value/100 ml				
		Average[b]		Percentile		Max.	Average[b]		Percentile		Max.
Alabama	Coastal						$\overline{X}$g	≤100			
	Other						$\overline{X}$g	≤200			
Alaska	All	Mean[1]	≤1000	80%	≤2400						
Connecticut	SA	Med	≤70	90%	≤230						
	SB	Med	≤700	90%	≤2300						
Florida	III	Ave[1]	≤1000	80%	≤1000	≤2400	Ave[1]	≤200	90%	≤400	≤800
Georgia	All						$\overline{X}$g	≤1000			≤4000
Los Angeles	Rec-1						LM[1]	≤200	90%	≤400	
Louisiana	A						LM[1]	≤200	90%	≤400	
Maine	SA	Max.	70	90%	≤230		Max.	15	90%	≤50	
	SB-1	Max.	240	90%	≤500		Max.	50	90%	≤150	
	SB-2	Max.	500	90%	≤1000		Max.	100	90%	≤200	
Maryland	I						LM	≤200			
Massachusetts	SA	Med.	≤70	90%	≤230						
	SB	Ave.	≤700	80%	≤1000						

TABLE 6 (continued)
Some U.S. Applied Standards for Primary Contact Recreational Marine Waters

Location	Water class[a]	Total coliform value/100 ml					Fecal coliform value/100 ml				
		Average[b]		Percentile		Max.	Average[b]		Percentile		Max.
Mississippi	All						$\overline{X}g$	≤200	90%	≤400	
New Hampshire	SB					≤240					
New Jersey	TW-1						Ave g	≤200			
	CW-1						Ave g	≤50			
New York	SA	Med.	≤70								
	SB	Med.	≤2400	80%	≤5000		$\overline{X}g$	≤200			
North Carolina	SB						LM	≤200	80%	≤400	
Oregon	All	Ave	≤240	80%	≤240						
Rhode Island	SA	Med.	≤70	90%	≤230		Med.	15	90%	≤50 (guide)	
	SB-2	Med.	≤700	90%	≤2300		Med.	≤50	90%	≤500 (guide)	
San Diego	Rec-I						LM[1]	≤200	90%	≤400	
South Carolina	SB						$\overline{X}g$	≤200	90%	≤400	
Virginia	Sub B						LM	≤200	90%	≤400	
							Ave[1]	≤2400	80%	≤2400	
Washington	A, fresh						Med.	≤50	90%	≤200	
	A, sea						Med.	≤14	90%	≤43	
Puerto Rico	SB						Ave[1]g	≤200	80%	≤400	
Virgin Islands	B						LM	≤70			

[a] Class — that designated in reference and relating to primary contract recreation; all — all waters with primary contact recreation.
[b] Average — $\overline{X}g$: geometric mean,
ave : average of ≮5 samples in ≯30 d,
ave[1]: monthly average,
ave g: geometric average,
LM : log mean,
LM[1]: log mean of ≮5 samples in 30 d,
mean[1]: mean of <5 samples in 1 month,
med.: median,
max.: maximum.

Compiled from USEPA 1978[73] as presented in McNeill, 1985.[1]

It was suggested that either *E. coli* or enterococci could be used to measure the potential for swimming-associated illness in fresh bathing waters. Furthermore, it was concluded that criteria developed from marine bathing waters are not applicable to fresh bathing waters, and with equivalent densities of enterococci the illness rate was approximately three times greater among swimmers using marine waters than those using freshwaters.

These recent U.S. studies[75,76] have led to the EPA recommending new criteria[63,64] (Table 5) and indicate that by using the existing criterion of 200 fecal coliforms per 100 ml a risk level of 15 gastrointestinal illnesses per 1000 population in marine waters and 6/1000 population in freshwaters has been unknowingly accepted.

While the U.S. school of thought considers microbiological guidelines and standards for primary contact recreation relevant to public health protection, others (primarily Moore)[77,78] have an opposing viewpoint and consider them merely useful for aesthetic considerations.

For many years it was considered unnecessary to have standards for bathing waters in the U.K.,[78] but in 1976 the European Economic Community (EEC) standards (Table 5) were introduced. Previously Moore,[77] in retrospective studies conducted in marine waters in England and Wales, showed that cases of enteric diseases were too infrequent to be used in the development of criteria for bacteriological standards.

Currently in the U.K.,[79] the debate continues as to the evidence for and potential magnitude of adverse health effects associated with the recreational use of natural surface waters. A scheme has been proposed as a basis to assess health hazards and to aid management of water recreation facilities. While the central thesis of the proposed guidelines is that a degree of risk exists from the recreational use of surface water, no attempt is made to assess the actual risks involved since this depends on the findings of sophisticated epidemiological appraisals which to date have not been undertaken. The guidelines, instead, establish a framework for assessing relative degrees of risk based on the nature of the intended use and the extent of foreseeable pollution and score some five different components. These are river classification, proportion of river volume which is sewage or trade effluent, proximity to sources of pollution, bacteriological quality (*E. coli* and fecal streptococci levels), and environmental health factors. A score ranging from 0 to 20 can be accumulated, and risk assessment strategy recommends:

Score	Acceptable recreational uses
0—10	Primary contact pursuits
0—16	Secondary contact pursuits
16—20	Use not advised

The criteria debate has been mainly centered around indicator bacteria derived from the intestinal tract, yet if the full spectrum of risk to swimmers is to be addressed, those indicators considered appropriate for swimming pools, which are associated with microorganisms derived from the mouth, nose, skin, and throat of bathers, may require attention. Canadian freshwater beach studies by Seyfried and co-workers,[80,81] in addition to documenting that illness associated with swimming can be related to the microbiological quality of water, found that the relationship between morbidity and enterococci was not as consistent as it was with staphylococci (bacteria more representive of microbial flora associated with body sites).

In 1983, an Israeli prospective epidemiological study[82] of marine beaches that were all within the WHO/UNEP guidelines for fecal coliforms found that symptoms of enteric morbidity among swimmers, particularly in the 0 to 4-year-old age group, were related to "high" density levels of enterococci (49), *E. coli* (49), and staphylococci (44), with medians shown in parentheses. Furthermore, swimmers had more morbidity symptoms (enteric, respiratory, other) than nonswimmers, regardless of the microbial quality of seawater. Again, total staphylococci were shown to be a consistent indicator in predicting total morbidity rates among swimmers.

South African studies[40] found coliform indicators to have shortcomings with regard to viruses and have recommended criteria to include limits for human viruses, fecal coliforms, fecal streptococci, and coliphages (Table 5). The test methods to be used in conjunction with these criteria are considered important since reoviruses should be able to be detected. Virus limit is similar to that recommended by the EEC (1976) and is considered realistic in terms of numbers of viruses in seawater with different levels of pollution. While the fecal coliform value is the same as EEC guidelines, the fecal streptococci limit of 40/100 ml is between that of the EEC (1976) 100 and EPA (1986) 3 or 35/100 ml. Both fecal streptococci and coliphage values are based on the local study findings of a relative number of indicators in seawater at sites with different pollution levels.

There is no evidence for adopting existing temperate criteria and, in fact, tropical studies (Tables 3 and 4) show a variety of results, many of which are conflicting. Others indicate possible deficiencies with utilization of traditional microbial indicators. Furthermore, there appears to be no emerging satisfactory indicator(s), and direct detection of pathogens is being recommended primarily because of the advancement of biotechnological techniques and because of the lack of correlation between indicators and pathogen presence.

Despite the doubts surrounding the appropriateness of temperate-derived criteria, i.e., using traditional indicators proven in temperate regions, the tropics are more in need of guideline values, given the diversity of waterborne disease organisms present, the under-developed nature of many countries and their large populations, poor medical services, and the known high morbidity and mortality rates. Furthermore, these around-the-year warm environments are receiving ever-increasing popularity as recreational resorts where water sports are the prime leisure activities. Clearly reevaluation of human health risks and means of assessing these risks is of immense importance, and the advancement of biotechnological techniques may soon see accurate determinations of the occurrence of waterborne and water-contact pathogens in recreational waters, followed by appropriate epidemiological studies in tropical regions.

REFERENCES

1. **McNeill, A. R.,** *Microbiological water quality criteria: a review for Australia,* Australian Water Resources Tech. Pap. No. 85, Australian Government Publishing Service, Canberra, 1985, 561.
2. **Hazen, T. C.,** Faecal coliforms as indicators in tropical waters: a review, *Int. J. Toxicol. Assess.*, 3, 461, 1988.
3. **Hazen, T. C., Santiago-Mercado, J., Toranzos, G. A., and Bermudez, M.,** What does the presence of fecal coliforms indicate in the waters of Puerto Rico? A review, *Bull. Puerto Rico Med. Assoc.*, 79, 189, 1987.
4. **Carrillo, M., Estrada, E., and Hazen, T. C.,** Survival and enumeration of the fecal indicators *Bifidobacterium adolescentis* and *Escherichia coli* in a tropical rain forest watershed, *Appl. Environ. Microbiol.*, 50, 468, 1985.
5. **Rivera, S. C., Hazen, T. C., and Toranzos, G. A.,** Isolation of fecal coliforms from pristine sites in a tropical rain forest, *Appl. Environ. Microbiol.*, 54, 513, 1988.
6. **Bermudez, M. and Hazen, T. C.,** Phenotypic and genotypic comparison of *Escherichia coli* from pristine tropical waters, *Appl. Environ. Microbiol.*, 54, 979, 1988.
7. **Lopez-Torres, A. J., Hazen, T. C., and Toranzos, G. A.,** Distribution and *in situ* survival and activity of *Klebsiella pneumoniae* and *Escherichia coli* in a tropical rain forest watershed, *Curr. Microbiol.*, 15, 213, 1987.
8. **Lopez-Torres, A. J., Prieto, L., and Hazen, T. C.,** Comparison of the *in situ* survival and activity of *Klebsiella pneumoniae* and *Escherichia coli* in tropical marine environments, *Microb. Ecol.*, 15, 41, 1988.
9. **Biamon, E. J. and Hazen, T. C.,** Survival and distribution of *Aeromonas hydrophila* in near-shore coastal waters of Puerto Rico receiving rum distillery effluent, *Water Res.*, 17, 319, 1983.

10. **Valdes-Collazo, L., Schultz, A. J., and Hazen, T. C.,** Survival of *Candida albicans* in tropical marine and fresh waters, *Appl. Environ. Microbiol.*, 53, 1762, 1987.
11. **Jimenez, L., Muniz, I., Toranzos, G. A., and Hazen, T. C.,** Survival and activity of *Salmonella typhimurium* and *Escherichia coli* in tropical freshwater, *J. Appl. Bacteriol.*, 67, 61, 1989.
12. **Perez-Rosas, N. and Hazen, T. C.,** *In situ* survival of *Vibrio cholerae* and *Escherichia coli* in tropical coral reefs, *Appl. Environ. Microbiol.*, 54, 1, 1988.
13. **Perez-Rosas, N. and Hazen, T. C.,** *In situ* survival of *Vibrio cholerae* and *Escherichia coli* in a tropical rain forest watershed, *Appl. Environ. Microbiol.*, 55, 495, 1989.
14. **Ortiz-Roque, C. M. and Hazen, T. C.,** Abundance and distribution of *Legionellaceae* in Puerto Rican waters, *Appl. Environ. Microbiol.*, 53, 2231, 1987.
15. **Elias-Montalvo, E. E., Calvo, A., and Hazen, T. C.,** Survival and distribution of *Yersinia enterocolitica* in a tropical rain forest stream, *Curr. Microbiol.*, 18, 119, 1989.
16. **Santiago-Mercado, J. and Hazen, T. C.,** Comparison of four membrane filter methods for fecal coliform enumeration in tropical waters, *Appl. Environ. Microbiol.*, 53, 2922, 1987.
17. **Hawkins, P. R., Runnegar, M. T. C., Jackson, A. R. B., and Falconer, I. R.,** Severe hepatotoxicity caused by the tropical cyanobacterium (blue-green alga) *Cylindrospermopsis raciborskii* (Woloszynska) Seenaya and Subba Raju isolated from a domestic water supply reservoir, *Appl. Environ. Microbiol.*, 50, 1292, 1985.
18. **Dufour, A. P.,** Diseases caused by water contact, in *Waterborne Diseases in the United States,* Craun, G. F., Ed., CRC Press, Boca Raton, FL, 1986, chap. 3.
19. **Briscou, J.,** Yeasts and fungi in marine environments, *Bull. Soc. Fr. Mycol. Med.*, 4, 159, 1975.
20. **Winner, H. I. and Hurley, R.,** *Candida albicans,* Little, Brown, Boston, 1964.
21. **Craun, G. F.,** Recent statistics of waterborne disease outbreaks (1981-1983), in *Waterborne Diseases in the United States,* Craun, G. F., Ed., CRC Press, Boca Raton, FL, 1986, chap. 6.
22. **Gerba, C. P., Rose, J. B., and Singh, S. N.,** Waterborne gastroenteritis and viral hepatitis, *Crit. Rev. Environ. Control,* 15, 213, 1985.
23. **Hoadley, A. W. and Dutka, T. J., Eds.,** *Bacterial Indicators/Health Hazards Associated with Water,* Spec. Tech. Publ. 635, ASTM, Philadelphia, 1977.
24. **James, A. and Evison, L., Eds.,** *Biological Indicators of Water Quality,* John Wiley & Sons, Chichester, 1979.
25. **Fujioka, R. S. and Shizumura, L. K.,** *Clostridium perfringens,* a reliable indicator of stream water quality, *J. Water Pollut. Control Fed.*, 57(10), 986, 1985.
26. **Fujioka, R. S., Tenno, K., and Kansako, S.,** Naturally occurring fecal coliforms and fecal streptococci in Hawaii's freshwater streams, *Int. J. Toxicol. Assess.*, 3, 613, 1988.
27. **Elliot, E. L. and Colwell, R. R.,** Indicator organisms for estuarine and marine waters, *FEMS Microbiol. Rev.*, 32, 61, 1985.
28. **Evison, L. M.,** Comparative studies on the survival of indicator organisms and pathogens in fresh and sea water, *Water Sci. Tech.*, 20, 309, 1988.
29. **Hagler, A. N., Mendonca-Hagler, L. C., Santos, E. A., Farage, S., Silva Filho, J. B., Schrank, A., and de Oliveria, R. B.,** Microbial pollution indicators in Brazilian tropical and subtropical marine surface waters, *Sci. Total Environ.*, 58, 151, 1986.
30. **Jones, K. and Betaieb, M.,** Nitrogen-fixing heterotrophic bacteria in coastal waters: a comparison of the contribution made by sewage effluent in a temperate and tropical environment, *Water Sci. Tech.*, 20, 323, 1988.
31. **Deetz, T. R., Smith, E. M., Goyal, S. M., Gerba, C. P., Vollet, J. J., Tsai, L., DuPont, H. L., and Keswick, B. H.,** Occurrence of rota- and enteroviruses in drinking and environmental water in a developing nation, *Water Res.*, 18, 567, 1984.
32. **Baron, R. C., Murphy, F. D., Greenberg, H. B., Davis, C. E., Bregman, D. J., Gary, G. W., Hughes, J. M., and Schonberger, L. B.,** Norwalk gastrointestinal illness. An outbreak associated with swimming in a recreational lake and secondary person-person transmission, *Am. J. Epidemiol.*, 115, 163, 1982.
33. **Koopman, J. S., Echert, E. A., Greenberg, H. B., Strohm, B. C., Isaacson, R. E., and Monto, A. S.,** Norwalk virus enteric illness acquired by swimming exposure, *Am. J. Epidemiol.*, 115, 173, 1982.
34. **Kaplan, J. E., Gary, G. W., Baron, R. C., Singh, N., Schonberger, L. B., Fieldman, R., and Greenberg, H. B.,** Epidemiology of Norwalk gastroenteritis and the role of Norwalk virus in outbreaks of acute nonbacterial gastroenteritis, *Am. Intern. Med.*, 96, 756, 1982.
35. **Divizia, M., De Filippis, P., Di Napoli, A., Venuti, A., Perez-Bercoff, R., and Para, A.,** Isolation of wild-type hepatitis A virus from the environment, *Water Res.*, 23, 1155, 1989.
36. **Sobsey, M. D., Oglesbee, S. E., Wait, D. A., and Cuenca, A. I.,** Detection of hepatitis A virus (HAV) in drinking water, *Water Sci. Technol.*, 17, 23, 1985.
37. **Jiang, X., Estes, M. K., and Metcalf, T. G.,** Detection of hepatitis A virus by hybridization with single-stranded RNA probes, *Appl. Environ. Microbiol.*, 53, 2487, 1987.

38. **Patti, A. M., Santi, A. L. Gabrieli, R., Fiamma, S., Cauletti, M., and Pana, A.,** Hepatitis A virus and poliovirus 1 inactivation in estuarine water, *Water Res.*, 21, 1335, 1987.
39. **Borrego, J. J. and Romero, P.,** Coliphage survival in seawater, *Water Res.*, 19, 557, 1985.
40. **Grabow, W. O. K., Idema, G. K., Coubrough, P., and Bateman, B. W.,** Selection of indicator systems for human viruses in polluted seawater and shellfish, *Water Sci. Tech.*, 21, 111, 1989.
41. **Dutka, B. J., El Shaarawi, A., Martins, M. T., and Sanchez, P. S.,** North and South American studies on the potential of coliphage as a water quality indicator, *Water Res.*, 21, 1127, 1987.
42. **Evison, L. M. and James, A.,** A comparison of the distribution of intestinal bacteria in British and East African water sources, *J. Appl. Bacteriol.*, 36, 109, 1973.
43. **Raghavachari, T. N. S. and Seetharama Iyer, P. U.,** Longevity of coliform organisms in water stored under natural conditions, *Indian J. Med. Res.*, 26, 877, 1939.
44. **Faechem, R.,** Fecal coliforms and fecal streptococci in streams in the New Guinea Highlands, *Water Res.*, 8, 367, 1974.
45. **Wright, R. C.,** A comparison of the levels of fecal indicator bacteria in water and human feces in a rural area of a tropical developing country (Sierra Leone), *J. Hyg. Camb.*, 89, 69, 1982.
46. **Wright, R. C.,** The seasonality of bacterial quality of water in a tropical developing country (Sierra Leone), *J. Hyg. Camb.*, 96, 75, 1986.
47. **Blum, D., Huttly, S. R. A., Okoro, J. J., Akujobi, C., Kirkwood, B. R., and Faechem, R. G.,** The bacteriological quality of traditional water sources in north-eastern Imo State, Nigeria, *Epidemiol. Inf.*, 99, 429, 1987.
48. **Thomson, J. A.,** Inadequacy of *Escherichia coli* as an indicator of water pollution in a tropical climate: a preliminary study in Botswana, *S. Afr. J. Sci.*, 77, 44, 1981.
49. **Lavoie, M. C.,** Identification of strains isolated as total and fecal coliforms and comparison of both groups as indicators of fecal pollution in tropical climates, *Can. J. Microbiol.*, 29, 268, 1983.
50. **Resnick, I. G. and Levin, M. A.,** Assessment of bifidobacteria as indicators of human fecal pollution, *Appl. Environ. Microbiol.*, 42, 433, 1981.
51. **Resnick, I. G. and Levin, M. A.,** Quantitative procedure for enumeration of bifidobacteria, *Appl. Environ. Microbiol.*, 42, 427, 1981.
52. **Bockemühl, J.,** Epidemiology, etiology and laboratory diagnosis of infectious diarrhea diseases in the tropics, *Immun. Infekt.*, 13, 269, 1985.
53. **Pagel, J. E., Qureshi, A. A., Young, D. M., and Vlassoff, L. T.,** Comparison of four membrane filter methods for fecal coliform enumeration, *Appl. Environ. Microbiol.*, 43, 787, 1982.
54. **Avila, M. J., Morinigo, M. A., Cornax, R., Romero, P., and Borrego, J. J.,** Comparative study of coliform-enumeration media from seawater samples, *J. Microbiol. Methods,* 9, 175, 1989.
55. **W.H.O.,** *Coastal Water Quality Control in the Mediterranean Sea,* World Health Organization, Copenhagen, 1982.
56. **Colwell, R. R.,** From counts to clones, *J. Appl. Bacteriol. Symp. Suppl.*, IS, 1987.
57. **Colwell, R. R.,** Plenary address — the need for interaction between microbiologists and engineers, *Water Sci. Tech.*, 20, 1, 1988.
58. **Hazen, T. C. and Jiménez, L.,** Enumeration and identification of bacteria from environmental samples using nucleic acid probes, *Microbiol. Sci.*, 5, 340, 1988.
59. **Gerba, C. P., Margolin, A. B., and Hewlett, M. J.,** Application of gene probes to virus detection in water, *Water Sci. Tech.*, 21, 147, 1989.
60. **Joret, J. C., Cervantes, P., Levi, Y., Dumoutier, N., Cognet, L., Hasley, C., Husson, M., and Leclerc, H.,** Rapid detection of *E. coli* in water using monoclonal antibodies, *Water Sci. Tech.*, 21, 161, 1989.
61. **Salas, H. J.,** History and application of microbiological water quality standards in the marine environment, *Water Sci. Tech.*, 18, 47, 1986.
62. **FWPCA,** *Rep. of the Natl. Tech. Advisory Committee on Water Quality Criteria,* Federal Water Pollution Control Administration, U.S. Department of Interior, Washington, D.C., 1968.
63. **E.P.A.,** Bacteriological ambient water quality criteria for marine and fresh recreational waters, *Fed. Reg.*, 51(45), 8012, 1986.
64. **Dufour, A. P. and Ballentine, R. K.,** *Bacteriological Ambient Water Criteria for Marine and Fresh Recreational Water — 1986,* EPA 440/5-84-002, National Technical Information Service, Springfield, VA, 1986.
65. **CCREM,** Recreational water quality and aesthetics, in *Canadian Water Quality Guidelines,* Canadian Council of Resources and Environment Ministers, Ontario, 1987, chap. 2.
66. **CEC,** Council directive of 8 Dec. 1975 concerning the quality of bathing water, 76/260/EEC, *Official J. Eur. Commun.*, No. L31/1-7, 1976.
67. **U.S.S.R.,** *All Union State Standard 17.1.5.02-80, State Standard of the USSR, Nature Conservation. The Hydrosphere. Hygiene Requirements for Recreation Areas Comprising Bodies of Water,* State Standards Committee of the U.S.S.R., Moscow, 1980.

68. **Grabow, W. O. K.,** South African experience on indicator bacteria, *Pseudomonas aeruginosa* and R+ coliforms in water quality control, in *Bacterial Indicators/Health Hazards Associated with Water,* Hoadley, A. W. and Dutka, B. J., Eds., ASTM, Philadelphia, 1977, 168.
69. **W.H.O.,** *Guidelines and Criteria for Recreational Quality of Beaches and Coastal Waters, Rep.* on a Working Group convened by the Regional Office of Europe, WHO, Bilthoven, 28th Oct-1 Nov 1974, Regional Office for Europe, World Health Organization, Copenhagen, 1975.
70. **WHO/UNEP,** *Health Criteria and Epidemiological Studies Related to Coastal Water Pollution,* Rep. of a Group of Experts jointly convened by WHO & UNEP, Athens 1-4 March 1977, Regional Office for Europe, World Health Organization, Copenhagen, 1977.
71. **NZ,** NZ Water and Soil Conservation Act of 1967, *Int. Dig. Health Leg.,* 16, 120, 1969.
72. **NH & MRC,** *Australian Guidelines for Recreational Use of Water,* National Health & Medical Research Council, Australian Government Publishing Service, Canberra, 1990.
73. **USEPA,** *A Compilation of State Water Quality Standards for Marine Waters,* U.S. Environmental Protection Agency, Washington, D.C., 1978.
74. **Geldreich, E. E.,** *Sanitary Significance of Fecal Coliforms in the Environment,* Water Pollut. Control Res. Service Pub. No. WP-20-3, U.S. Department of the Interior, Washington, D. C., 1966.
75. **Cabelli, V. J., Dufour, A. P., McCabe, L. J., and Levin, M. A.,** Swimming-associated gastroenteritis and water quality, *Am. J. Epidemiol.,* 115, 606, 1982.
76. **Dufour, A. P.,** *Health Effects Criteria for Fresh Recreational Waters,* EPA 600/1-84/004, U.S. Environmental Protection Agency, Cincinnati, OH, 1984.
77. **Moore, B.,** Sewage contamination of coastal bathing waters in England and Wales: a bacteriological and epidemiological study, *J. Hyg.,* 57, 435, 1959.
78. **Moore, B.,** The case against microbial standards for bathing beaches, in *Discharge of Sewage from Sea Outfalls,* Gamerson, A. L. H., Ed., Pergamon Press, London, 1975, 103.
79. **Jones, F. and Godfree, A. F.,** Recreational and amenity use of surface waters: the public health implications, *Water Sci. Tech.,* 21, 137, 1989.
80. **Seyfried, P. L., Tobin, R. S., Brown, N. E., and Ness, P. F.,** A prospective study of swimming-related illness. I. Swimming-associated health risk, *Am. J. Public Health,* 75, 1068, 1985.
81. **Seyfried, P. L., Tobin, R. S., Brown, N. E., and Ness, P. F.,** A prospective study of swimming-related illness. II. Morbidity and the microbiological quality of water, *Am. J. Public Health,* 75, 1071, 1985.
82. **Fattal, B., Peleg-Olevsky, E., Yoshpe-Purer, Y., and Shuval, H. I.,** The association between morbidity among bathers and microbial quality of seawater, *Water Sci. Tech.,* 18, 59, 1986.

Chapter 9

DISPOSAL OF WASTES AT SEA IN TROPICAL AREAS

Chih-Shin Shieh and Iver W. Duedall

TABLE OF CONTENTS

I. INTRODUCTION

Ocean disposal is a global issue. In tropical oceans (23.27°N to 23.27°S), like anywhere in the world oceans, ocean disposal requires both regional and international attention. In some tropical countries, the ocean has always been used for waste disposal. Unfortunately, very little published information is available to carry out a systematic evaluation of ocean disposal activity in tropical areas.

Since 1979, only one of the 27 Contracting Parties (nations) to The London Dumping Convention (LDC) in tropical areas (Figure 1) reported its activity on ocean disposal.[1,2] One may not assume that the absence of reports from other Contracting Parties indicates that ocean disposal is not practiced. Similarly, one may not rule out ocean disposal by non-Contracting Parties to the LDC. The purpose of this chapter is, therefore, to identify general problems associated with present and future ocean disposal in tropical areas and to provide recommendations.

While traditionally the practice of ocean disposal refers to the waste disposal from ships, barges, or aircraft,[3] we also consider outfall disharge as a form of disposal. Regardless of the disposal method, minimization of environmental degradation should be the single most important goal in considering the oceans for disposal of wastes.

Effects of ocean disposal of wastes on the marine environment depend on many complicated interactions among many factors or processes including: (1) quantities and kinds of waste, (2) chemical and physical properties of the waste, (3) rate of disharge of the waste into the ocean, (4) interaction of waste with seawater, (5) toxicity of the waste to organisms, and (6) oceanic processes in mixing and dispersing the wastes.[4,5] The ocean should only be used as the medium for waste disposal when the wastes have been well characterized, the oceanic processes at the dumpsite have been fully understood, the technology of waste disposal has been well developed, there is no permanent environmental degradation brought about by effects of disposal, and all other alternatives for waste disposal have been examined and compared with ocean disposal.

Serious marine pollution can occur from improper utilization of the ocean for waste disposal. More outbreaks of coastal water pollution are occurring all the time. In some countries disposal of wastes at sea conflicts with other appropriate activities involving use of the sea. Historically, the ocean has always been considered a resource. Recently (during the last 100 years), the ocean has been used as an easier and more accessible medium for disposal of major quantities of wastes. The conflict over using the ocean, either as a resource or a waste disposal site, will continue.

Although the problems of ocean pollution are recognized, the practice of ocean disposal may still be preferred in those areas where other options for waste disposal have significant limitations, such as groundwater contamination, the scarcity of vacant land for landfills, technological barriers, and economic and human health considerations. Under these limitations continuation of ocean disposal requires a comprehensive management program to minimize environmental impacts.

Our goals in this chapter are to evaluate current practices of ocean disposal in tropical areas and to provide recommendations for improving the management of ocean disposal. Information will be presented on the types and quantities of the wastes previously and currently disposed of in tropical ocean waters, some physical and chemical properties of these wastes, a discussion on oceanic processes that affect the behavior of wastes in the ocean, and recommendations.

II. CURRENT PRACTICES IN OCEAN DISPOSAL

A. OCEAN DUMPING AND TYPES OF WASTES

Ocean dumping of wastes (from ships, barges, or aircraft) in tropical oceans is currently

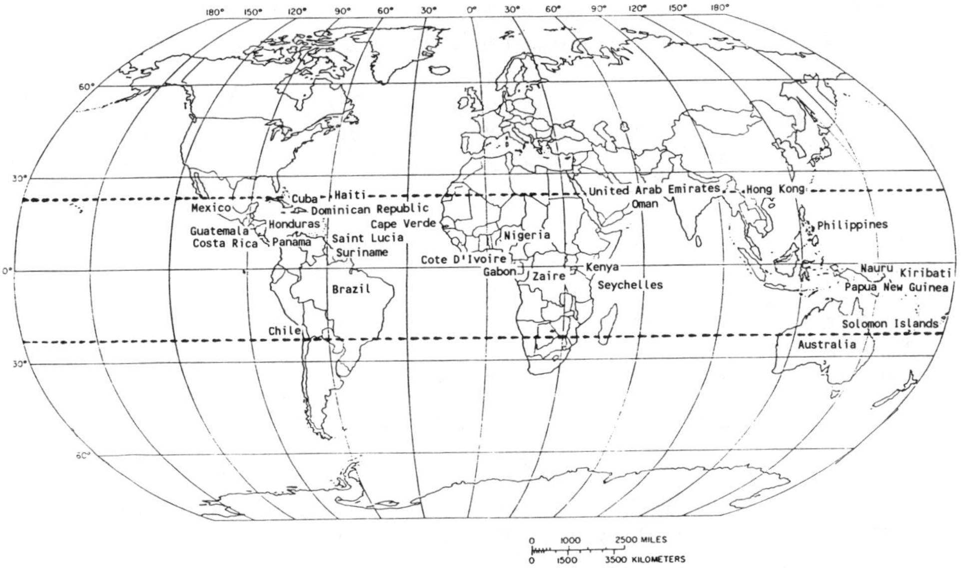

FIGURE 1. Contracting parties to the London Dumping Convention in tropical areas.

regulated by the LDC, an international organization. The LDC originated in June 1972 and entered into force on August 30, 1975. Each Contracting Party to the LDC is required to establish a national system to control ocean dumping, to maintain records of ocean dumping activities through a permit program, to monitor the conditions of the sea for the purposes of the LDC, and to report records of permits and information on monitoring activities to the International Maritime Organization (IMO).[5-7] At present, there are 64 Contracting Parties to the LDC.[2] The IMO (prior to 1982, IMO was called the Inter-Governmental Maritime Consultative Organization [IMCO]) has been receiving reports from the Contracting Parties to the LDC for a number of years. In annual reports prepared by the IMO, information on the number of permits, quantities and types of wastes to be discharged at sea, and location of the dumpsite are summarized and circulated to all Contracting Parties.[5] In the period 1976 to 1985, only two LDC Contracting Parties in tropical areas, Australia and the U.K. (Hong Kong), reported their ocean disposal activities to the convention. It should be noted that Australia had documented its ocean disposal activities even before becoming one of the Contracting Parties to the LDC in September 1985.

Wastes dumped in tropical oceans, according to IMO reports,[1,2] have been industrial wastes, sewage sludge, and dredged materials. Table 1 provides information on wastes dumped in tropical oceans over the period 1976 to 1985; the following describes briefly some properties of the dumped wastes.

1. Industrial Wastes

Industrial wastes include liquid and solid wastes. Table 2 lists the kinds of industrial wastes dumped at sea.[2] Historically, the liquid wastes dumped at sea have included acid-iron waste, fish-processing liquids, metal refinery wastes, and gas pipeline flushing wastes. Acid-iron waste is produced from the production of titanium dioxide and is characterized by its low pH (<1) and relatively high concentrations of dissolved iron and hydrochloric acid (Table 3). Its bulk density and suspended solids content are around 1.15 g cm^{-3} and 0.01 to 10 g l^{-1}, respectively. A number of trace metals of environmental concern are also contained in the acid-iron waste.[8,9]

According to the IMO[1,2] only Australia and Hong Kong provided information to the

TABLE 1
Quantities of Waste Dumped in Tropical Oceans, 1976—1985

	Tons (× 10^6)									
	1976	**1977**	**1978**	**1979**	**1980**	**1981**	**1982**	**1983**	**1984**	**1985**
Industrial wastes										
Liquid wastes										
Australia	0.03	0.2	0.2	—	—	—	—	—	0.3	0.3
Solid wastes										
Australia	0.02	0.2	0.2	—	—	—	—	—	—	0.02
Hong Kong	0.2	0.5	1.1	1.9	2.4	1.5	—	0.004	0.04	—
Sewage sludge										
Hong Kong	—	0.001	—	—	—	—	—	—	—	—
Dredged materials										
Australia	—	—	—	—	—	—	—	—	30	51
Hong Kong	—	6.3	7.3	8.3	1.4	7.6	19	7.6	5.5	—

Note: Data modified from References 1 and 2. The reader should be warned against overinterpretation of the data. Some of the data reflect amounts licensed for dumping, others reflect actual dumped amounts.

TABLE 2
Industrial Wastes Listed on Ocean Dumping Permits

Liquid wastes	Aqueous solutions from chemical industry
	Titanium dioxide industry wastes
	Drilling fluid and cuttings
	Inorganic sludges and phosphogypsum
	Paper mill wastes
Solid wastes	Filter cake
	Colliery wastes
	Fly ash
	Demolition material
	Cement
	Excavation material
	Barites
	Driftwood
	Rock
	Embers
Bulk solid wastes	Derelict vessels
	Drilling platforms
	Scrap metal
	Decommissioned arms and ammunition
	Unserviceable equipment
BOD wastes	Fish offal
	Food industry wastes
Other materials	Oil
	Oil dispersants
	Radioactive tracers for scientific experiments

LDC of their dumping of liquid wastes in the tropical oceans in the years 1976 to 1985. Before becoming a Contracting Party, Australia reported (relating to permitted quantities only) that chemical wastes licensed for ocean disposal were approximately 30,000 tons in 1976; 0.2 million tons in 1977 and 1978, respectively; and 0.3 million tons in 1984 and 1985, respectively.

Of the solid wastes dumpted at sea (Table 1), permits were issued by the U.K. for Hong Kong for the disposal of munitions: approximately 0.2 million tons in 1976; 0.5 million tons in 1977; 1 million tons in 1978; 2 million tons in 1979; and 1.5 million tons in 1981.[1,2]

TABLE 3
Elemental Composition ($\mu g\ l^{-1}$) of Acid-Iron Waste

Element	Allied Chemical Co. waste (U.S.)	DuPont-Edge Moor waste (U.S.)
Al	—	1.7×10^6
As	50	—
Cd	5.0	—
Cl	—	1.9×10^8
Cr	88	2.1×10^5
Cu	51	—
F	1.70×10^7	—
Fe	8.7×10^3	5×10^7
Pb	67	4.1×10^4
Mn	—	1.7×10^6
Hg	<1.0	—
Ni	64	—
SO_4^{2-}	—	7.9×10^5
Ti	—	3×10^6
V	—	2×10^5
Zn	330	—

Data modified from References 8 and 9.

Australia also issued permits in 1977 and 1978 for disposal of 0.2 million tons of iron calcine — a waste material produced when iron pyrite is roasted to obtain sulfur dioxide. In 1984 and 1985, about 23,000 ton of solid wastes were permitted by Australia for ocean disposal.

Australia was also the only country who submitted information on disposal of Annex I or Annex II liquid and solid wastes. This information only related to permitted quantities in 1977 and 1978.[1] The LDC categorizes waste substance into Annex I and Annex II for permit purposes (Table 4). Dumping of Annex I substances is prohibited; Annex II substances can be disposed of at sea only by special permit. All other substances that do not fall into either the Annex I or II category require a general permit to be dumped at sea; these permits must be obtained from either the flag state or the loading state.

2. Sewage Sludge

Sewage sludge is an anaerobic waste product from treatment of municipal wastewater. The sludge is in aqueous form containing about 3% suspended particles by weight. These particles span a wide range of sizes from large fragments of debris (>1 cm) to microscopic and colloidal particles.[10] The particles consist of organic detritus, fibers, food wastes, microorganisms, and inorganic matter.[5] Sewage sludge has an overall density of about 1.01 g cm^{-3}, which is less than that of seawater (1.025 g cm^{-3}) and is substantially less dense than heavier marine sediment (1.2 to 1.6 g cm^{-3}). The lower density of sludge particles is due to their predominantly organic composition. Sewage sludges contain trace elements in varying concentrations and can also contain a number of different types of synthetic organic compounds (Table 5).[11]

In tropical areas, over the period 1976 to 1985, the only Contracting Party to the LDC reporting its dumping of sewage sludge at sea was the U.K. (Hong Kong) in 1977 (Table 1).

3. Dredged Materials

Dredged materials range from clean sands to heavily contaminated fine grained materials with associated chemical and physical characteristics displaying significant variability.[10]

TABLE 4
London Dumping Convention Classification of Annex I and Annex II Waste Materials for Permit Purposes

Annex I Waste Substances

1. Organohalogen compounds
2. Mercury and mercury compounds
3. Cadmium and cadmium compounds
4. Persistent plastics and other persistent synthetic materials, for example, netting and ropes, which may float or may remain in suspension in the sea in such a manner as to interfere materially with fishing navigation or other legitimate uses of the sea
5. Crude oil, fuel oil, heavy diesel oil, and lubricating oils, hydraulic fluids, and any mixtures containing any of these, taken on board for the purpose of dumping
6. High-level radioactive wastes or other high-level radioactive matter, defined on public health, biological, or other grounds, by the competent international body in this field (at present the Internation Atomic Energy Agency) as unsuitable for dumping at sea
7. Materials in whatever form (e.g., solids, liquids, semiliquids, gases, or in a living state) produced for biological and chemical warfare
8. The preceding paragraphs of this annex do not apply to substances which are rapidly rendered harmless by physical, chemical, or biological processes in the sea provided they do not:

   (a) Make edible marine organisms unpalatable
   (b) Endanger human health or that of domestic animals

   The consultative procedure provided for under Artgicle XIV (of the LDC) should be followed by a Party if there is doubt about the harmlessness of the substance.
9. This Annex does not apply to wastes or other materials (e.g., sewage sludges and dredged spoils) containing the matters referred to in paragraphs 1—5 above as trace contaminants. Such wastes shall be subject to the provisions of Annexes II and III (wastes can be disposed of at sea by general permit) as appropriate.
10. Paragraphs 1 and 5 of this Annex do not apply to the disposal of wastes or other matter referred to in these paragraphs by means of incineration at sea. Incineration of such wastes or other matter at sea requires a prior special permit. In the issue of special permits for incineration the Contracting Parties shall apply the Regulations for the Control of Incineration of Wastes and Other Matter at Sea set forth in the Addendum to this Annex (which shall constitute an integral part of this annex) and take full account of the Technical Guidelines on the Control of Incineration of Wastes and Other Matter at sea adopted by the Contracting Parties in consultation.

Annex II Waste Substances

The following substances and materials requiring special care and only special permits are issued according to articles of the LDC.

1. Wastes containing significant amounts of the matters listed below: arsenic, lead, copper and their compounds, zinc, organosilicon compounds, cyanides, fluorides, pesticides and their by-products not covered in Annex I.
2. In the issue of permits for the dumping of large quantities of acids and alkalis, consideration shall be given to the possible presence in such wastes of the substances listed in paragraph 1 and to the following additional substances: beryllium, chromium, nickel and their compounds, vanadium.
3. Containers, scrap metal, and other bulky wastes liable to sink to the sea bottom which may present a serious obstacle to fishing or navigation.
4. Radioactive wastes or other radioactive matter not included in Annex I. In the issue of permits for the dumping of this matter, the contracting parties should take full account of the recommendations of the competent international body in this field, at present the Internation Atomic Energy Agency.
5. In the issue of special permits for the incineration of substances and materials listed in this Annex, the Contracting Parties shall apply the Regulations for the Control of Incineration of Wastes and Other Matter at Sea set forth in the Addendum to Annex I and take full account of the Technical Guidelines on the Control of Incineration of Wastes and Other Matter at At the Sea adopted by the Contracting Parties in consultation, to the extent specified in these Regulations and Guidelines.

From Office of the London Dumping Convention, International Maritime Organization, London.

TABLE 5
Concentrations ($\mu g\ l^{-1}$) of Constituents in New York City Sewage Sludge

Constituent	Oakwood Beach Treatment Plant	Newtown Creek Treatment Plant
Ag	4	38
As	18	14
Cd	<5	62
Cl	2.4	2.9
Cr	19	97
Cu	370	3220
Hg	3.8	6.9
Pb	180	640
Ni	57	520
Se	10	4
Zn	270	1370
PCBs	<2	0.5
Aldrin	<0.01	<0.01
Dieldrin	<0.04	0.35
DDT	0.03	0.58
Endosulfan	0.03	<0.03
Endrin	<0.05	<0.05
Heptachlor	<0.01	<0.01
Lindane	<0.02	<0.02
Toxaphene	<1	<1

Data from Reference 11.

TABLE 6
Trace Metal Concentration ($mol\ kg^{-1}$) of Dredged Material

Constituent	Dredged materials
Fe	0.02—0.90
Mn	$(0.4—10) \times 10^{-3}$
Zn	$(0.5—8) \times 10^{-3}$
Cu	$(0.8—9400) \times 10^{-6}$
Ni	$(0.2—2.6) \times 10^{-3}$
Cr	$(0.02—3.8) \times 10^{-3}$
Pb	$(5—1900) \times 10^{-6}$
Cd	$(0.4—600) \times 10^{-6}$
Hg	$(1—10) \times 10^{-6}$

Data modified from Reference 12.

Concentrations of major and trace elements as well as the concentration of the variety of synthetic organic compounds can vary over several orders of magnitude (Table 6).[12] Physical properties of dredged materials, including grain size, bulk density, water content, and geotechnical characteristics, are especially variable due to the kind or type of sediment being dredged, which is itself dependent on geological and watershed characteristics, as well as to the operational procedures used in dredging and disposal.[10]

In reporting the composition of dredged materials, specific comments are received by IMO from the Contracting Parties.[1] For example, samples of dredged materials for analysis are usually taken from the top 0.5 cm of harbor sediments. Below the surface layer the heavy metal content would normally be significantly lower in composition. Thus, the actual amount of heavy metals would always be lower than the amount reported to the LDC calculated on the basis of the sample. Also, comments are made on the reliability of calculating contents of metals and other substances in dredged spoils when calculations are based on a few small samples which may or may not be homogeneous from several tons. It is therefore, recommended[1] that in preparing a list of Annex I and II materials dumped in each dredged spoil site, an account should be taken of the way in which the sample was taken for analysis, the type of analyses undertaken (e.g., bulk sediment, chemical analysis of certain particle size fractions, elutriate analysis), and the origin of the spoil.

The U.K. (Hong Kong) was the only Contracting Party to LDC providing information on the quantities of dredged materials dumped in a tropical sea over the period 1977 to 1984 (Table 1). Permits from Hong Kong indicate that the quantity of dredged materials dumped was relatively consistent except in 1982. The significant increase in the amount of dredged materials during 1982 indicates that dredging is a sporadic activity. In Australia, permits were issued for disposal of 30 million tons of dredged spoils in 1984 and 51 million tons in 1985.

B. OCEAN OUTFALLS FOR SEWAGE DISCHARGE

Deep-ocean sewage outfall is another form of waste disposal. The primary goal in this form of disposal is to achieve a significant dilution of the sewage as quickly and as thoroughly as possible so as to minimize the impact on the marine environment.[13] Normally, the "outfall" consists of a pipe extended offshore; a manifold diffuser at the end of the pipe provides the best form of dilution. Sewage effluent passing through the diffuser is lighter than the more dense surrounding seawater and, thus, tends to rise toward the surface. The rising sewage plume mixes with the surrounding seawater and becomes diluted. Dilution will be dependent on the rate of discharge and oceanic conditions of the surrounding seawater. If currents around the diffuser are strong, the plume will tend to move with the current, forming a field of diluted sewage. Ultimately, sewage components will be diluted to conditions of ambient seawater. If the water body is shallow at the outfall and/or stratification in the water column is absent, the sewage plume will rise to the surface easily. This is obviously undesirable because winds and waves can lead to the transport of sewage toward the shore before it is diluted to ambient conditions or decomposed by marine organisms. If water column stratification is present near or at the diffuser of outfall, the sewage plume can become trapped well below the surface of the water. Subsurface trapping of the sewage is usually desirable since it tends to ensure that wastes will not wash up onto beaches.

Oahu, Hawaii, ocean outfalls have been used for the disposal of municipal wastewater. Monitoring to determine the effects of sewage effluent on benthic communities has been carried out since 1974,[14] and results show no major changes in the benthos associated with the discharged sewage effluent. Further studies by Dollar[15] indicate that at an ecosystem management level, the overall environmental effect of sewage effluent on benthic and pelagic ecosystems is insignificant. Nelson[16] also showed that the Sand Island outfall diffuser off the southern coast of Oahu, Hawaii, has a small effect on the benthic faunal community abundance and diversity. The reported insignificance of disposal effects is primarily due to effective dispersal and dilution of the sewage.[15] The most recent studies by Russo et al.[17] indicate that the abundance and species richness patterns of both mollusks and nonmollusks show little relationship to the location of the Sand Island outfall diffuser.

While island based outfalls appear to provide an environmentally acceptable means for disposal, similar discharge in enclosed regions can result in serious problems. Generally, effluent is enriched in nitrogen and phosphorus, and this can lead to eutrophication which will significantly alter the composition of the biological community.

III. FATE OF WASTES IN THE OCEAN

A. SOME GENERIC CONSIDERATIONS

The fate of a waste in the ocean results from the integration of various oceanic processes occurring both in the water column and in the sediments.[5] Once disposed of at sea, the waste may be dispersed in the water column or sink to the seafloor to form part of the sediment. The depth of penetration in the water column of a waste plume will depend on the kind of waste and its physical and chemical properties, density gradient in water column, and the currents present at the dumpsite.[18-20] When settling from surface water to the sediment, wastes interact with seawater through various physical, chemical, and biological processes. When the waste is discharged or dumped, some processes may occur immediately; others may occur later. The rate of interaction between the wastes and seawater ranges from seconds to years.

Physical processes are at all times acting to dilute wastes present in the water column either through dispersion or advection. Dilution of a waste will depend on the kinds of waste, the disposal site, and the time scales of physical processes. Ketchum and Ford[21] indicate that the mixing coefficient for ocean disposal of acid-iron waste is proportional to the rate of waste discharge from the transport vessel. Initially, dilution processes are usually rapid (a few seconds), but complete replacement of a water body containing waste can take a few years or possibly decades. Csanady[20] estimates that actual replacement of the entire water mass containing the diluted waste will depend on the extent of large-scale circulation processes and may take months to years.

Many chemical processes can occur almost simultaneously and all the way through the water column to the seafloor. The rate of chemical interaction (waste with seawater) in the water column ranges from seconds to hours and depends on physical and chemical processes. These processes include acid-base neutralization, dissolution or precipitation, flocculation of solid phases, surface exchange reactions between particles and seawater (adsorption, desorption), volatilization at the sea surface, and changes in oxidation state.[5] On mixing with seawater, the physical and chemical characteristics of a waste can be substantially modified by oceanic physical processes. The chemical and physical forms of an element in a liquid waste, rather than an element's total concentration, will determine a waste's chemical and biological behavior in the ocean.[5] Chemical speciation of an element includes consideration of its oxidation state, its formation of ion pairs and organic complexes, and its incorporation into solid phases.

Biological processes and effects occurring in the water column will be time dependent and will be a function of physical and chemical processes. The effect of a biological process may be short-term (toxicity to individual or groups of organism) or long-term (sublethal effects, food chain effects, community and ecosystem effects). Waste materials may be degraded by microorganisms, and, thus, their physical and chemical characteristics are changed. Some wastes containing organic matter may be ingested by marine organisms and incorporated into their body tissues, or they may be metabolized by the marine organisms resulting either in a change of the character of the compound or its complete digestion. Some waste components may be accumulated and passed upward to higher trophic levels in the food chain.

When the waste reaches the seafloor, it will be further changed in its physical and chemical characteristics. Once on the seafloor, waste materials will be concentrated rather than diluted due to sedimentation, diagenesis (changes that take place within a sediment during and after burial), and bioaccumulation processes. The ultimate fate of wastes reaching the sediment is determined by those processes occurring in the benthic region. The mobility of waste components in the sediment is mainly determined by geological (sediment transport), geochemical, and biological processes. Waste components may be chemically immobilized, resulting from the formation of insoluble compounds, such as sulfide salts, in sediment. Some of the immobilized waste components may later be remobilized due to the change of redox potential of sediment. The subsequent chemical mobility of these particles in the sediment will depend on the physicochemical properties of the sediment.[22,23]

The activity of benthic organisms, e.g., bioturbation, also affects the mobility and modification of waste components in the sediment. Organisms may ingest some waste particles and convert them to biomass or fecal material which will then be consumed by other organisms. Thus, waste materials can be recycled many times before they become permanently buried in sediment. Geochemical and biological processes taking place on the seafloor occur on much longer time scales than those in the water column.

Sediment transport, a process resulting from the interaction between sediment particles and bottom currents, will also determine the fate of wastes in the sediment. Waste materials may become permanent sediment components or remain at the surface of the sediment ready for further interaction with seawater.

B. FATE OF INDUSTRIAL WASTES

When dumped at sea, the acid portion of the acid-iron waste (a common industrial waste) is neutralized because of the alkaline nature of seawater; ferric hydroxide particles are immediately formed when the waste comes in contact with the seawater because of the relatively high concentrations of iron (Fe^{2+}) present in the wastes. Thus, during and immediately after disposal, a plume of ferric hydroxide particles clearly identifies the path of the acid-iron waste disposed of in the ocean.

The fate of the newly formed particles as well as the dissolved components will depend on the oceanographic conditions at the dumpsite, the dumping operation, and season of the year.[10] Waste particles can accumulate in the pycnocline during summer dumping and ride an internal wave for several hours after disposal. Therefore, density structure in the water column and the presence of internal waves are important factors in determining the short-term fate of the waste. Generally, oceanographic conditions at selected dumpsites favor significant dispersion, typically resulting in a several 1000-fold dilution soon after the dumping event.[24]

C. FATE OF SEWAGE SLUDGE

When sewage sludge is dumped into the ocean, a partitioning occurs between the dissolved constituents and the particulate matter. Factors affecting partitioning include the type of disposal (spot dump vs. line dump), oceanic conditions, depth of the pycnocline, and settling rate of the particles. The dissolved constituents mix fairly rapidly into surface waters. The dense particles can settle through the water column rapidly, while the lighter particles may accumulate above the pycnocline.[25] Stratification in the water column is an important barrier to some of the settling sludge particles. Proni and Hansen[25] also detected sludge particles spreading horizontally at a rate of about 50 cm s^{-1} on the thermocline from the apex of a plume after a dumping event.

Some particulate matter in sewage sludge may settle to the bottom of the ocean. Once on the seafloor, sewage sludge will undergo further decomposition by bottom-dwelling microorganisms; some particulates are transformed to their dissolved form and may release important plant nutrients. If these nutrients are mixed upward into the photic zone of the water column, they can be utilized by marine plants.[5]

D. FATE OF DREDGED MATERIALS

The philosophy of disposing of dredged materials is different from that of industrial liquid waste, sewage effluent, and sewage sludge. Usually, dispersion is preferred for the disposal of industrial and sewage wastes. For dredged materials, the containment philosophy has to be considered. Although the majority of the material is clean and uncontaminated, a smaller percentage of the dredged materials is severely contaminated, making it unsuitable for ocean disposal. Thus, a containment technique should be employed to minimize environmental impacts.

For uncontaminated materials dispersion of the materials is of minimal concern with regard to long-term effects. Hydraulic dredging techniques are employed, and the materials are disposed of via hopper dredge, pipeline, or side-casting techniques at a designated site. Sites are often selected to favor dispersal so as to avoid unnecessary mounding of dredged materials and potential development of associated navigational hazards. In areas where the dispersal of sediment associated with contaminants is of concern, clamshell buckets or similar confined removel techniques are favored. Sediments are generally placed in scows for transport to the designated offshore disposal sites. These disposal areas have to be carefully selected to maximize containment in order to minimize long-term environmental effects. In this regard, capping of contaminated dredged spoils with clean spoils is being utilized by the U.S. Army Corps of Engineers in New England waters in the U.S.

E. SPECIAL CONSIDERATIONS IN TROPICAL OCEANS

Tropical oceans comprise the major, central portion of the world oceans, representing about 37% of the area of all seawater.[26] In general, world oceans (major water masses) are considered chemically stable with little variation in chemical characteristics within a water mass situated over a relatively large area. The tropical oceans, in considering ocean disposal of wastes, are unique in their physical characteristics, i.e., ocean circulations and currents.

The tropical oceans are characterized by a well-mixed warm surface layer (100 to 300 m) and a sharp thermocline, which separates the warm surface layer from the cold water beneath. The major ocean currents of the tropics are a reflection of the wind system in the area. Beneath the trade winds are the westward-flowing Equatorial Currents confined mostly to the surface layer, with average speeds of 25 to 75 cm/s.[27] Between these two broad westward flows is a relatively narrow (300 to 500 km wide) eastward-flowing Equatorial Countercurrent. At the equator, there is a subsurface eastward-flowing Equatorial Undercurrent (200 m thick and 300 km wide). In addition, the tropical oceans have a specific ability to respond rapidly to a change in wind stress.[28] The direction of currents in the tropics may be reversed in a few weeks in response to monsoon alternation.[26] Longhurst and Pauly[26] attribute this unique ability to the occurrence of the El Niño-Southern Oscillation (ENSO) phenomenon.

Since the thermocline is a density interface where some waste particles from dumping may accumulate, extensive stratification in the water column of the tropical oceans is an important barrier to some of the settling particles. One may expect that the dumped wastes may be spread and mixed horizontally rather than vertically owing to stratification and strong currents. Therefore, wastes which are not dispersive preferred, e.g., severely contaminated dredged materials, may not be disposed of at this area. Also, rapid changes in current directions must be taken into account when selecting dumpsites.

IV. CONCLUSIONS

Twenty-seven tropical countries are members of the London Dumping Convention, yet only two countries actively report on their dumping practices. Therefore, lack of published information on the disposal activities has prevented us from providing a comprehensive review of the subject for most tropical regions. Rather, in this chapter we have focused on generic aspects because much of what we already know for other regions would apply to tropical areas.

It is probable that tropical regions will be used increasingly in the future for waste disposal in one way or another because a number of countries already belong to the LDC and also because efforts to strengthen agricultural and industrial activities are likely to occur in the area. Development of environmentally acceptable waste management practices should run in parallel with all activities involving improvement in the overall well-being of the tropical area. In this regard we make the following recommendations:

1. *Use the holistic approach to waste management.* It is likely that no single medium (land, air, or water) would satisfy the disposal requirements for all forms of wastes generated. Rather, each medium needs to be evaluated based on its suitability for assimilating waste in an environmentally acceptable manner. This approach will necessarily involve an analysis of cost-benefit by considering the evaluation of a number of factors, including minimization of waste production, characteristics and quantities of wastes, cost of disposal, health and environmental effects, the demonstrated need for ocean disposal in light of other disposal alternatives, and beneficial use of the wastes in order to avoid disposal in the first place.
2. *Provide rapid turnaround of information on current disposal activities.* Evaluation of effects of ocean dumping starts with information on the kinds and amounts of wastes permitted for disposal. While 27 of the 64 countries belonging to the LDC are found in tropical regions, only 2 nations reported on their ocean dumping activities in the period 1976 to 1985. Improvement in information on ocean disposal activities can lead to a better understanding of the fate and effects of waste, which, in turn, can be used for better management of waste disposal.
3. *Develop, use, and report on monitoring activities.* Concerns about the disposal of wastes in the marine environment stem from the worldwide desire to prevent the occurrence of irreversible degradation of marine environments. The purpose of monitoring is to determine if environmental quality objectives have been met. Each Contracting Party to the LDC is required to report information on monitoring activities.
4. *Participate in international workshops and symposia devoted to aspects of marine waste disposal.* There is an abundance of information continually being made available on waste disposal at sea. The information deals with the technology of waste disposal and environmental monitoring, scientific findings on the fates and effects of waste, and the special needs of some countries. Countries in tropical regions need to participate actively in the gathering and sharing of such information in order to make use of appropriate findings and also to contribute to the knowledge in this field.

ACKNOWLEDGMENTS

We thank Dr. Manfred Nauke, Office of the London Dumping Convention, International Maritime Organization, for providing information on dumping practices by Contracting Parties to the LDC. We also thank Annette Bernard for her work in preparing the manuscript.

REFERENCES

1. International Maritime Organization (IMO), Report on the Nature and Quantities of Wastes Dumped at Sea, IMO, London, 1985, 50.
2. International Maritime Organization (IMO), Amounts of Wastes and Other Matter Dumped and Incinerated at Sea 1976—1985, IMO, London, 1990, 15.
3. **Nauke, M. K.,** Obligations of contracting parties to the London Dumping Convention, in *Oceanic Processes in Marine Pollution,* Vol. 3, Champ, M. A. and Park, P. K., Eds., Krieger, Melbourne, FL, 1989, 123.

4. **Johnston, R.,** *Marine Pollution,* Johnston, R., Ed., Academic Press, London, 1976, 729.
5. **Duedall, I. W., Ketchum, B. H., Park, P. K., and Kester, D. R.,** Global inputs, characteristics, and fates of ocean-dumped industrial and sewage wastes: an overview, in *Wastes in the Ocean,* Vol. 1, Duedall, I. W., Ketchum, B. H., Park, P. K., and Kester, D. R., Eds., Wiley-Interscience, New York, 1983, 3.
6. **Moore, G.,** Legal aspects of marine pollution control, in *Marine Pollution,* Johnston, R., Ed., Academic Press, London, 1976, 589.
7. **Timagenis, G. J.,** *International Control of Marine Pollution,* Vol. 1, Oceana Publications, New York, 1980, 404.
8. **Brown, M. F., Kester, D. R., and Dowd, J. M.,** Automated iron measurements after acid-iron waste disposal, in *Wastes in the Ocean,* Vol. 1, Duedall, I. W., Ketchum, B. H., Park, P. K., and Kester, D. R., Eds., Wiley-Interscience, New York, 1983, 157.
9. ERCO, Physical, Chemical, and Biological Monitoring Program at the Acid Waste Disposal Site in New York Bight — Elizabeth Plant Waste, Second Quarter 1986 Dump, ERCO, Cambridge, MA, 1986.
10. U.S. National Research Council, *Monitoring Particulate Wastes in the Oceans,* Marine Board, National Research Council, Washington, D.C., 1988.
11. EA Engineering, Science, and Technology, Inc., Response to EPA's Comments on a Special Permit Application for the Disposal of Sewage Sludge from Fourteen New York City Water Pollution Control Plants at the Deepwater Municipal Sludge Dump Site, EA Engineering, Science, and Technology, Inc., Sparks, MD, 1988.
12. U.S. National Research Council, *Dredging Coastal Ports,* National Academy Press, Washington, D.C., 1985, 212.
13. **Bishop, P. L.,** *Marine Pollution and Its Control,* McGraw-Hill, New York, 1983, 357.
14. **Kay, E. A. and Kawamoto, R.,** Micromolluscan Assemblages in Mamala Bay, Oahu, 1974—1982, Tech. Rep. No. 158, Water Resources Research Center, University of Hawaii at Manoa, Honolulu, 1983, 73.
15. **Dollar, S. J.,** Response of Benthic Ecosystems to Deep Ocean Sewage Outfalls in Hawaii: Nutrient Fluxes at the Sediment-Water Interface, Ph.D. thesis, University of Hawaii, Honolulu, 1986, 216.
16. **Nelson, W. G.,** Benthic Infaunal Sampling in Vicinity of Sand Island Ocean Outfall, Oahu, Hawaii, Spec. Rep. 6:20:86, Water Resources Research Center, University of Hawaii at Manoa, 1986, 117.
17. **Russo, A. R., Kay, E. A., Bailey-Brock, J. H., and Cooke, W. J.,** Benthic Faunal Sampling Adjacent to Sand Island Ocean Outfall, Oahu, Hawaii, Spec. Rep. 06:12:88, Water Resources Research Center, University of Hawaii at Manoa, Honolulu, 1988, 95.
18. **Fischer, H. B., List, E. J., Koh, R. C. Y., Imberger, J., and Brooks, N. H.,** *Mixing in Inland and Coastal Waters,* Academic Press, New York, 1979, 483.
19. **Kullenberg, G. E. B.,** Physical oceanography studies related to waste disposal in the sea, in *Wastes in the Ocean,* Vol. 1, Duedall, I. W., Ketchum, B. H., Park, P. K., and Kester, D. R., Eds., Wiley-Interscience, New York, 1983, 87.
20. **Csanady, G. T.,** Long-term mixing processes in slopewater, in *Wastes in the Ocean,* Vol. 3, Duedall, I. W., Ketchum, B. H., Park, P. K., and Kester, D. R., Eds., Wiley-Interscience, New York, 1983, 103.
21. **Ketchum, B. H. and Ford, W. L.,** Waste Disposal at Sea, A Preliminary Report on Acid-Iron Waste Disposal, Woods Hole Oceanographic Institution, Woods Hole, MA, 1948, 18.
22. **Berner, R. A.,** *Principles of Chemical Sedimentology,* McGraw-Hill, New York, 1971, 240.
23. **Aller, R. C.,** The Influence of Macrobenthos on Chemical Diagenesis of Marine Sediments, Ph.D. thesis, Yale University, New Haven, CT, 1977, 600.
24. **Mukherji, P. and Kester, D. R.,** Acid-iron disposal experiments in summer and winter at deepwater dumpsite-106, in *Wastes in the Ocean,* Vol. 1, Duedall, I. W., Ketchum, B. H., Park, P. K., and Kester, D. R., Eds., Wiley-Interscience, New York, 1983, 141.
25. **Proni, J. R. and Hansen, D. V.,** Dispersion of particulates in the ocean studied acoustically: the importance of gradient surfaces in the ocean, in *Ocean Dumping of Industrial Wastes,* Ketchum, B. H., Kester, D. R., and Park, P. K., Eds., Plenum Press, New York, 1981, 161.
26. **Longhurst, A. R. and Pauly, D.,** *Ecology of Tropical Oceans,* Academic Press, New York, 1987, 407.
27. **Knauss, J. A.,** *Introduction To Physical Oceanography,* Prentice-Hall, Englewood Cliffs, NJ, 1978, 338.
28. **Philander, S. G.,** Tropical oceanography, *Adv. Geophys.,* 28A, 461, 1985.

Chapter 10

CONTROL OF METAL POLLUTION IN TROPICAL RIVERS IN AUSTRALIA

Douglas A. Holdway

TABLE OF CONTENTS

I. INTRODUCTION

The majority of potential sources of metal pollution to Australian tropical rivers are from mines and mineral refining operations. Materials that are extracted and refined in northern Australia include uranium, gold, manganese, bauxite, copper, zinc, and lead. All of these operations must manage waste and runoff waters to some extent. In the wet-dry monsoonal climate of the top end of northern Australia, practically all of the annual rain falls during the 4 months of the Southern Hemisphere summer, the remainder of the year being effectively dry. Thus, during the wet season (summer), large quantities of excess water falling onto a mining operation need to be managed, either through storage in large retention ponds (which, in turn, provide increased catchment for water) or release into the local streams and rivers.

During the dry season, large storage (or groundwater and pumping) capacity must be available to mining and milling operations to ensure sufficient water for the extraction and processing of minerals is available. River and stream flow during the dry season is low or nonexistent, and the safe release of any wastewater directly into these tropical rivers and streams is severely restricted by the extremely low dilution capacity. As a result, any operational water releases need to be made during the high flow-high dilution periods in the wet season. This highly restricted period of time in the year during which safe effluent release into the natural environment can occur is unique to the tropics and represents a considerable engineering, water, and wastewater management problem.

II. DESCRIPTION OF WASTEWATERS

The complex wastewaters and effluents produced by the mining and ore-processing industry generally involve mixtures of metals found in the ore, but also may contain many other contaminants such as petroleum products and organic chemicals used on site both in the processing of the minerals (such as cyanide used in gold extraction) as well as in the management of the mine site with the widespread use of pesticides and herbicides to control insect and noxious weed populations, especially in the tropics.[1] The impact of such heterogeneous wastes on the freshwater ecosystem cannot be accurately predicted by only measuring the chemical constituents within the mixture and estimating their toxicity, nor by determining the toxicity of the individual components (which typically change with time and are not fully known) and adding the toxic units.[2]

The complex chemical and biochemical interactions involved when biological organisms are exposed to known chemical mixtures of only two or three individual metals have not been fully understood. Unknown mixtures of chemicals are present in most wastewaters from industrial or mining sources which can involve hundreds of different chemicals in varying concentrations. Thus, the simple chemical measurement of every chemical with absolute certainty is next to impossible and extremely expensive since most analyses require *a priori* knowledge of the chemicals being looked for. The combination of synergistic, antagonistic, and additive effects of chemicals on living organisms, and the changes in concentrations and chemical constituents with time, preclude the prediction of toxicological effects of complex wastewaters on aquatic organisms.[3] The toxicity of Ranger Uranium Mine's Retention Pond No. 4 (RP4) wastewater discussed later on in this chapter is an excellent example of a wastewater toxicity that greatly exceeds the additive toxicity of its known chemical components.

III. TOXICITY TESTING

The only way to determine the real toxicity of complex wastewaters to aquatic organisms is to test the effluents directly for toxicity to whole aquatic organisms.[2] By using whole organisms and the actual wastewaters to be released, the biological and biochemical effects of the exposure are integrated by the exposed organisms, which act like experimental "black boxes" in this respect. Since the size, feeding status, and life stage of aquatic organisms can greatly modify their sensitivity to toxicants,[4-6] and since larval and juvenile life stages as well as reproduction are among the most toxicant sensitive,[7] it is important that toxicity tests be performed on embryo, larval, and early juvenile life stages of aquatic organisms whenever full chronic or reproduction toxicity tests cannot be run. Environmentally "safe" concentrations and dilutions can be estimated from laboratory-determined no-observed-effect concentrations (NOEC) by applying an appropriate safety factor based on both the nature of the toxicants within the wastewater (such as persistence and biomagnification potential) and on the scientific judgment of the environmental toxicologist regarding the overall risk that release of the wastewater represents to the aquatic environment. Safety factors are required owing to the possibility of significant deleterious effects that were not looked for in the laboratory, unknown differences between species, more sensitive species in the wild that were not tested, the restricted exposure times and experimental durations in the laboratory tests, and overall experimental error, both systematic and statistical.[1,2] Irrespective of the quality of the laboratory data, a scientifically defensible safety factor would not normally be less than a factor of ten applied to the lowest NOEC or calculated threshold concentration (defined as the geometric or arithmetic mean of the NOEC and the lowest-observed-effect concentration (LOEC) for the same laboratory test criteria) observed for a variety of test organisms representing different trophic levels within the aquatic ecosystem.

The validity of this "safe" concentration can then only be verified as having no detectable effect on the aquatic ecosystem by monitoring the aquatic ecosystem into which the wastewater is being released.[8] This monitoring must as a matter of design include assessment of impact both within and outside of the mixing zone, an arbitrarily defined area outside of which the wastewater will be at or below the defined "safe" concentration. Since the question is uncertain of whether an effect can be observed in the field before serious irreversible damage occurs outside of the mixing zone, it is essential that laboratory tests be as rigorous and applicable as possible.

Only detection of serious impact of a wastewater before release into the aquatic environment will ensure that the protection of aquatic ecosystems from pollution is possible. Laboratory toxicity testing of wastewaters also enables concentration-response relationships to be determined under controlled conditions, thus allowing for the mechanisms of toxicity, and the possible risk of persistence and biomagnification, to be understood.

IV. TOXICITY TESTING APPROACH

The toxicity of mine wastewaters containing mixtures of metals and other chemicals should be initially assessed at the whole-organism level of biological organization. Higher levels of organization such as population and community are not easily examined in the laboratory and generally require large and expensive equipment when such tests are attempted on other than microscopic ecosystems. Lower levels of biological organization such as organ, tissue, cellular, and molecular levels provide important information regarding mechanisms of toxicity and detoxication processes, but are of value in predicting safe levels only when their information can be related back to whole-body effects.[9] Individual organisms act as biological integrators of their environmental conditions, and any changes in those conditions which lead to significant whole-organism effects such as increased mortality, impaired growth, or impaired reproduction represent significant threats to aquatic ecosystems.

Thus, the three primary types of information necessary to evaluate the potential of any metal or mixture of metals to affect tropical rivers adversely are

1. Increased mortality
2. Decreased growth
3. Impaired reproduction

Associated with the above, but generally more difficult to determine biologically, is the risk of delayed effects as a consequence of persistence in the ecosystem and biomagnification of chemicals through the various trophic levels of the ecosystem to toxic levels. Notable examples of metals of this type are mercury, cadmium, and lead, and uptake-depuration studies can generally indicate the risk of biomagnification by generating specific bioconcentration factors and a biological half-life for each chemical of concern. These types of metals, as with all persistent toxicants, must be regulated based on total loadings into the environment as well as overall concentration within the water at any one time.[8,10] Ambient chemical concentrations within selected tissues (particularly, known target organs such as liver and kidney as well as important food tissues such as muscle) of local organisms representing the various trophic levels within an ecosystem (and especially the tertiary carnivores) should be measured before any major new source of persistent metals is allowed to be released into the environment. These values will establish the background levels against which future tissue analyses for chemical content will be assessed and which will, thus, provide the earliest warning system for the presence of unexpected persistent chemicals.

V. SPECIFIC EXAMPLES OF EVALUATING MINE WASTEWATERS FOR TOXICITY

Ranger Uranium Mines Pty. Limited is located on a lease area completely surrounded by Kakadu National Park in northern Australia. It is situated very close to the Magela Creek which runs into the Magela Floodplain, an extensive area of wetlands that are one of the major ecological assets of Kakadu National Park and which are a significant habitat for thousands of birds. The retention ponds of Ranger are designed to retain the excess water which falls on the mine site and pit during the wet season and which are slightly to moderately contaminated with metals and some organic materials.

The dominant metal of concern in these wastewaters is uranium, but manganese and zinc levels are also elevated relative to the natural Magela waters, as are sulfate levels of the retention pond waters. During years of high rainfall, it is sometimes necessary for the mine to release excess wastewater contained in these retention ponds into the Magela Creek to prevent having to stop mining operations. In order to ensure that any water releases will not result in significant deleterious effects to the aquatic ecosystems of Kakadu and specifically the Magela Creek and Floodplain, the Commonwealth Government of Australia established a statutory authority, the Office of the Supervising Scientist (OSS), as an environmental protector of the region.[11]

As part of this role as ''protector'', the OSS adopted the strategy of prerelease biological testing of any wastewaters with potential for release from the mine site.[12] These tests have been conducted for a number of years on the two major retention pond waters: Retention Pond Number 2 (RP2) located within the restricted release zone (RRZ), and Retention Pond Number 4 (RP4) located outside of the RRZ. At Ranger, the RRZ has been defined to include all areas within the project area in which significant mineralization (uranium concentrations greater than 0.02%) occurs. The zone includes the tailings dam, the mine pit, ore stockpiles, the mill, and RP2 and Retention Pond 3. All water entering the RRZ from rainfall, mine pit seepage, and groudwater bores is retained and may only be discharged with the prior approval of the supervising authority. In addition, no process water may be transferred from the mill-tailings circuit to retention ponds other than the tailings dam.[13] Up to 1990, permission to discharge RRZ waters to Magela Creek has not been granted even though this restriction has caused interruptions to mining. Authorization has been given for the disposal of some RRZ water by land irrigation within the RRZ during the dry season: this authorization excludes process water and tailings water. RP2 receives runoff water from the ore stockpiles and mill area as well as water pumped out of the mine pit. RP4 receives runoff water from the waste rock stockpile. Neither of these retention ponds contains process water from the mill, which is pumped to the tailings dam and evaporated. Waters within the RRZ could only be released from the mine site under very specific and limited conditions, based on social as well as environmental considerations.

The organisms used in these tests were selected from a group of 19 tropical freshwater organisms assessed by the Alligator Rivers Institute for their suitability as laboratory test organisms (Table 1) and included several species of local fish, water fleas (Cladoceran), shrimp, freshwater snails, two species of hydra (Phylum Cnidaria), tadpoles, larval freshwater mussels, and a species of duckweed. Test criteria included mortality, reproduction, behavior, and physiology and sought the most sensitive response for each species.[14-18] After the first season of method development and application (Tables 2 and 4), a series of standardized test methods (in draft form) were published.[19-27] These methods were subsequently applied in the following wet season's prerelease biological tests (Tables 3 and 5 to 8) as well as in the postrelease biological toxicity assessment of naturally diluted RP4 wastewaters following release via a siphon into a billabong (small pond) located on the mine site (Tables 9 to 11).

The results of these tests conducted over 2 years provided a number of insights into complex wastewater toxicity. The first and most important finding was that no one species was always the most sensitive to wastewater exposure. For example, although larval purple-spotted gudgeons *(Mogurnda mogurnda),* green and pink hydra (*Hydra* sp. A and sp. B), and cladocerans (*Moinodaphnia macleayi*) tended to be among the most sensitive organisms, there were some tests where a wastewater (RP4 collected February 1988) was extremely toxic to larval gudgeons while being essentially nontoxic to both species of hydra and only moderately toxic to cladocerans (Table 4). However, exposure to wastewater collected from the same retention pond a year later showed the greatest toxicity to pink hydra and cladocerans, with larval gudgeons less sensitive to exposure (Table 8).

Similar differences were noted in the toxic effects of another major retention pond wastewater (RP2). Water collected from RP2 in March 1988 was extremely toxic to larval gudgeons and cladocerans, while only of moderate toxicity to both species of hydra (Table 2), whereas water collected from the same pond in March 1989 was moderately toxic to cladoceran and pink hydra but effectively non-toxic to larval gudgeons, checkered rainbow fish (*Melanotaenia splendida inornata*), and dainty blue-eyes (*Pseudomugil tenellus*) after 14 d of exposure (Table 3).

Differences in metal toxicity have been previously reported between Australian tropical freshwater organisms. Checkered rainbow fish were noted to be more numerous in metal- and acid-polluted regions than in unpolluted zones of the Finniss River, NT below Rum Jungle, while atyid shrimp (*Caridina* sp) and eel-tailed catfish (*Porochilus rendahli*) were reported absent from the metals-polluted regions.[28] Similarly, laboratory tests assessing the acute toxicity of copper in Sydney tap water to 15 species of Australian freshwater animals revealed a 510-fold difference in copper toxicity between the most and least sensitive organisms.[29]

Another finding was that field dilution of the wastewater by releasing it via a siphon into a billabong (Tables 9 to 11) did not reduce or alter the toxicity of the water from that predicted by the laboratory tests (Table 8). This was true even for water collected from the billabong some 2 weeks after release.[30] There was no obvious biological filter effect operating at this time, and the predictions of toxicity made from laboratory experiments were supported by the observed toxicities of field diluted waters. In effect, a type of field validation was

TABLE 1
Organisms from the Alligator Rivers Region Assessed for Prerelease Biological Testing

Species	Bred	Reared	Potential: Type of tests	Comments
			Vertebrates	
Fish				
1. Fly-specked hardyhead *(Craterocephalus stercusmuscarum)*	Yes	Yes	Moderate Hatchability Larval survival	Obtaining large number of eggs difficult, early rearing difficult
2. Mariana's hardyhead *(C. marianae)*[a]	Yes	No	Poor Hatchability	Larvae not yet successfully reared: low fecundity
3. Dainty blue-eye *(Pseudomugil tenellus)*[a]	Yes	Yes	Good Hatchability Larval survival Reproduction	Low number of eggs; larvae difficult to rear in quantity
4. Spotted blue-eye *(P. gertudae)*	Yes	Yes	Moderate Hatchability Larval survival Reproduction	As for *P. tenellus*
5. Checkered rainbow fish *(Melanotaenia splendida inornata)*	Yes	Yes	Good Hatchability Larval survival Reproduction	Large quantities of eggs; difficult to raise larvae in first week–requires passive aquaculture approach
6. Black-banded rainbow fish *(M. nigrans)*	Yes	Yes	Good Hatchability Larval survival Reproduction	As for *M. splendida inornata*
7. Reticulated perchlet *(Ambassis macleayl)*	Yes	Yes	Moderate Juvenile growth Juvenile survival	Very small eggs; larvae difficult to rear, sexes not dimorphic; low fecundity
8. Purple-spotted gudgeon *(Morgunda morgunda)*	Yes	Yes	Excellent Hatchability Larval survival Reproduction Growth Full life cycle	Large number of eggs; young hatch as embryos; larvae feed on brine shrimp immediately; rapid growth
Frogs				
9. Brown tree frog *(Litoria rothit)*	No	Yes	Good Hatchability Larval survival Growth	Available in large numbers during West season only; robust in captivity
			Invertebrates	
Clam shrimps				
10. *Cyclestheria hislopt*	Yes	Yes	Excellent Larval survival Reproduction Life cycle	Short life cycle; robust; high fecundity; requires flow-through system
Water fleas				
11. *Moinodaphnia macleayt*	Yes	Yes	Excellent Larval survival Reproduction Life cycle	Very short life cycle of 5—7 d; 4-d 3-brood bioassay developed; require careful maintenance of main culture

TABLE 1 (continued)
Organisms from the Alligator Rivers Region Assessed for Prerelease Biological Testing

Species	Bred	Reared	Potential: Type of tests	Comments
Shrimps				
12. *Caridina graciltrostris*	Yes	Yes	Poor Larval survival	Breeding and rearing difficult; larvae not robust
13. *Caridinides wilkinsi*	Yes	Yes	Poor Larval survival	As for *C. graciltrostris*
Snails				
14. *Amerianna carinata*[b]	Yes	Yes	Excellent Hatchability Larval survival Reproduction Life cycle	6-month life cycle test protocol, 14-d test protocol developed
15. *Physastra* sp.[b]	Yes	Yes	Excellent Hatchability Larval survival Reproduction Life cycle	As for *A. carinata*
Mussels				
16. *Velesunio angasi*	No	No	Moderate Larval survival Larval snap rate	Behavioral assay developed; labor intensive, adults must be taken from wild
Hydra				
17. Green hydra *(Hydra* sp. 1)[b]	Yes	Yes	Excellent Reproduction Asexual/sexual	Appear very sensitive to pollutants; very fast reproduction test (3 or 4 d)
18. Pink hydra *(Hydra* sp. 2)[b]	Yes	Yes	As above	As above
Plants				
Duckweed				
19. *Lemna aequinoctialts*	Yes	Yes	Excellent Frond growth Root growth	Appear very sensitive to presence of nutrients and some chemicals; very easy to run tests

[a] Endemic to region.
[b] May be endemic to region.

provided by these results indicating that this particular water was toxic and should not have been released, even indirectly via a billabong, into the waters flowing into Kakadu National Park if a policy of no-observed-effect was to be upheld.

A third concept that this work gives credence to is that a battery of toxicity tests must be performed on a complex and continuously changing wastewater using a number of organisms representing different trophic levels if a dilution (concentration) is to be proved "safe" for release into a natural ecosystem. Any one species or life stage may or may not be the most sensitive organism tested at any one time, and a battery of tests will increase the likelihood of any deleterious effects being observed in the laboratory.

This is borne out by the results of toxicity tests on seepage water of Rockhole Mine

TABLE 2
Toxicity Test Results for RP2 Water Collected March 16, 1988 Given as % RP2 Water ($p \leq 0.1$)

Species tested	Test time (d)	Endpoint	LOEC (%)	NOEC (%)
Purple spotted gudgeon (*Mogurnda mogurnda*)	14	Larval mortality	0.3	<0.3
Checkered rainbow fish (*Melanotaenia splendida*)	14	Hatchability Larval mortality	>32	32
Dainty blue-eye (*Pseudomugil tenellus*)	14	Adult mortality	>32	32
Reticulated perchlet (*Ambassis macleayi*)	14	Growth Adult mortality	>32	32
Freshwater snail (*Amerianna carinata*)	14	Larval mortality Reproduction (underlined) Larval heartrate (underlined)	0.3	<0.3
Green hydra (*Hydra* sp. A)	6	Morphology (underlined) Reproduction Mortality	10	3.2
Pink hydra (*Hydra* sp. B)	6	Morphology Reproduction (underlined) Mortality	32	10
Conchostracan (*Cyclestheria* sp.)	4	Larval mortality	>32	32
Cladoceran (*Moinodaphnia macleayi*)	6	Larval mortality (underlined) Adult mortality Reproduction	0.3	<0.3
Duckweed (*Lemna aequinoctialis*)	14	Frond propagation Root growth (underlined)	0.3	<0.3

Note: When endpoints differed in sensitivity, the most sensitive endpoint (underlined) was used to determine the respective lowest-observed-effect concentration (LOEC) and no-observed-effect concentration (NOEC).

Adit No. 1 collected in September of 1988. This was an old uranium mine from the 1950s which was abandoned in 1962 and has water flowing out of adit No. 1 into Rockhole Mine Creek. Rockhole Mine Creek flows into the South Alligator River, an important river and catchment within Kakadu National Park. This water, containing high levels of uranium, copper, manganese, and radium,[31] proved to be extremely toxic to green hydra reproduction (NOEC <0.1%) and highly toxic to purple-spotted gudgeon larvae (NOEC = 0.1%), while being of only slight toxicity to cladocerans (Tables 12 and 13). If only cladocerans had been tested, this water would not have revealed the extent of toxicity which it possessed and which represents a toxicological threat to the aquatic ecosystem of the South Alligator River below the creek entrance, at least at periods of low river flow.

VI. SUMMARY

1. Protection of tropical rivers from metal pollution requires that industrial and mining wastewaters be biologically tested for aquatic toxicity before release from the site into natural ecosystems occurs and that a "safe" dilution which incorporates a minimum tenfold safety factor applied to the lowest NOEC be determined and utilized.
2. Nineteen species of tropical freshwater organisms from the Alligator Rivers Region

TABLE 3
Toxicity Test Results for RP2 Water Collected March 29, 1989 Given as % RP2 Water

Test	Test time (d)	Endpoint	LOEC (%)	NOEC (%)
Larval gudgeon Test (OFR No. 52)	14	Mortality Length, weight	>32	32
Rainbow fish test	14	Mortality Length, weight	>32	32
Dainty blue-eye test (*Pseudomugil tenellus*)	14	Mortality Reproduction Hatchability	>32	32
Cladoceran test (OFR No. 56)	5	Adult mortality Reproduction Juvenile mortality	10	3.2
Pink hydra test (OFR No. 58)	6	Tentacle morphology Population reproduction	10	3.2
Duckweed test (OFR No. 59)	14	Frond propagation Root growth Frond condition	32 (1.0	10 0.3)[a]

Note: When endpoints differed in sensitivity, the most sensitive endpoint (underlined) was used to determine the respective LOEC and NOEC values. Standard test methods published as Open File Records (OFRs)[19-27] have the OFR number given in parentheses; 32 and 0.3% RP2 water were the maximum and minimum concentrations tested, respectively.

[a] Improved plant growth based on frond propagation and root growth.

of Australia's Northern Territory have been assessed for their applicability in biological toxicity tests, and nine standardized toxicity test methods for eight of these organisms have been developed and published in draft form.

3. Application of these test methods to wastewaters from an operating uranium mine has shown that prerelease toxicity testing provides accurate information on the toxicity of metal-containing wastewaters with a high degree of confidence. Field validation of the laboratory results was obtained when wastewaters which were field diluted through a release into a billabong gave similar results to laboratory-diluted wastewaters.
4. No one species is always the most sensitive to exposure to complex wastewaters. Changes with time in wastewater chemistry, in toxicity, and in the physiological capacity of specific organisms to survive in a contaminated environment (tolerance) can result in different species having varying sensitivities over time to exposure to complex wastewaters collected from the same location.
5. As a result of the unlikelihood of finding the "most sensitive species", it is necessary to test the toxicity of complex wastewaters to a battery of organisms representing different trophic levels of the ecosystem under physical conditions representative of the specific environment needing protection.
6. Use of natural billabongs as "biological filters" for releasing mine wastewaters does not necessarily result in reduced toxicity of the effluent.

Protection of an aquatic ecosystem is an extremely difficult goal to achieve if any policy other than "no detectable impact" is pursued. It is essential that prerelease testing be undertaken before any wastewater or any other waste-material release into aquatic environ-

TABLE 4
Toxicity Test Results for RP4 Water Collected February 25—28, 1988 Given as % RP4 Water ($p \leq 0.1$)

Species tested	Test time (d)	Endpoint	LOEC (%)	NOEC (%)
Purple spotted gudgeon (*Mogurnda mogurnda*)	14	Larval mortality Hatchability Heartrate	0.3	<0.3
Chequered rainbow-fish (*Melanotaenia splendida*)	14	Hatchability Larval mortality	0.3	<0.3
Dainty blue-eye (*Pseudomugil tenellus*)	14	Adult mortality	32	10
Reticulated perchlet (*Ambassis macleayi*)	14	Growth Adult mortality	1.0	0.3
Brown tree frog (*Litoria rothii*)	13	Larval mortality	>32	32
Freshwater snail (*Amerianna carinata*)	14	Reproduction Larval heartrate	3.2	1.0
Green hydra (*Hydra* sp. A)	6	Morphology Reproduction	>32	32
Pink hydra (*Hydra* sp. B)	6	Morphology Reproduction	>32	32
Conchostracan (*Cyclestheria* sp.)	4	Larval mortality	>32	32
Cladoceran (*Moinodaphnia macleayi*)	6	Larval mortality Adult mortality Reproduction	3.2	1.0
Duckweed (*Lemna aequinoctialis*)	14	Frond propagation Root growth	10	3.2

Note: When endpoints differed in sensitivity, the most sensitive endpoint (underlined) was used to determine the respective LOEC and NOEC values.

ments is permitted. Should a wastewater release be approved as being "safe", then a biological monitoring program must be implemented to ensure that unlooked for or unobserved deleterious effects do not occur and that the biological integrity and health of the receiving aquatic ecosystem is preserved.

ACKNOWLEDGMENTS

The prerelease biological testing research discussed in this chapter was undertaken during my tenure as a Senior Research Scientist with the Office of the Supervising Scientist. I am particularly grateful to Ms. Helen Allison, Ms. Magda Wiecek, and Mr. Michael Mannion for technical support. I also wish to thank the now retired Deputy-Director of the Institute, Mr. Vince M. Brown, and the Supervising Scientist of the past 10 years, Mr. Robert M. Fry, for their help and constructive advice. I hope this excellent research organization will continue its valuable research which is essential for the protection of the Alligator Rivers Region and which is slowly revealing the numerous scientific mysteries of this unique World Heritage Area.

TABLE 5
Toxicity Test Results for RP4 Water Collected December 12, 1988 Given as % RP4 Water ($p \leq 0.1$)

Test	Test time (d)	Endpoint	LOEC (%)	NOEC (%)
Embryo gudgeon test (OFR No. 51)	4	Hatchability	1.0	0.3
		Embryo heartrate		
Larval gudgeon test (OFR No. 52)	14	Mortality	>32	32
		Length, weight		
Cladoceran test (OFR No. 56)	5	Adult mortality	3.2	1.0
		Reproduction		
		Juvenile mortality		
		Juvenile length		
Green hydra test (OFR No. 57)	6	Tentacle morphology	32	10
		Population reproduction		
Conchostrachan test (*Cyclestheria*)	4	Larval mortality	>32	32
		Larval length		

Note: When endpoints differed in sensitivity, the most sensitive endpoint (underlined) was used to determine the respective LOEC and NOEC values. Standard test methods published as Open File Records (OFRs)[19-27] have the OFR number given in parentheses. 32 and 0.3% RP4 water were the maximum and minimum concentrations tested, reespectively.

TABLE 6
Toxicity Test Results for RP4 Water Collected January 16, 1989 Given as % RP4 Water ($p \leq 0.1$)

Test	Test time (d)	Endpoint	LOEC (%)	NOEC (%)
Larval gudgeon test (OFR No. 52)	14	Mortality	0.3	<0.3
		Length, weight		
Freshwater snail test (OFR No. 55)	14	Mortality	0.3	<0.3
		Reproduction		
		Juvenile heartrate		
Cladoceran test (OFR No. 56)	5	Adult mortality	3.2	1.0
		Reproduction		
		Juvenile mortality		
Green hydra test (OFR No. 57)	6	Tentacle morphology	0.3	<0.3
		Population reproduction		
Duckweed test (OFR No. 59)	14	Frond propagation	0.3	<0.3
		Root Growth		
		Plant dry weight		
		Frond condition		
Tadpole test	14	Mortality	>32	32

Note: When endpoints differed in sensitivity, the most sensitive endpoint (underlined) was used to determine the respective LOEC and NOEC values. 32 and 0.3% RP4 water were the maximum and minimum concentrations tested, respectively.

TABLE 7
Toxicity Test Results for RP4 Water Collected February 7, 1989 Given as % RP4 Water ($p \leq 0.1$)

Test	Test time (d)	Endpoint	LOEC (%)	NOEC (%)
Embryo gudgeon test (OFR No. 51)	4	Hatchability Embryo heartrate	0.3	<0.3
Green hydra test (OFR No. 57)	6	Tentacle morphology Population reproduction	1.0	0.3

Note: When endpoints differed in sensitivity, the most sensitive endpoint (underlined) was used to determine the respective LOEC and NOEC values. 32 and 0.3% RP4 water were the maximum and minimum concentrations tested, respectively.

TABLE 8
Toxicity Test Results for RP4 Water Collected February 21, 1989 given as % RP4 Water ($p \leq 0.1$)

Test	Test time (d)	Endpoint	LOEC (%)	NOEC (%)
Embryo gudgeon test (mortality only)	4	Mortality	1.0	0.3
Larval gudgeon test (OFR No. 52)	14	Mortality Length, weight	1.0	0.3
12-d-old Gudgeon test (mortality only)	14	Mortality	>32	32
Juvenile perchlet test (OFR No. 54)	14	Mortality Weight	>32	32
Cladoceran test (OFR No. 56)	6	Adult mortality Reproduction Juvenile mortality	0.3	<0.3
Pink hydra test (OFR No. 58)	6	Tentacle morphology Population reproduction	0.3	<0.3
Duckweed test (OFR No. 59)	14	Frond propagation Root growth Plant dry weight Frond condition	0.3	<0.3

Note: When endpoints differed in sensitivity, the most sensitive endpoint (underlined) was used to determine the respective LOEC and NOEC values. 32 and 0.3% RP4 water were the maximum and minimum concentrations tested, respectively.

TABLE 9
Djalkmara Billabong Toxicity Test Results for Water Collected February 20, 1989, 1 h before a Water Release from Retention Pond Number 4 into the Billabong Began

Test	Test time (d)	Endpoint	Toxic water	Nontoxic water
Cladoceran Test (OFR No. 56)	5	Adult mortality Reproduction Juvenile mortality	M	U
Green Hydra Test (OFR No. 57)	6	Tentacle morphology Population reproduction Mortality	M	U
Duckweed Test (OFR No. 59)	14	Frond propagation Root growth Plant dry weight Frond condition	M	U

Note: When endpoints differed in sensitivity, the most sensitive endpoint (underlined) was used to determine the presence or absence of toxicity to each test species ($p \leq 0.1$). Waters tested were middle Djalkmara Billabong water (M) during backfilling by Magela water (pH = 6.56, conductivity = 20.0 μS/cm) and upper Djalkmara Billabong water (U) containing "original" billabong water (pH = 7.49, conductivity = 94.7 μS/cm).

TABLE 10
Djalkmara Billabong Toxicity Test Results for Water Collected February 22, 1989, 2 d after a Water Release from Retention Pond Number 4 into the Billabong Began

Test	Test time (d)	Endpoint	Toxic water	Nontoxic water
Cladoceran test (OFR No. 56)	5	Adult mortality Reproduction Juvenile mortality	C,MU,U	UM,009 M
Green hydra test (OFR No. 57)	6	Tentacle morphology Population reproduction	C,M,MU U	UM,009
Pink hydra test (OFR No. 58)	6	Tentacle morphology Population reproduction	C[a] U	UM,009 M,MU
Duckweed test (OFR No. 59)	14	Frond propagation Root growth Plant dry weight Frond condition	C, M,MU,U	UM,009

Note: When endpoints differed in sensitivity, the most sensitive endpoint (underlined) was used to determine the presence or absence of toxicity to each test species ($p \leq 0.1$). Waters tested were upstream Magela (UM), downstream Magela (009), the confluence of Djalkmara billabong and the Magela Creek (C), middle Djalkmara Billabong (M), midupper Djalkmara Billabong (MU), and upper Djalkmara Billabong water (U).

[a] = stimulation of reproduction

TABLE 11
Water Chemistry for Toxicity Tests Using Djalkmara Billabong Water Collected February 22, 1989

Water	pH	Conductivity (μS/cm)
Upstream Magela (UM)	6.36	24
Downstream Magela (009)	6.40	16
Djalkmara/Magela confluence (C)	6.83	60
Middle Djalkmara (M)	7.52	220
Midupper Djalkmara (MU)	7.45	380
Upper Djalkmara (U)	7.52	520

TABLE 12
Toxicity Test Results for Rockhole Mine Adit Seepage Water Collected September 6, 1988 (% seepage water)

Test	Test time (d)	Endpoint	LOEC (%)	NOEC (%)
Embryo gudgeon test (OFR No. 51)	7	Hatchability Embryo heartrate	1.0	0.1
17-d-old Gudgeon test (OFR No. 52)	14	Mortality Length, weight	1.0	0.1
Cladoceran test (OFR No. 56)	6	Adult mortality Reproduction Juvenile mortality	100	10
Green hydra test (OFR No. 57)	6	Tentacle morphology Population reproduction	0.1	<0.1

Note: When endpoints differed in sensitivity, the most sensitive endpoint (underlined) was used to determine the respective LOEC and NOEC values. 100 and 0.1% Rockhole mine adit seepage water were the maximum and minimum concentrations tested, respectively. Control and dilution water was upstream Rockhole Creek water collected September 6, 1988 100 m above first seepage.

TABLE 13
Water Chemistry for Toxicity Tests Using Rockhole Mine Adit Seepage Water Collected September 6, 1988

Water	pH	Conductivity (μS/cm)
Upstream Rockhole Creek control water	7.10	30
0.1% Rockhole mine adit seepage water	7.10	30
1.0% Rockhole mine adit seepage water	7.00	40
10% Rockhole mine adit seepage water	5.70	60
100% Rockhole mine adit seepage water	3.00	630

REFERENCES

1. **Holdway, D. A.,** Mining and tropical freshwater environments: laboratory studies, in *Proc. Joint Australian Institute of Biology/Australian and New Zealand Association for the Advancement of Science (ANZAAS) Symp. on Aquatic Toxicology in Australia* Sydney, May 17, 1988, Skidmore, J. F., Ed., Australian Institute of Biology Special Publication, in press.
2. **Brown, V. M.** Development of water quality criteria from toxicological data, in *Water Quality Management Freshwater Ecotoxicity in Australia,* Hart, B. T., Ed., Water Studies Centre, Chisholm Institute of Technology, Melbourne, 1986, chap. 11.
3. **USEPA,** Water Quality Criteria Documents, U.S. Environmental Protection Agency, Washington, D.C., *Fed. Reg.,* 1980, 45, 231, 79318-79323.
4. **Sprague, J. B.,** Factors that modify toxicity, in *Fundamentals of Aquatic Toxicology,* Rand, G. M. and Petrocelli, S. R., Eds., Hemisphere, New York, 1985, chap. 6.
5. **Holdway, D. A. and Dixon, D. G.,** Acute toxicity of pulse-dosed methoxychlor to juvenile American flagfish (*Jordanella floridae* Goode and Bean) as modified by age and feeding, *Aquat. Toxicol.,* 6, 243, 1985.
6. **Holdway, D. A., Kempe, E. J., and Dixon, D. G.,** Acute toxicity of methoxychlor to larval white sucker (*Catostomus commersoni*) as modified by age, food availability and egg preexposure, *Can. J. Fish. Aquat. Sci.,* 44, 227, 1987.
7. **Woltering, D. M.,** The growth response in fish chronic and early life stage toxicity tests: a critical review, *Aquat. Toxicol.,* 5, 1, 1984.
8. **Hodson, P. V.,** Water quality criteria and the need for biochemical monitoring of contaminant effects on aquatic ecosystems, in *Water Quality Management Freshwater Ecotoxicity in Australia,* Hart, B. T., Ed., Water Studies Centre, Chisholm Institute of Technology, Melbourne, 1986, chap. 2.
9. **Dixon, D. G., Hodson, P. V., Klaverkamp, J. F., Lloyd, K., and Roberts, R.,** *The Role of Biochemical Indicators in the Assessment of Aquatic Ecosystem Health — their Development and Validation,* Publ. No. NRCC 24371, Environmental Secretariate, National Research Council of Canada, 1985.
10. **Hodson, P. V., Blunt, B. R., and Whittle, D. M.,** Monitoring lead exposure of fish, in *Contaminant Effects on Fisheries,* Advances in Environmental Science and Technology, Vol. 16, Cairns, V. W., Hodson, P. V., and Nriagu, J. O., Eds., Wiley Interscience, New York, 1984, 87.
11. Supervising Scientist for the Alligator Rivers Region, *1st Annual Report 1978-79,* Australian Government Publishing Service, Canberra, 1979, 1.
12. Supervising Scientist for the Alligator Rivers Region, *Annual Report 1987-88,* Australian Government Publishing Service, Canberra, 1988, 25.
13. Supervising Scientist for the Alligator Rivers Region, *Annual Report 1986-87,* Australian Government Publishing Service, Canberra, 1987, 36.

14. **Holdway, D. A., Humphrey, C., Mcbride P., and Allison, H.,** Toxicity of retention pond waters, in *Alligator Rivers Region Research Institute Annual Research Summary 1986-87,* Supervising Scientist for the Alligator Rivers Region, Australian Government Publishing Service, Canberra, 1987, 99.
15. **Allison, H. E. and Holdway, D. A.,** Toxicity of uranium to hydra, in *Alligator Rivers Region Research Institute Annual Research Summary 1987-88,* Supervising Scientist for the Alligator Rivers Region, Australian Government Publishing Service, Canberra, 1988, 68.
16. **Holdway, D. A. and Wiecek, M. M.,** Toxicity of uranium to fish, in *Alligator Rivers Region Research Institute Annual Research Summary 1987-88,* Supervising Scientist for the Alligator Rivers Region, Australian Government Publishing Service, Canberra, 1988, 63.
17. **Holdway, D. A., Allison, H. E., Mannion, M. M. D., Wiecek, M. M., and Templeman, M. A.,** Development of toxicity tests, in *Alligator Rivers Region Research Institute Annual Research Summary 1987-88,* Supervising Scientist for the Alligator Rivers Region, Australian Government Publishing Service, Canberra, 1988, 59.
18. **Holdway, D. A., Allison, H. E., Mannion, M. M. D., Wiecek, M. M., and Templeman, M. A.,** Toxicity of Ranger retention pond waters, in *Alligator Rivers Region Research Institute Annual Research Summary 1987-88,* Supervising Scientist for the Alligator Rivers Region, Australian Government Publishing Service, Canberra, 1988, 69.
19. **Holdway, D. A. and Wiecek, M. M.,** OSS test protocols for the biological testing of waste waters for release into Magela Creek. I. Embryo gudgeon test *(Mogurnda mogurnda),* Open File Rec. No. 51, Supervising Scientist for the Alligator Rivers Region, 1988.
20. **Holdway, D. A. and Wiecek, M. M.,** OSS test protocols for the biological testing of waste waters for release into Magela Creek. II. Larval gudgeon test *(Mogurnda mogurnda),* Open File Rec. No. 52, Supervising Scientist for the Alligator Rivers Region, 1988.
21. **Holdway, D. A., Wiecek, M. M., and Brown, V. M.,** OSS test protocols for the biological testing of waste waters for release into Magela Creek. III. Larval rainbow-fish test *(Melanotaenia splendida),* Open File Rec. No. 53, Supervising Scientist for the Alligator Rivers Region, 1988.
22. **Holdway, D. A. and Wiecek, M. M.,** OSS test protocols for the biological testing of waste waters for release into Magela Creek. IV. Juvenile perchlet test *(Ambassis macleayi),* Open File Rec. No. 54, Su-
23. **Allison, H. E., Mannion, M. M. D., and Holdway, D. A.,** OSS test protocols for the biological testing of waste waters for release into Magela Creek. V. Freshwater snail test *(Amerianna carinata),* Open File Rec. No. 55, Supervising Scientist for the Alligator Rivers Region, 1988.
24. **McBride, P., Allison, H. E., Templeman, M. A., Brown, V. M., and Holdway, D. A.,** OSS test protocols for the biological testing of waste waters for release into Magela Creek. VI. Cladoceran test *(Moinodaphnia macleayi),* Open File Rec. No. 56, Supervising Scientist for the Alligator Rivers Region, 1988.
25. **Allison, H. E. and Holdway, D. A.,** OSS test protocols for the biological testing of waste waters for release into Magela Creek. VII. Hydra test (*Hydra* Sp. A), Open File Rec. No. 57, Supervising Scientist for the Alligator Rivers Region, 1988.
26. **Allison, H. E. and Holdway, D. A.,** OSS test protocols for the biological testing of waste waters for release into Magela Creek. VIII. Hydra test (*Hydra* Sp. B), Open File Rec. No. 58, Supervising Scientist for the Alligator Rivers Region, 1988.
27. **Allison, H. E. and Holdway, D. A.,** OSS test protocols for the biological testing of waste waters for release into Magela Creek. IX. Duckweed test *(Lemna aequinoctialis),* Open File Rec. No. 59, Supervising Scientist for the Alligator Rivers Region, 1988.
28. **Jeffree, R. A. and Williams, N. J.,** Biological indications of pollution of the Finniss River system, especially fish diversity and abundance, in *Rum Jungle Environmental Studies,* Davey, D. R., Ed., Rep. No. E365, Australian Atomic Energy Commission, Sydney, 1975, chap. 7.
29. **Skidmore, J. F.,** Toxicity of copper to selected Australian freshwater animals, in *Water Quality Management Freshwater Ecotoxicity in Australia,* Hart, B. T., Ed., Water Studies Centre, Chisholm Institute of Technology, Melbourne, 1986, chap. 4.
30. **Holdway, D. A., Mannion, M. M. D., Rossendell, J., and Brown, K.** Toxicity of Djalkmara Billabong water before and after two RP4 water releases into the billabong via siphon, in *Alligator Rivers Region Research Institute Annual Research Summary 1988-89,* Supervising Scientist for the Alligator Rivers Region, Australian Government Publishing Service, Canberra, 1991, 62.
31. Supervising Scientist for the Alligator Rivers Region, *Annual Report 1988-89,* Australian Government Publishing Service, Canberra, 1989, 19.

Index

INDEX

T

U

V

W

Z